Flax

Medicinal and Aromatic Plants – Industrial Profiles

Individual volumes in this series provide both industry and academia with in-depth coverage of one major genus of industrial importance.

Edited by Dr Roland Hardman

Flax
The genus *Linum*

Edited by

Alister D. Muir and Neil D. Westcott

Agriculture and Agri-Food Canada, Saskatoon,
Saskatchewan, Canada

CRC Press
Taylor & Francis Group
Boca Raton London New York

CRC Press is an imprint of the
Taylor & Francis Group, an **informa** business
A TAYLOR & FRANCIS BOOK

CRC Press
Taylor & Francis Group
6000 Broken Sound Parkway NW, Suite 300
Boca Raton, FL 33487-2742

First issued in paperback 2019

© 2003 by Taylor & Francis Group, LLC
CRC Press is an imprint of Taylor & Francis Group, an Informa business
Typeset in 11/12pt Garamond by
Graphicraft Limited, Hong Kong

No claim to original U.S. Government works

ISBN-13: 978-0-415-30807-6 (hbk)
ISBN-13: 978-0-367-39519-3 (pbk)

British Library Cataloguing in Publication Data
A catalogue record for this book is available from the British Library

Library of Congress Cataloging in Publication Data
Flax : the genus linum / edited by Alister D. Muir and Neil D. Westcott.
        p. cm. — (Medicinal and aromatic plants—industrial profiles;
v. 34)
Includes bibliographical references (p.).
   1. Flax. 2. Flax—Therapeutic use. I. Muir, Alister D. II. Westcott,
Neil D. III. Series.
      SB253 .F58 2003
      583'.79—dc21
                                                    2002155288

Visit the Taylor & Francis Web site at
http://www.taylorandfrancis.com

and the CRC Press Web site at
http://www.crcpress.com

Contents

Contributors

William F. Clark
Department of Medicine
The University of Western Ontario
London Health Sciences Center
800 Commissioners Road East
London, Ontario N6A 4G5
Canada

Stephen C. Cunnane
Department of Nutritional Sciences
Faculty of Medicine
FitzGerald Building
University of Toronto
150 College Street, Toronto
Ontario M5S 1A8
Canada

John R. Dean
Agricore United
Box 6600, 201 Portage Avenue
Winnipeg, Manitoba R3C 3A7
Canada

Axel Diederichsen
PGRC, Saskatoon Research Center
Agriculture & Agri-Food Canada
107 Science Place, Saskatoon
Saskatchewan S7N 0X2
Canada

Anatoly Marchenkov
All-Russian Flax Research Institute
(VNIIL), 35, Lunacharsgogo Str.
172060 Torzhok
Russian Federation

Diane H. Morris
President, Mainstream Nutrition
904-130 Carlton St, Toronto
Ontario M5H 4K3
Canada

Alister D. Muir
Bioproducts and Processing
Saskatoon Research Center
Agriculture & Agri-Food Canada
107 Science Place, Saskatoon
Saskatchewan S7N 0X2
Canada

Malcolm Ogborn
Department of Pediatrics
Winnipeg Children's Hospital
The University of Manitoba
Winnipeg, Manitoba R3A 1S1
Canada

N. Lee Pengilly
Saskatchewan Flax Development
Commission
A5A 116 103rd Street, Saskatoon
Saskatchewan S7N 1Y7
Canada

Kailash Prasad
Professor Emeritus
Department of Physiology
College of Medicine
University of Saskatchewan
107 Wiggins Rd, Saskatoon,
Saskatchewan S7N E5E
Canada

Khalid Y. Rashid
Morden Research Center
Agriculture & Agri-Food Canada
Morden, Manitoba R6M 1Y5
Canada

Ken Richards
PGRC, Saskatoon Research Center
Agriculture & Agri-Food Canada
107 Science Place, Saskatoon
Saskatchewan S7N 0X2
Canada

Sharon E. Rickard-Bon*
Department of Nutritional Sciences
Faculty of Medicine
University of Toronto
150 College St., Toronto
Ontario M5S 3E2
Canada

Tatiana Rozhmina
All-Russian Flax Research Institute
(VNIIL), 35, Lunacharsgogo Str.
172060 Torzhok
Russian Federation

Sheila E. Scheideler
Professor of Animal Science
Institute of Agriculture and
Natural Resources
University of Nebraska Lincoln
Lincoln, Nebraska 68588-0908
USA

Juliana J. Soroka
Ecological Pest Management
Saskatoon Research Center
Agriculture & Agri-Food Canada
107 Science Place, Saskatoon
Saskatchewan S7N 0X2
Canada

Lilian U. Thompson
Department of Nutritional Sciences
Faculty of Medicine
University of Toronto
150 College St., Toronto
Ontario M5S 3E2
Canada

Igor Uschapovsky
All-Russian Flax Research Institute
(VNIIL), 35, Lunacharsgogo Str.
172060 Torzhok
Russian Federation

Marion Vaisey-Genser
Professor and Senior Scholar
Department of Human Nutritional
Sciences
Faculty of Human Ecology
University of Manitoba, Winnipeg
Manitoba R3T 2N2
Canada

Neil D. Westcott
Bioproducts and Processing
Saskatoon Research Center
Agriculture & Agri-Food Canada
107 Science Place, Saskatoon
Saskatchewan S7N 0X2
Canada

Ian L. Wise
Breeding and Genetics
Cereals Research Center
Agriculture & Agri-Food Canada
195 Dafoe Road, Winnipeg
Manitoba R3T-2M9
Canada

*Current Address
St. Joseph's Care Group
35 Algoma Street North
P.O. Box 3251
Thunder Bay, Ontario P7B 5G7
Canada

Preface to the series

There is increasing interest in industry, academia and the health sciences in medicinal and aromatic plants. In passing from plant production to the eventual product used by the public, many sciences are involved. This series brings together information which is currently scattered through an ever increasing number of journals. Each volume gives an in-depth look at one plant genus, about which an area specialist has assembled information ranging from the production of the plant to market trends and quality control.

Many industries are involved such as forestry, agriculture, chemical, food, flavour, beverage, pharmaceutical, cosmetic and fragrance. The plant's raw materials are roots, rhizomes, bulbs, leaves, stems, barks, wood, flowers, fruits and seeds. These yield gums, resins, essential (volatile) oils, fixed oils, waxes, juices, extracts and spices for medicinal and aromatic purposes. All these commodities are traded worldwide. A dealer's market report for an item may say 'Drought in the country of origin has forced up prices'.

Natural products do not mean safe products and account of this has to be taken by the above industries, which are subject to regulation. For example, a number of plants which are approved for use in medicine must not be used in cosmetic products.

The assessment of safe-to-use starts with the harvested plant material which has to comply with an official monograph. This may require absence of, or prescribed limits of, radioactive material, heavy metals, aflatoxin, pesticide residue, as well as the required level of active principle. This analytical control is costly and tends to exclude small batches of plant material. Large scale contracted mechanised cultivation with designated seed or plantlets is now preferable.

Today, plant selection is not only for the yield of active principle, but for the plant's ability to overcome disease, climatic stress and the hazards caused by mankind. Such methods as *in vitro* fertilization, meristem cultures and somatic embryogenesis are used. The transfer of sections of DNA is giving rise to controversy in the case of some end-uses of the plant material.

Some suppliers of plant raw material are now able to certify that they are supplying organically-farmed medicinal plants, herbs and spices. The European Union directive (CVO/EU No 2092/91) details the specifications for the *obligatory* quality controls to be carried out at all stages of production and processing of organic products.

Fascinating plant folklore and ethnopharmacology leads to medicinal potential. Examples are the muscle relaxants based on the arrow poison, curare, from species of *Chondrodendron*, and the anti-malarials derived from species of *Cinchona* and *Artemisia*. The methods of detection of pharmacological activity have become increasingly reliable and specific, frequently involving enzymes in bioassays and avoiding the use of laboratory animals. By using bioassay-linked fractionation of crude plant juices or extracts, compounds can be specifically targeted which, for example, inhibit blood platelet aggregation, or have anti-tumour, or anti-viral, or any other required activity. With the assistance of robotic devices, all the members of a genus may be readily screened. However, the plant material must be *fully* authenticated by a specialist.

The medicinal traditions of ancient civilizations such as those of China and India have a large armamentaria of plants in their pharmacopoeias which are used throughout South-East Asia. A similar situation exists in Africa and South America. Thus, a very high percentage of the world's population relies on medicinal and aromatic plants for their medicine. Western medicine is also responding. Already in Germany all medical practitioners have to pass an examination in phytotherapy before being allowed to practise. It is noticeable that throughout Europe and the USA, medical, pharmacy and health related schools are increasingly offering training in phytotherapy.

Multinational pharmaceutical companies have become less enamoured of the single compound magic bullet cure. The high costs of such ventures and the endless competition from "me too" compounds from rival companies often discourage the attempt. Independent phytomedicine companies have been very strong in Germany. However, by the end of 1995, eleven (almost all) had been acquired by the multinational pharmaceutical firms, acknowledging the lay public's growing demand for phytomedicines in the Western World.

The business of dietary supplements in the Western World has expanded from the health store to the pharmacy. Alternative medicine includes plant-based products. Appropriate measures to ensure the quality, safety and efficacy of these either already exist or are being answered by greater legislative control by such bodies as the Food and Drug Administration of the USA and the recently created European Agency for the Evaluation of Medicinal Products, based in London.

In the USA, the Dietary Supplement and Health Education Act of 1994 recognized the class of phytotherapeutic agents derived from medicinal and aromatic plants. Furthermore, under public pressure, the US Congress set up an Office of Alternative Medicine and this office in 1994 assisted the filing of several Investigational New Drug (IND) applications, required for clinical trials of some Chinese herbal preparations. The significance of these applications was that each Chinese preparation involved several plants and yet was handled as a *single* IND. A demonstration of the contribution to efficacy, of *each* ingredient of *each* plant, was not required. This was a major step forward towards more sensible regulations in regard to phytomedicines.

My thanks are due to the staff of Taylor & Francis publishers who have made this series possible and especially to the volume editors and their chapter contributors for the authoritative information.

Roland Hardman, 1997

Preface

Linum usitatissimum is a widely distributed Mediterranean and temperate climate zone oilseed that has a long history of traditional use both as an industrial oil and a fiber crop. However in North America and in Western Europe, its use as a food and medicinal crop is less well known in spite of an equally long history of traditional use for these purposes. Known as linseed in the United Kingdom and many of its former colonies, or flax in North America, the oilseed cultivars of this species are now the predominant forms that are grown around the world, while fiber cultivars have declined in importance as synthetic fibers have displaced linen fibers from many of their traditional uses. Although the primary focus of this book is on the medicinal and nutritional constituents of the seed of *Linum usitatissimum*, several other *Linum* spp. contain podophyllotoxin derivatives that have biological activity or potential for use in the semisynthesis of anticancer drugs (Chapter 3).

Beginning in the mid-1980s there has been a steadily growing interest in the medicinal and nutraceutical value of flax. At first, interest was concentrated on the biological activity of the omega-3 fatty acid α-linolenic acid (Chapter 7). As time passed and more investigators focused on the biological activity of flax, other biologically active molecules were identified including the flax lignans and soluble dietary fiber (Chapter 12). The experimental evidence for the role of flax lignans in preventing cancer is described in considerable detail in Chapter 8. The role of lignans in the prevention of cardiovascular diseases and the associated reduction in risk factors is described in Chapter 9. The potential for flaxseed and flax lignans to delay the progression and severity of lupus nephritis and other kidney diseases is documented in Chapter 10. Even the cyanogenic glycosides, long perceived to be anti-nutritional, may have a protective effect from selenium toxicity (Chapter 12). Flax lignans are converted to a number of different compounds in the mammalian GI tract. These compounds, including the mammalian lignans, undergo enterohepatic circulation after initial absorption (Chapter 11). The lignan secoisolariciresinol diglucoside and several of its metabolites also possess significant antioxidant activity resulting in a series of oxidized metabolites.

While the primary focus of this book is on the human use of flaxseed, increasing attention is being placed on modification of animal diets to modify the fatty acid profile of meat and dairy products (Chapter 14).

In keeping with other books in this series, Chapters 1 and 13 focus on the history of the traditional food and medicinal uses of flaxseed and provide a fascinating compilation of the diverse uses humans have found for this crop. The correct identification of medicinal plants has long been a problem for the medicinal plant industry. While correct identification of *Linum usitatissimum* as it applies to the commercial species is rarely a problem, the taxonomy of this genus is complicated and there are few reliable English language resources available on this subject. We are very pleased therefore to be able to present in Chapter 2 a detailed analysis of the taxonomy and genetic resources of this economically important species. As we become more aware of the biological activity of compounds such as the lignans, knowledge and conservation of the genetic resources of this genus will become increasingly important in the future as the few remaining wild populations of *Linum* species come under increasing pressure from human activities.

Fortunately the commercial sources of flaxseed are not in any danger of extinction as flax is cultivated in many countries around the world. The main growing areas for flax (both fiber and oilseed) are documented in Chapter 4, along with descriptions of the typical agronomic practices for flax production in each region. The diseases (Chapter 5) and insect pests (Chapter 6) encountered by flax growers around the world are described along with the preferred strategies for their control.

The commercial trade in flaxseed and its oil are described in Chapter 15, while the current regulatory status of flax products for human consumption is described in Chapter 16.

This book is intended to give the reader a comprehensive overview of the present knowledge of this genus, both from the perspective of producing the crop as well as its many uses, both old and new. As a result of the rapidly growing interest in flax and the wide range of biological activities associated with constituents of this plant, flax is emerging as a strong rival for soybean as a major nutraceutical crop.

We wish to thank Mr Ralph Underwood for his many contributions to the art work in this book, to the library staff of the Saskatoon Research Center for locating many obscure references and to our colleagues at the Center for their many helpful suggestions.

1 Introduction

History of the cultivation and uses of flaxseed

Marion Vaisey-Genser and Diane H. Morris

What department is to be found in active life in which flax is not employed? – Pliny, the Roman natural historian[1]

Introduction

Linum usitatissimum, the specific name for flax within the family Linaceae, aptly describes its usefulness and versatility. In fact, the name Linum originated from the Celtic word *lin* or "thread," and the name *usitatissimum* is Latin for "most useful" (Kolodziejczyk and Fedec, 1995). According to the Compact Oxford Dictionary, the old French word "linnet" means "the bird which hovers feeding on the seeds over flax fields." The word "line" is derived from a Latin or Greek ancestor, *linum*, meaning "flax"; other words such as linen, lining, linear, and lineage are all derived from the word "line" (Judd, 1995). These ancient linguistic origins underscore the importance of flaxseed or linseed, as it is sometimes called, to the economic and social development of humans.

The terms *flaxseed* and *linseed* have particular meanings, depending on the region. In Europe, flax refers to the seed grown for fiber (linen) production, while linseed refers to oilseed flax grown for industrial and nutritional uses. In North America, the terms are used interchangeably to describe the linum species, *Linum usitatissimum*, although there is a slight preference for using the term flaxseed when referring to the flax used for human consumption. Traditional flaxseed varieties consumed by humans are rich in alpha-linolenic acid (ALA), the essential omega(ω)-3 fatty acid, which constitutes about 57 percent of the total fatty acids in flax. Solin is a generic term for flax varieties that contain less than 3 percent ALA (Oomah and Mazza, 1988).[2] Flax grown for human consumption and flax grown to manufacture linen belong to the same species but are different varieties, and the two products are not usually obtained from the same crop (BeMiller, 1973).

Flaxseed has historically been consumed as a cereal and valued for its medicinal qualities, while the oil from the seed has served variously as a frying medium for food, a lamp oil and a preservative in products such as paint and flooring. However, fabric made from the fibers of flax stalks is the best documented example of the earliest use of flax, likely because linen artifacts have survived over time better than perishable oils and foods. For example, the weaving of linen is credited with having a role in some of the major landmarks in the evolution of Western Civilization. Today, Canada is a world leader in the production and export of flax. In 1996 and 1997, Canada produced about 860,000 tonnes (or 40 percent) of the world's total flax. Canada exports flax mainly to Europe, Japan, South Korea and the United States (Anonymous, 1999b).

Table 1.1 Landmarks in the chronology of flax from ancient to contemporary times

Years B.C.E.*

8000   Wild flaxseed (*L. bienne*) dated in the Fertile Crescent (Syria, Turkey and Iran) (Helbaek, 1969; van Zeist, 1970; van Zeist, 1972)

7000   Agriculture fully established in the Fertile Crescent; flax among the first crops domesticated (van Zeist and Bakker-Heeres, 1975; Smith, 1995)

6000   Remains of linen artifacts identified in the Dead Sea area (Schick, 1988)

5000   Earliest dated Egyptian linen cloth (Judd, 1995)

4000   European use of flax: artifacts of the Swiss Lake dwellers (Zohary and Hopf, 1993)

2000   The first industry? Babylonians twisted flax fibers into thread for weaving (Harris, 1993)

1400   Egyptians used linseed oil for embalming and linen for binding mummies (Judd, 1995); linen was the primary fabric for Egyptian cloth (Barber, 1994)

1000   Flaxseed used in breads in Jordan and Greece (Stitt, 1994)

500   Flaxseed used as a laxative and, by Hippocrates, as a poultice (Judd, 1995); Phoenicians' linen sails may have introduced flax to Flanders and Britain (Wilson, 1979)

Years C.E.

800   Charlemagne prescribed flaxseed production for all subjects in the Roman Empire (Anonymous, 1999b)

1000   Flanders was a leading center of the linen industry (Wilson, 1979)

1400   The Renaissance: Van Eyck pioneered linseed oil in oil painting preservation (Judd, 1995)

1500   The Reformation: Huguenots took linen-producing skills to the British Isles (Baines, 1985)

1600   Colonization: French immigrants took flax to North America (Atton, 1989)

1800   Industrial Revolution began. Linoleum flooring patented in Britain (Judd, 1995); invention of the cotton gin in 1793 forecast the end of the dominance of linen and the 1920s' collapse of the flax industry (Duder, 1992)

1980   Consumer emphasis on environmentally friendly, natural products renewed producer interest in flax in North America and Europe (Prentice, 1990)

1995   Emergence of flaxseed as a functional food

* B.C.E. = Before the Common Era; C.E. = Common Era. The Common Era is understood to begin about 1 A.D.

This chapter describes the biological origins of flax, its importance in the agricultural economies of the Fertile Crescent of about 8000 B.C.E.,[3] early methods of producing flax fiber, and ancient and contemporary uses of flaxseed as a food, medicine, industrial agent and source of fiber and oil. Although the specific points in time when each of the functional qualities of flax emerged remains somewhat imprecise, the chronological landmarks in Table 1.1 provide a guidepost for exploring the history of flax from early times to the present day.

Biological origins

As is the case with several other oilseed crops, cultivated annual flax appears to have evolved from wild or weedy forms that are either perennial or annual. A likely

Figure 1.1 Flowering and fruiting branches, flower (without petal), petal, fruiting calyx with capsule (Plate 374, Zohary, 1972, used with permission of The Israel Academy of Sciences and Humanities, Jerusalem).

progenitor is *L. bienne* Mill. (syn. *L. angustifolium* Huds.), which has the same chromosome number ($2n = 30$), typical strong branches, periwinkle blue flowers and splitting (dehiscent) bolls or capsules of seed (Figure 1.1). This form of wild flax is widely distributed across western Europe, the Mediterranean Basin, North Africa and the Caucasus (Zohary and Hopf, 1993). Perennial wild flax also grows in China (Pan, 1990).

The Indian subcontinent seems the most likely geographic area for the botanical origin of flax, as this region still claims the greatest biological diversity within the genus *Linum*. This was demonstrated by the Russian plant breeders, Vavilov and Dorofeyev, who collected a worldwide database of flax varieties, among other species, as a reference for plant breeding (Judd, 1995). The English translation of their treatise on the *Origin and Geography of Cultivated Plants* was published in 1992 (Vavilov and Dorofeyev, 1992). The thesis of India's position as the source of flax is consistent with early trade routes linking India with the Middle East through the Indus valley. Edible flaxseed rather than fiber flax dominated India's production because other fiber species, such as hemp, were already in wide use (Judd, 1995).

Ancient cultivation and processing of flax

Flaxseed is considered a founding crop because it was among the first domesticated plants. Its cultivation likely began in the fertile valleys of the Tigris and Euphrates rivers in Mesopotamia – the so-called Fertile Crescent – given that these were the sites of the earliest agrarian societies (Zohary and Hopf, 1993; Smith, 1995). Archeological evidence suggests that flax probably has been used by humans for about 10,000 years and cultivated for some 8,000 years. Linseed, apparently wild *L. bienne*, has been found in excavation sites in Syria, Turkey and Iran and dated between 8000 and 6750 B.C.E. (Helbaek, 1969; van Zeist, 1970; van Zeist, 1972). Larger seeds recovered in Syria, Iraq and other parts of the Mesopotamian basin point to flax cultivation as part of an evolving irrigation-based agricultural system before 6000 B.C.E. (Zohary and Hopf, 1993). Over time, conquest and trade spread flax cultivation from the Near East to Europe, the Nile Valley and West Asia (Zohary and Hopf, 1993). It was reportedly used in Scandinavia for spinning during the Iron Age (900–400 B.C.E.) (Geijer, 1979). The Chinese heritage of flax use, predominantly as oil, goes back at least 2,000 years and possibly 5,000 years, based on archeological records (Pan, 1990). Later, colonization took flax cultivation as far afield as North America (Atton, 1988) and Australia (Frost, 1985).

With domestication, some of the characteristic traits of the weedy forms of flax evolved through random selection via the consistent practice of sowing, reaping and threshing. For example, today's domestic flax bolls no longer split or dehisce to distribute seed, and seed size, number and oil yield have increased (Zohary and Hopf, 1993). Different flax varieties were cultivated in different geographic regions according to local growing conditions and local techniques. In Egypt, where flax was prized for its use in weaving fine linens, varieties that grew tall with minimal branching were encouraged because of the longer fibers available from tall stalks. Also, fiber plants were sown close together to further discourage branching. In contrast, along the southern Nile, the Abyssinians fostered a shorter variety of flax that was less susceptible to windstorms and lodging and bore larger seeds. They used flaxseed both as a source of edible oil and as a cereal and flour (Judd, 1995).

Flax harvesting practices developed in keeping with the variety and its designated use. Where flax was cultivated for its fiber, the stalks were pulled up by the roots before the seeds matured and stored upright to dry so that the fibers could be separated from the stem more easily (Wilson, 1979). However, Heinrich (1992) described a three-stage harvesting system apparently used by early Egyptians to achieve a range of qualities in their linens, as evidenced by the microscopic examination of cloth artifacts. The earliest harvest was of green stalks just after the plant flowered, yielding the fine soft fibers that were used for the delicate fabrics reserved for the aristocracy. Most flax crops were pulled about thirty days after flowering and yielded a stronger fiber. The final harvest of mature plants took place several weeks later, yielding the coarse fibers used in ropes and mats, as well as the seeds for future planting and for linseed oil. Where flaxseed was used primarily for oil or cereal, the seed was always allowed to mature fully before harvesting took place (Atton, 1989).

Post-harvest handling techniques also evolved according to utility. Dry storage of whole flaxseed was undoubtedly practiced early and effectively. Despite its high oil content and the highly unsaturated nature of linseed oil, mature whole flaxseed has demonstrated impressive oxidative stability in contemporary studies at ambient temperatures (Chen *et al.*, 1994).

Information is sparse concerning the technologies for extracting linseed oil and flax fibers before or during the early periods of flax cultivation. Zohary and Hopf (1993) proposed that edible flaxseed oil was decanted simply from crushed seeds soaked in water, a form of cold pressing. In contrast, the extraction of linseed oil for paints and varnishes in later times involved hot pressing the seed (> 35°C). Concerning the fiber, Judd (1995) speculated that the separation of flax fiber from stems, a process known as retting, may have been discovered as a result of flooding. People might have noted and gathered the long lines of strong white fibers left tangled along the riverbanks in the wake of receding flood waters. Wilson (1979) reported that ancient Egyptians took advantage of the nighttime dew for retting linseed stalks. Pulled flax was simply spread on rooftops or in dry fields for several weeks and turned daily to spread the moisture that was needed for the microbial decomposition of the pectins and gums binding the fibers. Later, Pliny's *Natural History* (77 B.C.E.) described how, in the Roman Empire, flax stems were retted by weighting them with stones in tanks of water (Clarke, 1986). Today retting is done using tanks and modern controls.

Ancient and contemporary uses of flax

The historical record provides convincing evidence that flax was valued in Ancient and Early Modern times as a food, medicine, preservative and source of fiber for linen. Today, flax is valued for many of the same – and some new – reasons. This section describes the many ancient and contemporary uses of flax.

Flaxseed as a food crop

Flax appears to be the earliest and best documented oil and fiber crop cultivated in the earliest agricultural stages of the Old World some 8,000 years ago (Zohary and Hopf, 1993; Smith, 1995). Crop cultivation was accompanied by the storage of surplus grains and oilseeds. These practices brought the reliable food supplies necessary for the development of Western Civilization (Harlan, 1975). As flax and other crops were domesticated, there was a shift from the nomadic hunter/gatherer lifestyle to agricultural societies where people put down roots alongside the crops they tended. Villages grew up around farms, providing farm owners and their families with a central location for trading services and goods. The bartering of farm produce included flaxseed and linens made at home from flax fiber (Barber, 1994).

Flaxseed cereal dishes

At about 1000 B.C.E., Jordanian farmers saved flaxseeds harvested at the end of the growing season for next year's crop and for adding to mixed grain breads. The Greeks preferred to mix roasted flax, barley and coriander with salt, which could be stored in earthenware jars until needed for making bread (Stitt, 1994).

In Ethiopia, flax was particularly well-suited for planting in the highlands at elevations of 2,500–3,000 meters. It was a key ingredient in stews, porridges and drinks. Roasted flaxseeds, for example, were crushed and mixed with pulses to make a stew called "w'et." In the Begemedhir region, w'et was made from roasted, crushed flaxseeds and split peas. A popular porridge was also made from roasted, crushed and cooked flaxseeds, to which salt and red pepper were added. Stone-ground flax was occasionally used to make porridge. Sometimes a drink called "ch'ilk'a" was made by mixing lightly roasted, milled flax with water, salt and honey. Ch'ilk'a was consumed during periods of fasting, when consumption of alcoholic beverages was not allowed (Siegenthaler, 1994).

In today's North American food market, consumers likewise enjoy the pleasant, nutty taste of flaxseed and can buy it in a variety of forms: whole flaxseed, flaxseed oil, capsules, powders and food products made with added flaxseed; all are available from supermarkets and health food stores. Whole or milled flaxseed is a popular addition to baked goods such as bagels, muffins, sports nutrition bars and multigrain breads. For consumers who enjoy baking at home, there are muffin, cookie, bun, dinner roll, focaccia and bread machine recipes that contain added milled or whole flaxseed (Anonymous, 1999a). Red River™ hot cereal, a popular breakfast item that originated in Manitoba's Red River valley in 1924, contains cracked and whole flaxseed (Sarah Caplan, personal communication).

Flaxseed as an edible oil source

The use of flaxseed oil in food preparation by the ancients is seldom dealt with in detail apart from acknowledging its long culinary use in Egypt, India and China (Judd, 1995). Pan (1990) cites archeological studies supporting the production of flaxseed in China for possibly 5,000 years, and at least since the beginning of the Common Era. He maintains that flaxseed oil has been the major edible oil used in peasant communities from the earliest times to the present day in those areas of China where flax is grown. The oil is extracted locally by pressing and consumed in home-prepared foods. He points out, however, that the consumption of fats in China is modest.

In the first half of the twentieth century, interest turned to flaxseed as a source of edible oil in areas of Europe and North America where the climate precluded olive and peanut production. Normal growth and reproduction in laboratory animals fed flaxseed oil was demonstrated by Molotkow (1932) in the Soviet Union, by Maynard and colleagues (1942) in the United States and by Crampton *et al.* (1951) in Canada. Prompted by import restrictions on edible oils during the 1939–1945 World War and the legalization in some provinces of margarine as a tablespread, the National Research Council of Canada launched studies on linseed and rapeseed for their potential as domestic oilseed crops. The Council benefitted from experiences in the North and West of Europe where the linseed crushing industry originated (Hunt, 1969).

In the 1950s, the limiting factor for the use of flaxseed oil in food was its susceptibility to "flavor reversion," an off-flavor that developed in industrial prototypes of both salad oil and shortenings. This defect appeared to be associated with flaxseed oil's high ALA content (Lemon, 1947; Armstrong and McFarlane, 1944). Experimental high temperature processing of flaxseed oil to polymerize ALA (275°C for 8–12 hr.) succeeded in averting flavor reversion, but animal feeding trials showed that it significantly reduced the oil's nutritive value (Crampton *et al.*, 1951; Lips and Crampton, 1952). As a result, the commercial processing of flaxseed oil did not proceed at that time. Today, the high-ALA flaxseed oil sold in health food stores is cold pressed at < 35°C, processed in a low-oxygen environment and packaged in light-proof containers to maintain stability (Carter, 1993). The increasing consumer demand for cold-pressed flaxseed oil appears promising (Prentice, 1990).

Flax use in the pet and livestock feed industries

Two emerging areas for flax include the pet food industry and animal feeds for livestock. For instance, Hill's Pet Nutrition introduced a pet food product formulated with added flax to meet the needs of dogs undergoing chemotherapy. The new food product is rich in ALA and the amino acid arginine, both of which appear to reduce the painful side-effects of cancer treatment and increase survival time by more than 54 percent (Anonymous, 1998a). Anecdotal evidence indicates that dog and other animal breeders add flaxseed to animal diets to produce a glossy, shiny coat – an important characteristic for show animals (Barry Hall, personal communication).

Flax is added to animal feeds as a means of increasing the ω-3 fatty acid content of animal products. In one study of broiler chicks, diet supplementation with linseed (at levels of 1.0, 2.5 and 5.0%) significantly increased the 3-fatty acid content of thigh muscle (Chanmugam *et al.*, 1992). When flaxseed was added to poultry rations at a level of 10 or 20 percent, the ALA content of egg yolk lipid increased about 5 and 9 percent, respectively (Caston and Leeson, 1990). Men who ate eggs derived from flaxseed-fed hens showed significant increases in the total ω-3 fatty acid content of platelet phospholipids (Ferrier *et al.*, 1995), thus making this approach an effective strategy for increasing the ω-3 fatty acid content of human tissues. Flaxseed has also been added to rations of dairy cattle, hogs and farm-raised fish, all with the goal of increasing the ω-3 fatty acid content of animal tissues and products derived from them (Vaisey-Genser and Morris, 1997).

Flaxseed as a functional food

Ancient peoples likely valued flaxseed as a functional food, although they would not have labeled it as such. The term "functional food" is thoroughly modern, and it is used to describe the nutritional and disease prevention value of flaxseed and other foods.

Definition of functional food

In one sense, all foods are functional, because they provide nutrients, taste or aroma (Hasler, 1996). Within the last decade, however, the term *functional food* emerged and has been used interchangeably with the terms "nutraceutical," "designer food," "pharmafood" and "foodaceutical" to describe foods or components of foods that may enhance health or prevent disease (Hasler, 1996).

Recent efforts have been made to standardize the language in this area. The Institute of Medicine of the U.S. National Academy of Sciences defines *functional foods* as those that "encompass potentially healthful products," including any food ingredient or modified food product that may provide a health benefit over and above the traditional nutrients the food contains (Thomas and Earl, 1994). Health Canada proposes to differentiate the terms "functional food" and "nutraceutical." It proposes to define a *functional food* as one that is "similar in appearance to, or may be, a conventional food, is consumed as part of a usual diet, and is demonstrated to have physiological benefits and/or reduce the risk of chronic disease beyond basic nutritional functions." It proposes to define a *nutraceutical* as "a product isolated or purified from foods that is generally sold in medicinal forms not usually associated with food . . . [and that] also has a physiological benefit or provides some protection against chronic disease"

(Anonymous, 1998b). In other words, nutraceuticals include pills, powders and capsules that contain nutrients and other ingredients derived from functional foods.

The functional food aspects of flaxseed

In Ancient times, around 500 B.C.E., Hippocrates wrote about the value of flax in relieving abdominal distress. Flaxseed has also been used as a laxative and poultice. In preparing a laxative, roasted flaxseeds were ground and mixed with water. A harsher stool softener could be made by boiling raw flaxseed in water and then drinking the solution (Siegenthaler, 1994). During the Early Modern era (around the 8th century C.E.), Charlemagne, the Frankish king and emperor of the West, considered flaxseed so important for health that he passed laws and regulations requiring its consumption by his subjects (Anonymous, 1999b). In Ethiopia, a flaxseed-containing drink was prepared to help relieve itching. The rationale was that the oils in the flax drink eventually permeated the affected skin from the inside out and reduced itching (Siegenthaler, 1994).

Flax appeared on the North American continent about 400 years ago when a European farmer brought it to New France (Anonymous, 1999b). Letters written in the early 1600s extolled the virtues of flax as a crop. Samuel Argall wrote in 1617 that flax would likely grow well in Virginia (Anonymous, 1896/97). Instructions written on behalf of King Charles I of England to Sir William Berkeley, a gentleman and governor of Virginia, admonished the governor to encourage "ye people there" to raise staple commodities such as flax (Anonymous, 1895). In 1691, Virginia enacted laws "enjoyning ye planting and dressing of flax" and levied penalties for not "mak[ing] a pound of Flax . . . every year" (Anonymous, 1903).

Pioneers in the American West at the beginning of the nineteenth century often used homemade remedies for common ills such as chills, fevers, snakebites and indigestion. Herbs, wildflowers, tree bark and other indigenous plants were staples in the frontier doctor's medicine box. Flax, along with slippery elm and pokeberry leaves, was fashioned into poultices for treating cuts and burns (Buley, 1933/34). Perennial flax (*L. perenne*) was used for eyewashes by the Paiutes and Shoshones of the Great Plain Basin in North America. Other flax varieties were used in the Missouri River region as folk remedies for coughs, gallstones, lung disorders and digestive disorders, including constipation (Haggerty, 1999).

Today, flaxseed is considered a functional food because it contains nutrients and other components, listed in Table 1.2, that promote health and may help prevent certain diseases. Flaxseed's main components of interest include dietary fiber, ALA and lignans.

Dietary fiber

Flaxseed averages about 28 percent dietary fiber on a dry-weight basis. The proportion of soluble to insoluble fiber in flaxseed varies between 20:80 and 40:60.[4] The soluble fiber component of flaxseed is mainly mucilage gums (Vaisey-Genser and Morris, 1997).

The soluble fiber fraction of flaxseed is likely the fraction responsible for its blood lipid-lowering effects. In a randomized, crossover study, hyperlipidemic men and postmenopausal women ate muffins containing either partially defatted flaxseed (about 50 g defatted flaxseed/day) or wheat bran daily for three weeks and then switched to

Table 1.2 Functional food components in flaxseed

Component in flaxseed	Health benefit
Dietary fiber	
Soluble fiber	Blood total cholesterol reduction
	Blood glucose reduction
Insoluble fiber	Laxation
Alpha-linolenic acid*	Blood total cholesterol reduction
	Reduction in production of arachidonic acid-derived eicosanoids
	Reduction of inflammation
	Reduction in risk of coronary heart disease
	Reduction in risk of stroke
Lignans	Prevention of cancer[†]

Notes
* Alpha-linolenic acid is the essential ω-3 fatty acid.
[†] Data on the anticancer effects of lignans are available at present only in animals.

the alternate muffin type for three weeks. Defatted flaxseed was used because it is rich in the seed coat mucilage gums and low in ALA, an ω-3 fatty acid that may also lower blood cholesterol levels. Using defatted flaxseed allowed for a more precise test of the lipid-lowering effects of flaxseed's mucilage gums. When subjects consumed flaxseed muffins, there were significant reductions in total cholesterol, LDL-cholesterol, apolipo-protein B and apolipoprotein A-I compared with the wheat bran group (Jenkins *et al.*, 1999).

Flaxseed also lowers blood glucose. In a study of six healthy volunteers, the glycemic response of subjects to 50 g of a test carbohydrate was compared with 50 g of a standard carbohydrate after an overnight fast. Flaxseed flour rich in flax mucilage was the test carbohydrate, while bread made from wheat flour was the standard. Consumption of a flaxseed bread test meal produced a 28 percent reduction in the area under the blood glucose curve compared with consumption of the standard white bread test meal (Cunnane *et al.*, 1993).

Finally, flaxseed promotes laxation because it contains both soluble fiber, which absorbs water from the gastrointestinal tract and increases stool bulk, and insoluble fiber, which increases stool transit time. Cunnane *et al.* (1995) reported a 30 percent increase in bowel movements among young, healthy adults who ate 50 g flaxseed per day for four weeks.

Alpha-linolenic acid

Flaxseed averages about 41 percent fat on a dry-weight basis. It is low in saturated fat (9% of total fatty acids), moderate in monounsaturated fat (18%) and rich in polyun-saturated fat (73%). Alpha-linolenic acid (ALA), an 18-carbon polyunsaturated ω-3 fatty acid, constitutes about 57 percent of the total fatty acids in flaxseed, making it the leading source of plant-based ω-3s. The other polyunsaturated fatty acid in flax is linoleic acid, an essential ω-6 fatty acid that constitutes about 16 percent of the total fatty acids (Daun and DeClercq, 1994).

ALA is the parent fatty acid of the ω-3 family; it is essential in the human diet because our bodies cannot manufacture it. The conversion of ALA to the long-chain ω-3 fatty acids, eicosapentaenoic acid (EPA) and docosahexaenoic acid (DHA), and the

Metabolic pathways of the omega-3 and omega-6 fatty acids[*]

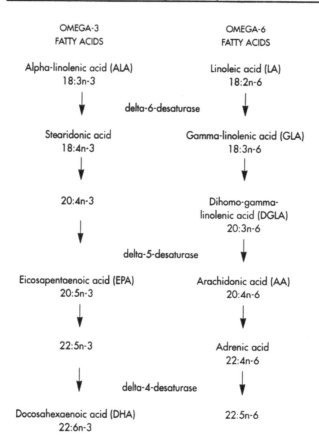

[*]Only the names of major fatty acids are shown in this figure.

Figure 1.2 Alpha-linolenic acid is the parent fatty acid of the ω-3 family. Linoleic acid is the parent fatty acid of the ω-6 family. Both are essential fatty acids, required in the diet because our bodies cannot manufacture them. Alpha-linolenic acid and linoleic acid are desaturated and elongated to longer-chain fatty acids. The main metabolites of α-linolenic acid are eicosapentaenoic acid and docosahexaenoic acid; the main metabolite of linoleic acid is arachidonic acid. Because both families compete for the same desaturase enzymes, the ω-3 fatty acids can interfere with the activities of the ω-6 fatty acids and vice versa. (Used with permission of the Flax Council of Canada).

metabolism of the ω-6 fatty acids are shown in Figure 1.2. ALA and the other ω-3 fatty acids compete with the ω-6 fatty acids for the same desaturase enzymes involved in *their* metabolism. Consequently, the ω-3 fatty acids can interfere with the activities of the ω-6 fatty acids, and vice versa.

An important biological action of ALA is that, like EPA and DHA, it is incorporated into cell membrane phospholipids. Supplementing the diet with traditional flaxseed,

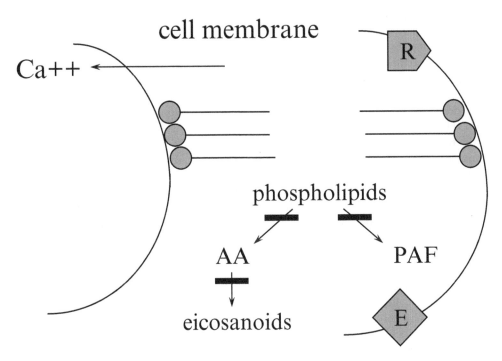

Figure 1.3 Flaxseed consumption increases the ω-3 fatty acid content of cell membrane
phospholipids two- to eight-fold, thus influencing membrane fluidity and function.
ω-3 fatty acids, including alpha-linolenic acid, inhibit the conversion of linoleic
acid to arachidonic acid, decrease the amount of arachidonic acid released from
cell membrane phospholipids, and block the formation of eicosanoids derived from
arachidonic acid that promote inflammation and platelet aggregation. Lignans, also
present in flaxseed, inhibit inflammatory responses induced by platelet-activating
factor, a lipid mediator derived from arachidonic acid. Reactions blocked by ω-3
fatty acids or lignans are shown with a solid bar (━). Abbreviations: AA, arachidonic acid; Ca++, calcium ion; E, membrane-bound enzyme; R, membrane-bound
receptor; PAF, platelet-activating factor.

for example, increases the ALA content of blood phospholipids, triglycerides and/or
cholesteryl esters two- to eight-fold (Cunnane *et al.*, 1993; Mantzioris *et al.*, 1994).

Increased ALA, EPA and DHA concentrations in cell membranes affect the activities of membrane-bound phospholipids, including their regulation of membrane-bound
enzymes, receptors and ion channels (Nair *et al.*, 1997). (A highly stylized diagram of
a cell membrane is shown in Figure 1.3.) The ω-3 fatty acids also affect membrane
fluidity and elasticity. For example, consumption of a high ALA diet, in which the
ALA was derived from a flaxseed oil-based margarine, enhanced the elasticity of arterial membranes in 15 overweight adults (Nestel *et al.*, 1997).

Moreover, because they interfere with the metabolism and biological actions of the
ω-6 fatty acids, ALA and the other ω-3 fatty acids block the production of arachidonic
acid and the formation of potent eicosanoids and other lipid mediators derived from
it. Eicosanoids are 20-carbon fatty acids derived from specific ω-3 and ω-6 fatty
acids; they include the prostaglandins, leukotrienes and thromboxanes. Eicosanoids are
lipid mediators of inflammatory reactions (Heller *et al.*, 1998). Several eicosanoids

derived from arachidonic acid, particularly thromboxane A_2, promote inflammation and platelet aggregation – two pathological processes involved in the development of atherosclerosis (Ross, 1999) and general inflammatory reactions associated with rheumatoid arthritis (Ariza-Ariza *et al.*, 1998), systemic lupus erythematosus (SLE) (Das, 1994) and lupus nephritis (Clark *et al.*, 1995). By comparison, the eicosanoids derived from the ω-3 fatty acid EPA are not as biologically active as the arachidonic acid-derived eicosanoids and tend not to promote inflammation and platelet aggregation (Heller *et al.*, 1998).

Platelet-activating factor (PAF), also shown in Figure 1.3, is a biologically active compound implicated in the pathology of chronic inflammatory diseases such as SLE and rheumatoid arthritis. PAF is derived from arachidonic acid. Feeding a 15 percent flaxseed diet to mice resulted in a significant inhibition of PAF-induced platelet aggregation (Hall *et al.*, 1993). In a study of nine patients diagnosed with SLE, consumption of 15 g or 30 g flaxseed daily for four weeks also inhibited PAF-induced platelet aggregation (Clark *et al.*, 1995).

These biological actions of ALA and the other ω-3 fatty acids have important implications for health and the prevention of chronic diseases that involve inflammatory processes, such as cardiovascular diseases, including myocardial infarction and stroke; cancer; lupus nephritis; and rheumatoid arthritis. For example, the ω-3 fatty acids may help prevent arrhythmia by affecting cell signaling or calcium channels and by blocking eicosanoid formation from arachidonic acid in cardiomyocytes (Nair *et al.*, 1997). Indeed, populations with high intakes of ALA from all sources have a lower risk of certain types of cancer (de Lorgeril *et al.*, 1998), myocardial infarction and stroke (de Lorgeril *et al.*, 1994; Simon *et al.*, 1995; Ascherio *et al.*, 1996; Hu *et al.*, 1999).

Given the emerging importance of ω-3 fatty acids in helping prevent chronic diseases, some experts have called for an increased dietary intake of ω-3s to offset the current high intakes of ω-6 fatty acid in Western diets. Whereas the human diet during the Paleolithic era (20,000–8000 B.C.E.) provided roughly equal amounts of ω-6 and ω-3 fatty acids, giving an ω-6/ω-3 ratio of about 1:1, the current ω-6/ω-3 ratio in the North American diet ranges from about 10:1 to 25:1 (Simopoulos, 1991). Although the United States presently has no specific recommended ω-6/ω-3 ratio, Health Canada recommends an ω-6/ω-3 ratio of 4:1 to 10:1, particularly for pregnant and lactating women and infants (Anonymous, 1990). The Food and Agriculture Organization/World Health Organization (FAO/WHO) (FAO, 1997) recommends a ratio of between 5:1 to 10:1 and advises individuals consuming diets with a higher ratio to consume more foods containing ω-3 fatty acids, such as green leafy vegetables, legumes and fish and other seafood (Anonymous, 1995). Because flaxseed is a rich source of ALA, it makes a significant contribution to the recommended Adequate Intake of ALA established by the Institute of Medicine (2002).

Lignans

Lignans are phytoestrogens, compounds that occur naturally in plants and exhibit weak estrogenic activity. For example, lignans can bind to estrogen receptors on cell membranes (Zava *et al.*, 1997). Other common phytoestrogens include the isoflavonoids, diadzein and genistein, found in soybeans, and the coumestans found in red clover and alfalfa.

Specifically, lignans are diphenolic compounds found in higher plants and the biological fluids of animals and humans. Flaxseed is rich in the plant lignan precursor, secoisolariciresinol diglucoside (SDG), which is acted on by bacteria in the colon to produce the mammalian lignans enterolactone and enterodiol. Mammalian lignans undergo enterohepatic circulation and are excreted in the urine, mainly as glucuronide conjugates. The greater the intake of plant lignans from foods such as flaxseed, the greater the urinary excretion of mammalian lignans (Thompson, 1995).

Lignans inhibit cell proliferation and growth, making them potential anticancer agents. They appear to protect against certain cancers, particularly hormone-sensitive cancers such as those of the breast, prostate and endometrium. Lignans stimulate the hepatic synthesis of sex hormone-binding globulin, thus increasing the clearance of circulating estrogen, and they bind to estrogen receptors, thus interfering with estrogen-mediated tumorigenic processes (Adlercreutz *et al.*, 1986). Studies in laboratory animals confirm the anticancer properties of flaxseed lignans. Rats fed defatted flaxseed meal for four weeks, for instance, had decreased epithelial cell proliferation and nuclear aberrations in mammary gland cells (early markers of carcinogenesis), compared with rats fed the basal diet alone (Serraino and Thompson, 1991). A clinical trial of flaxseed's effect on tumor development in women with breast cancer is underway at the University of Toronto (L.U. Thompson, personal communication).

The functional food trend, driven by health-conscious baby boomers, was the leading food trend in 1998 (Hasler, 1998). The trend is destined to continue, as baby boomers seek foods that help fight disease, enhance longevity and promote quality of life (Hollingsworth, 1999). Consumers and food manufacturers alike are increasingly identifying flaxseed, which is rich in ω-3 fatty acids, antioxidants (lignans) and fiber, as a functional food.

Flax fiber for fabric

The utility of flax for fabric lies in the structure of the plant's stalk. Flax is a dicotyledonous plant. Flax fibers are part of the inner bark or the walls of phloem cells within the stem. The fibers are hollow tubes that are primarily cellulose. They grow as bundles held together by complex carbohydrates such as pectins, gums and waxes. These fiber bundles support the plant and, because the fibers are hollow, likely transport photosynthesized material from the leaves by capillary action (Atton, 1989; Judd, 1995).

There are two primary advantages to flax fiber for linen cloth, both of which were well-recognized by early civilizations. One is the strength of the yarns spun from long line flax fibers, which are more than twice as strong as cotton. The other flax advantage is that the hollow flax fibers are very absorbent due to the property of *wicking*, the movement of moisture along the surface (Atton, 1989). This makes linen fabrics comfortable to wear in hot climates because sweat evaporates quickly, producing a cooling effect. Moreover, wicking makes cloth made from flax even stronger when wet, giving a desirable durability in products like sails, tents, rugs and heavy bagging. A disadvantage of wicking is that flax fibers resist dyeing (Wilson, 1979).

A further advantage of the hollow flax fibers is that they break sound effectively. This feature led to the contemporary use of linen in the interior construction of Roy Thomson Hall, a major concert hall in Toronto, Ontario, inaugurated in 1982 (Atton, 1989).

Ten centuries later, the Belgian master artist Jan Van Eyck capitalized on the susceptibility of linseed oil to oxidative polymerization in preserving the quality of his magnificent oil paintings. Like others before him from early Grecian times, Van Eyck used linseed oil as a vehicle for his pigments. However, he was the first to apply a final coat of linseed oil that, on drying, oxidized and polymerized to provide a permanent translucent glaze, preserving the original beauty of his artwork (Judd, 1995). The application of Van Eyck's invention spread rapidly throughout the European art community of the Renaissance, yielding masterpieces that are today's treasures. In many of these works, the canvases were linen, giving flax a dual role in art history (Eastlake, 1960; Mayer, 1975).

Linseed oil also played a role in the Reformation. Boiled linseed oil served as the drying agent for the lamp black used as the pigment by Johannes Gutenberg in his movable metal type printing press (Eastlake, 1960). By preserving the printed word, linseed oil contributed to literacy and communication throughout the world (Judd, 1995).

Other uses of flax

The process for making the easy-to-clean, resilient floor covering, linoleum, was invented by Frederick Walton in Britain in the latter part of the nineteenth century. The process he patented in 1860 was a technique for oxidizing and polymerizing linseed oil, which was mixed with rosin and ground cork, pressed onto a backing of burlap or canvas, rolled to an even thickness, baked in large ovens and cut into sheets (Panati, 1987; Heinrich, 1992). This inexpensive and durable material was widely adopted for home and institutional use, but it was later supplanted by synthetic floorings. Current enthusiasm for natural products has brought renewed interest in linoleum.

Oilcloth table coverings were standard items in North American kitchens in the first half of the twentieth century and are still used today. These "just wipe to clean" tablecloths consisted of linen canvas coated with several layers of linseed oil-based paint (Heinrich, 1992) in decorative designs and colors. Over time, these cloths became brittle, making cracking a common sign of age and excess wear.

Present-day research is focusing on the application of linseed oil as a weathering agent on concrete roads and structures (Xie *et al.*, 1999). Dairy farmers in Ontario have reported that coating new cement floors in their barns with linseed oil alone or in combination with diesel oil not only preserves them but also helps prevent hoof damage, which is apparently related to the caustic alkalinity of new concrete (Cumming, 1999).

Flax provides a good example of the efficient use of crops by today's agribusiness. The flax straw left in the field after threshing, for example, is collected and processed into a fiber used for cigarette and other high-quality papers. Industrial linseed oil is used in manufacturing paint, printing inks, soap, patent leather, core oils and brake linings (Carter, 1993).

Summary

Flaxseed has been used by humans for thousands of years. It is a founding crop, being among the first plants domesticated for human use. In ancient times, flaxseed was valued as a food and medicine, for the versatility of its oil and for its long fibrous stalks which were made into fine linen. At the end of the twentieth century, flaxseed continues to be valued for its contribution to linen production and industrial processes. And with the dawn of the new millennium, flaxseed has emerged as a functional food, valued for its soluble and insoluble dietary fiber, alpha-linolenic acid content and lignans – all constituents that promote health and may help prevent disease, particularly the chronic diseases of Western countries.

Acknowledgments

The authors appreciate the generous aid of Don Frith and Barry Hall at the Flax Council of Canada; Sarah Caplan at Robin Hood Multifoods, Inc.; Drs. Robert Hill (Plant Science), Sami Rizkalla (Civil and Geological Engineering), Jennifer Shay (Botany) and Thomas Shay (Anthropology) at the University of Manitoba; and Dr. Lilian U. Thompson at the University of Toronto in providing information and identifying helpful reference material.

Notes

1 As cited in Haggerty (1999).
2 Traditional flaxseed varieties rich in alpha-linolenic acid (ALA) have a brown seed coat. Solin varieties, which contain less than 3 percent ALA, have a yellow seed coat. The exception is the Omega variety, which is high in ALA and has a yellow seed coat (Barry Hall, personal communication, 1999).
3 B.C.E. = Before the Common Era; C.E. = Common Era. The Common Era is understood to begin in about 1 A.D.
4 A Joint FAO/WHO Expert Consultation report has proposed phasing out the terms soluble and insoluble fiber (Anonymus, 1995). We elected to continue using these terms because they are widely recognized by consumers and health professionals.

References

Adlercreutz, H., Fotsis, T., Bannwart, C., Wähälä, K., Mäkelä, T., Brunow, G., and Hase, T. (1986). Determination of urinary lignans and phytoestrogens metabolites, potential antiestrogens and anticarcinogens, in urine of women on various habitual diets. *J. Steroid Biochem.* 25, 791–797.
Anonymous (1895). Instructions to Berkeley, 1642. *Virginia Hist. Mag.* 2, 281–288.
Anonymous (1896/97). Letter and proclamation of Argall. *Virginia Hist. Mag.* 4, 28–29.
Anonymous (1903). Abridgment of Virginia laws, 1694: Flax. *Virginia Hist. Mag.* 10, 51–52.
Anonymous (1990). *Nutrition Recommendations. The Report of the Scientific Review Committee.* (Department of Supply and Services, Cat. No. H49-42/1990E), Health and Welfare Canada, Ottawa, Canada.
Anonymous (1995). WHO and FAO Joint Consultation. Fats and oils in human nutrition. *Nutr. Rev.* 53, 202–205.
Anonymous (1998a). Anticancer diet. *Petfood Industry*, Nov/Dec, p. 71.
Anonymous (1998b). Policy paper: Nutraceuticals/functional foods and health claims on foods. Health Canada, www.hc-sc.gc.ca
Anonymous (1999a). *Flax Family Favourites: Recipes and Healthful Tips.* Flax Council of Canada and Saskatchewan Flax Development Commission. Winnipeg, Canada.
Anonymous (1999b). *Flax history.* Flax Council of Canada, www.flaxcouncil.ca.
Ariza-Ariza, R., Mestanza-Peralta, M., and Cardiel, M.H. (1998). Omega-3 fatty acids in rheumatoid arthritis: An overview. *Semin. Arthritis Rheum.* 27, 366–370.
Armstrong, J.G., and McFarlane, W.D. (1994). Flavour reversion in linseed shortening. *Oil and Soap* 21, 322–327.

Ascherio, A., Rimm, E.B., Giovannucci, E.L., Spiegelman, D., Stampfer, M., and Willett, W.C. (1996). Dietary fat and risk of coronary heart disease in men: Cohort follow up study in the United States. *Brit. Med. J.* 313, 84–90.

Atton, M. (1989). *Flax Culture from Flower to Fabric*. The Ginger Press, Owen Sound, ON.

Baines, P. (1985). Weaving linen cloth. In *Flax and Linen*, Shire Publications, North Bridge, Dyfed, UK, pp. 17–20.

Barber, E.W. (1994). Land of Linen. In E. Wayland (ed.), *Women's Work: The First 20,000 Years-Women, Cloth and Society in Early Times*, W.W. Norton and Co., New York, pp. 185–206.

BeMiller, J.N. (1973). Quince seed, psyllium seed, flaxseed and okra gums. In Whistler, R.L., and BeMiller, J.N. (eds), *Industrial Gums*, Academic Press, New York, pp. 339–367.

Buley, R.C. (1933/34). Pioneer health and medical practices in the old northwest prior to 1840. *Mississippi Valley Hist. Rev.* 20, 497–520.

Carter, J.F. (1993). Potential of flaxseed and flaxseed oil in baked goods and other products in human nutrition. *Cereal Foods World* 38, 753–759.

Caston, L., and Leeson, S. (1990). Research note: Dietary flaxseed and egg composition. *Poultry Sci.* 69, 1617–1620.

Chanmugam, P., Boudreau, M., Boutte, T., Park, R.S., Hebert, J., Berrio, L., and Hwang, D.H. (1992). Incorporation of different types of n-3 fatty acids into tissue lipids of poultry. *Poultry Sci.* 71, 516–521.

Chen, Z.-Y., Ratnayake, W.M.N., and Cunnane, S.C. (1994). Oxidative stability of flaxseed lipids during baking. *J. Am. Oil Chem. Soc.* 71, 629–632.

Clark, W.F., Parbtani, A., Huff, M.W., Spanner, E., De Salis, H., Chin-Yee, I., Philbrick, D.J., and Holub, B.J. (1995). Flaxseed: A potential treatment for lupus nephritis. *Kidney Int.* 48, 475–480.

Clarke, L.J. (1986). The weaver's raw materials. In *The Craftsman in Textiles*, G. Bell and Sons, London, p. 87.

Coles, B., and Coles, J. (1989). Pioneers of the inland waters. In *People of the Wetlands*, Thames and Hudson, London, pp. 86–116.

Crampton, E.W., Farmer, F.A., and Berryhill, F.M. (1951). The effect of heat treatment on the nutritional value of some vegetable oils. *J. Nutr.* 43, 431–440.

Cumming, I. (1999). *Linseed oil and diesel combination helps on new floors*. Manitoba Cooperator, Winnipeg. Feb. 4, p. 30.

Cunnane, S.C., Ganguli, S., Menard, C., Liede, A.C., Hamadeh, M.J., Chen, Z.-Y., Wolever, T.M.S., and Jenkins, D.J.A. (1993). High α-linolenic acid flaxseed (*Linum usitatissimum*): Some nutritional properties in humans. *Br. J. Nutr.* 49, 443–453.

Cunnane, S.C., Hamadeh, M.J., Liede, A.C., Thompson, L.U., Wolever, T.M.S., and Jenkins, D.J.A. (1994). Nutritional attributes of traditional flaxseed in healthy young adults. *Am. J. Clin. Nutr.* 61, 62–68.

Das, U.N. (1994). Beneficial effect of eicosapentaenoic and docosahexaenoic acids in the management of systemic lupus erythematosus and its relationship to the cytokine network. *Prost. Leuk. Essent. Fatty Acids* 51, 207–213.

Daun, J.K., and DeClercq, D.R. (1994). Sixty years of Canadian flaxseed quality surveys at the Grain Research Laboratory. *Proc. of the Flax Institute of the United States*, Flax Institute of the United States, Fargo, ND, pp. 192–200.

de Lorgeril, M., Renaud, S., Mamelle, N., Salen, P., Martin, J.-L., Monjaud, I., Guidollet, J., Touboul, P., and Delaye, J. (1994). Mediterranean alpha-linolenic acid-rich diet in secondary prevention of coronary heart disease. *The Lancet* 343, 1454–1459.

de Lorgeril, M., Salen, P., Martin, J.-L., Monjaud, I., Boucher, P., and Mamelle, N. (1998). Mediterranean dietary pattern in a randomized trial: Prolonged survival and possible reduced cancer rate. *Arch. Intern. Med.* 158, 1181–1187.

Duder, C.J.D. (1992). Beadoc-the British East Africa disabled officers' colony and the white frontier in Kenya. *Agricultural Hist. Rev.* 40, 142–150.

Eastlake, C.L. (1960). *Methods and Materials of Painting of the Great Schools and Masters*, Vol. I and II, Dover, New York, pp. 182–268. (cited by Judd, 1995).

FAO (1997). FAO/WHO Expert Consultation on Carbohydrates in Human Nutrition, Food and Agriculture Organization. www.fao.org/WAICENT/FAOINFO/ECONOMIC/ESN/ carboweb/carbo.htm

Ferrier, L.K., Caston, L.J., Leeson, S., Squires, J., Weaver, B.J., and Holub, B.J. (1995). α-Linolenic acid- and docosahexaenoic acid-enriched eggs from hens fed flaxseed: Influence on blood lipids and platelet phospholipid fatty acids in humans. *Am. J. Clin. Nutr.* 62, 81–86.

Frost, A. (1985). Botany Bay: An imperial venture of the 1780's. *English Historical Rev.* 100, 309–330.

Geijer, M. (1979). Materials. In *A History of Textile Art*, Sotheby Parke Bernet, London, pp. 6–8.

Haggerty, W.J. (1999). Flax-Ancient herb and modern medicine. *HerbalGram* 45, 51–57.

Hall, A.V., Parbtani, A., Clark, W.F., Spanner, E., Keeney, M., Chin-Yee, I., Philbrick, D.J., and Holub, B.J. (1993). Abrogation of MRL/lpr lupus nephritis by dietary flaxseed. *Am. J. Kidney Dis.* 22, 326–332.

Hall, R. (1986). Woven fabrics and dyeing. In *Egyptian Textiles*, Shire Publications, Aylesbury, UK, pp. 9–12.

Harlan, J.R. (1975). The Near Eastern Center. In *Crops and Man*, American Soc. Agron. & Crop Science Soc. America, Madison, WI, pp. 171–189.

Harris, J. (1993). The Ancient World. In Harris, J. (ed.) *Textiles, 5,000 Years: An International History and Illustrated Survey*, Harry N. Abrams, New York, pp. 54–74.

Hasler, C.M. (1996). Functional foods: The Western perspective. *Nutr. Rev.* 54, S6–S10.

Hasler, C.M. (1998). Functional foods: Their role in disease prevention and health promotion. *Food Tech.* 52, 63–70.

Heinrich, L. (1992). Linen: Cloth of the Ancient Egyptians. In *The Magic of Linen: Flaxseed to Woven Cloth*, Orca Book, Victoria, B.C., pp. 173–190.

Helbaek, H. (1969). Plant collecting, dry-farming and irrigation agriculture in prehistoric Deh Luran. In Hole, F., Flannery, K.V., and Neely, J.F. (eds), *Prehistory and Human Ecology of the Deh Lurah Plain, Memoirs of the Museum of Anthropology*, University of Michigan, Ann Arbor, MI, pp. 386–426. (Cited by Zohary and Hopf, 1993).

Heller, A., Koch, T., Schmeck, J., and van Ackern, K. (1998). Lipid mediators in inflammatory disorders. *Drugs* 55, 487–496.

Hollingsworth, P. (1999). Food priorities for an aging America. *Food Tech.* 53, 38–40.

Hopf, M. (1961). Pflanzenfunde aus Lerna/Argolis. *Zuchter* 31, 239–247. (Cited by Zohary and Hopf, 1993).

Hu, F.B., Stampfer, M.J., Manson, J.E., Rimm, E.B., Wolk, A., Colditz, G.A., Hennekens, C.H., and Willett, W.C. (1999). Dietary intake of α-linolenic acid and risk of fatal ischemic heart disease among women. *Am. J. Clin. Nutr.* 69, 890–897.

Hunt, K.E. (1969). Raw materials. In van Stuyvenberg, J.H. (ed.) *Margarine-An Economic, Social and Scientific History 1969–1969*, University of Toronto Press, Toronto, pp. 37–82.

Institute of Medicine (2002). *Dietary Reference Intakes for Energy, Carbohydrate, Fiber, Fat, Fatty Acids, Cholesterol, Protein, and Amino Acids*, National Academies Press, Washington, DC, pp. 8-1 to 8-97.

Jenkins, D.J., Kendall, C.W., Vidgen, E., Agarwal, S., Rao, A.V., Rosenberg, R.S., Diamandis, E.P., Novokmet, R., Mehling, C.C., Perera, T., Griffin, L.C., and Cunnane, S.C. (1999). Health aspects of partially defatted flaxseed, including effects on serum lipids, oxidative measures, and ex vivo androgen and progestin activity: a controlled crossover trial. *Am. J. Clin. Nutr.* 69, 395–402.

Judd, A. (1995). Flax-Some historical considerations. In Cunnane, S.C., and Thompson, L.U. (eds), *Flaxseed in Human Nutrition*, AOCS Press, Champaign, IL, pp. 1–10.

Kolodziejczyk, P.P., and Fedec, P. (1995). Processing flaxseed for human consumption. In Cunnance, S.C., and Thompson, L.U. (eds), *Flaxseed in Human Nutrition*, AOCS Press, Champaign, IL, pp. 261–280.

Lemon, H.W. (1947). Flavor reversion in hydrogenated linseed oil. III. The relations of isolinolenic acid to flavor deterioration. *Can. J. Res.* **25**, 34–43.

Lips, H.J., and Crampton, E.W. (1952). Linseed oil problems in food use. *Can. Chem. Proc.* **36**, 66–67.

Mantzioris, E., James, M.J., Gibson, R.A., and Cleland, L.G. (1994). Dietary substitution with an a-linolenic acid-rich vegetable oil increases eicosapentaenoic acid concentrations in tissues. *Am. J. Clin. Nutr.* **59**, 1304–1309.

Mayer, R. (1981). Oil painting. In *The Painter's Craft – An Introduction to Artists' Methods and Materials*, Viking Press, New York, pp. 84–114.

Maynard, L.A., McKay, C.M., Loosli, J.K., Lingenfetter, J.F., Barrentine, B., and Sperling, G. (1942). *Physiology of lactation*. Cornell Agr. Expt. Sta., Ithaca (Cited in *Chem. Abstr.*, 1945, **39**, 4996).

Molotkow, W.G. (1932). Experimental morphology after single and repeated doses of linseed oil. *Nutr. Abstracts Rev.* **2**, 797.

Nair, S.S.D., Leitch, J.W., Falconer, J., and Garg, M.L. (1997). Prevention of cardiac arrhythmia by dietary (n-3) polyunsaturated fatty acids and their mechanism of action. *J. Nutr.* **127**, 383–393.

Nestel, P.J., Pomeroy, S.E., Sasahara, T., Yamashita, T., Liang, Y.L., Dart, A.M., Jennings, G.L., Abbey, M., and Cameron, J.D. (1997). Arterial compliance in obese subjects is improved with dietary plant n-3 fatty acid from flaxseed oil despite increased LDL oxidizability. *Arterioscler. Thromb. Vasc. Biol.* **17**, 1163–1170.

Oomah, B.D., and Mazza, G. (1998). Flaxseed products for disease prevention. In Mazza, G. (ed.) *Functional Foods: Biochemical and Processing Aspects*, Technomic Publishing Company, Lancaster, PA, pp. 91–138.

Pan, Q. (1990). Flax production, utilization and research in China. *Proc. of the Flax Institute of the United States*, Flax Institute of the United States, Fargo, ND, pp. 59–63.

Panati, C. (1987). In and around the house. In *Extraordinary Origins of Everyday Things*, Harper and Row, New York, pp. 131–167.

Prentice, B.E. (1990). Whither the end-product markets for flax? The emerging picture. *Proc of the Flax Institute of the United States*, Flax Institute of the United States, Fargo, ND, 45–54.

Ross, R. (1999). Atherosclerosis-An inflammatory disease. *New England J. Med.* **340**, 115–126.

Schick, T. (1988). Nahal Hemar cave: Cordage, basketry and fabrics. *Atiqot* **38**, 31–43.

Serraino, M., and Thompson, L.U. (1991). The effect of flaxseed supplementation on early risk markers for mammary carcinogenesis. *Cancer Lett.* **60**, 135–142.

Siegenthaler, I.E. (1994). The use of flaxseed in Ethiopia. *Proc. of the Flax Institute of the United States*, Flax Institute of the United States. Fargo, ND, pp. 143–149.

Simon, J.A., Fong, J., Bernert Jr., J.T., and Browner, W.S. (1995). Serum fatty acids and the risk of stroke. *Stroke* **26**, 778–782.

Simopoulos, A.P. (1991). Omega-3 fatty acids in health and disease and in growth and development. *Am. J. Clin. Nutr.* **54**, 438–463.

Smith, B.D. (1995). *The Emergence of Agriculture*. Scientific American Library, New York, pp. 11–13.

Stitt, P.A. (1994). History of flax: 9000 years ago to 1986. *55th Flax Institute of the United States*, Flax Institute of the United States, Fargo, ND, pp. 152–153.

Stout, E.E. (1970). Linen. In *Introduction to Textiles*, John Wiley & Sons, Toronto, pp. 75–77.

Thomas, P.R., and Earl, R. (1994). Enhancing the food supply. In *Opportunities in the Nutrition and Food Sciences: Research Challenges and the next Generation of Investigators*, National Academy Press, Washington, DC, pp. 98–142.

Thompson, L.U. (1995). Flaxseed, lignans and cancer. In Cunnane, S.C., and Thompson, L.U. (eds), *Flaxseed in Human Nutrition*, AOCS Press, Champaign, IL, pp. 219–236.

Trinder, B. (1992). Ditherington flax mill, Shrewsbury-A re-evaluation. *Textile History* **23**, 189–223.

Turner, W.H.K. (1982). The development of flax-spinning mills in Scotland 1787–1840. *Scottish Geographical Mag* **98**, p. 4–15.

Vaisey-Genser, M., and Morris, D.H. (1997). *Flaxseed: Health, Nutrition and Functionality*. Flax Council of Canada, Winnipeg, MB.

van Zeist, W. (1970). The Oriental Institute excavations at Mureybit, Syria: Preliminary report on the 1965 campaign. Part III. Palaeobotany. *J. Near East Studies* **29**, 67–176. (Cited by Zohary and Hopf, 1993).

van Zeist, W. (1972). Palaeobotanical results in the 1970 seasons at Cayonu, Turkey. *Helinium* **12**, 3–19. (Cited by Zohary and Hopf, 1993).

van Zeist, W., and Bakker-Heeres, J.A.H. (1975). Evidence for linseed cultivation before 6000 BC. *J. Archaeological Sci.* **2**, 215–219.

Vavilov, N.I., and Dorofeyev, V.F. (1992). *Origin and Geography of Cultivated Plants*. Löne, D. (Trans). Cambridge University Press. Cambridge, pp. 67 and 313.

Wilford, J.N. (1993). *Site in Turkey yields oldest cloth ever found*. New York Times, New York. July 13, pp. C1, C8.

Wilson, K. (1979). Spinning and raw materials. In *A History of Textiles*, Westview Press, Boulder, CO, pp. 11–16.

Xie, Y., Rizkalla, S.H., Chan, H.C., and Kwan, A.K.H. (1999). Linseed oil-based preservatives for concrete structures and highways in Hong Kong. Unpublished data, Department of Civil and Geological Engineering, University of Manitoba, Winnipeg, Canada

Zava, D.T., Blen, M., and Duwe, G. (1997). Estrogenic activity of natural and synthetic estrogens in human breast cancer cells in culture. *Environ. Health Perspect.* **105 (Suppl 3)**, S637–S645.

Zohary, D., and Hopf, M. (1993). Oil and fiber crops. In *Domestication of Plants in the Old World-The Origin and Spread of Cultivated Plants in West Asia, Europe, and the Nile Valley*, Clarendon Press, Oxford, pp. 118–126.

Zohary, M. (1972). Plate 374. In *Flora Palaestina, Part 2. Plantanaceae to Umbelliferae*, Israel Academy of Sciences and Humanities, Jerusalem.

2 Cultivated flax and the genus *Linum* L.

Taxonomy and germplasm conservation

Axel Diederichsen and Ken Richards

Introduction

Genetic resources of a given crop species cover the entire range of genetic diversity available, even if only particular genotypes are relevant and significant in industrialized agriculture. For breeding purposes, the entire range of different genotypes of the species has to be considered as a resource for crop improvement. Describing and analyzing the structure of a genepool is primarily the concern of systematics and taxonomy. Thus, systematics and taxonomy, if applied and understood properly, are effective tools to assist activities in germplasm conservation and utilization. Modern molecular techniques, which allow for the description of the genetic structure at the molecular level, are a powerful complement to traditional phenotypic approaches for plant classification.

The most important criterion of the taxonomic species concept is the ability to cross hybridize and produce fertile offspring. This criterion has also been applied by Harlan and de Wet (1971) for defining the primary, secondary and tertiary genepools for a given crop species. Therefore, taxonomy, plant breeding, and evolutionary studies have a common interest in the study of plant species diversity, in spite of differing scientific and practical considerations. Some scientists have clearly stated the usefulness and economic relevance of taxonomy in considering plant genetic resources for food and agriculture (Hanelt, 1988; Small, 1993). However, taxonomy is sometimes also looked upon as being a classical and obsolete branch of biological science, based on inflexible rules of nomenclature. Another criticism is that taxonomic classifications are usually based on phenotypic observations, which are influenced by the environment. The taxonomic rules for nomenclature were initially developed for wild plants. Difficulties in applying these rules to cultivated plants have resulted in suggestions to reject any formal taxonomic treatment of cultivated plants (Hetterscheid and Brandenburg, 1995). However, the first published scientific observations on genetic diversity within a species were made on domesticated plants and animals (Darwin, 1868; Candolle, 1883). The striking phenotypic diversity within a given species of domesticated plants or animals initiated the research into biodiversity in the nineteenth century. Taxonomic classifications of particular groups within a given species, i.e. at the infraspecific (= intraspecific) level, were identified and named.

In order to understand and benefit from the use of classical research on the diversity of flax and its wild relatives, it is essential to be familiar with the history and terminology of taxonomy. As gene erosion at the species and at the infraspecific levels continues, it is also important to consider historical studies on genotypes that are no longer cultivated. Flax provides a good example for this, as several genotypes studied and described by N.I. Vavilov and his co-workers in the first half of the twentieth century no longer exist (Elladi, 1940). Knowledge of the systematic and taxonomic issues at the species, infraspecific, and above species levels facilitates an understanding of the classical literature and provides for effective communication among present-day genebank curators, plant scientists and breeders.

To consider phenotypic studies in genetic diversity as obsolete due to modern molecular methods is not justified. The final use of a plant is always aimed at the entire plant or parts of it, which is the phenotype. Plant breeding works on the improvement of phenotypic traits. Molecular characterizations will add valuable information to germplasm description and provide new insights into biodiversity (Fu *et al.*, 2002a).

This chapter attempts to provide a better understanding of classical and modern characterization of flax germplasm. It includes a description of the genus *Linum* L. and pays particular attention to the infraspecific groups of cultivated flax (*Linum usitatissimum* L.). A review of the world's flax germplasm collections is also provided.

Description of cultivated flax (*Linum usitatissimum* L.)

General description of the plant

Cultivated flax (Figures 2.1 and 2.2) is a summer annual crop in temperate climates. The time from seeding to harvest varies between 90 and 150 days. In subtropical areas, flax is also cultivated under short-day environments. In the past, winter-annual flax was grown in some Mediterranean, south-, central- and east-European countries with mild winters. Recently, winter hardy cultivars of flax have been bred in the UK. However, these cultivars and landraces are not obligatory winter crops, as they will grow normally and also produce seeds when sown in the spring.

Flax has a tap root. The leading shoot is erect and lateral branching at the stem base occurs. If the leading shoot of the young plant is injured, the plant will develop several secondary basal prostrate-ascending sprouts. The number of basal shoots also increases with high levels of soil fertility (Dillman and Brinsmade, 1938). Lateral branching is suppressed by dense planting, which is practiced in fiber flax cultivation.

Plant height varies considerably and depends largely on the genotype (Figure 2.2). Plant height ranges from 20 to 150 cm (Hegi, 1925). Large-seeded flax is much shorter than typical fiber flax. Branching in the upper parts of the stem is also determined by the genotype. Large-seeded flax typically has secondary branches from the middle of the stem, while fiber flax only branches in the upper quarter of the stem. The three-nerved leaves alternate. Smaller leaves are linear; larger leaves are linear-lanceolate. The leaves vary from 3 to 13 mm in width and 15 to 55 mm in length. During the ripening of the plant, the leaves senesce and drop off.

The flowers are arranged panicle-like and all branches of the shoot terminate in flowers. The five sepals are ovate, carinate, acuminate, and shorter than the ripe fruit. The corolla forms into a tube, funnel or bowl and has a diameter of 10–32 mm (Figure 2.3). The five petals are inversely egg-shaped and some genotypes have petals with a longitudinal fold. Genotypes with inward folded petal margins occur. Petal color varies from white, light blue, blue, dark blue, pink, violet, to red-violet (Figure 2.2). Further differentiations in petal color are possible (Tammes, 1928; Dillman, 1936). The most frequent color is blue, followed by white. The petals are veined and the veins also vary

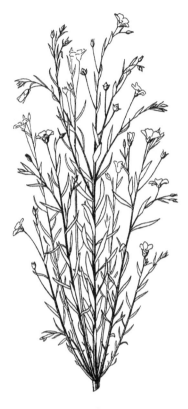

Figure 2.1 Flax, flowering plant (Howard, 1924).

Figure 2.2 Flax accessions of a genebank grown for rejuvenation and characterization. The diversity in flower color, plant height, and development stage is obvious. (Photo: A. Diederichsen, Saskatoon, Saskatchewan, Canada, 1995).

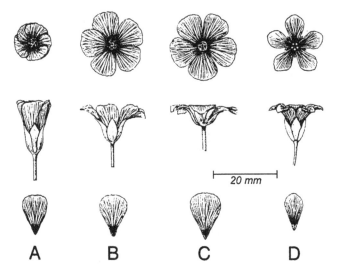

Figure 2.3 Shape of flowers in flax. A: tube; B: funnel; C: bowl. Petals overlap in A, B, and C; petals overlap less than 50% in D. (Adapted from Kulpa and Danert, 1962).

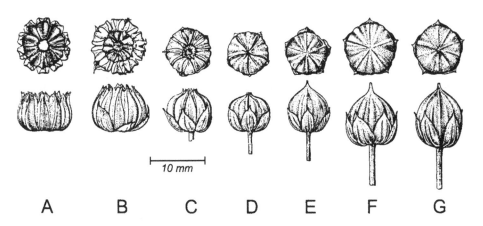

Figure 2.4 Degrees of spontaneous opening in mature flax capsules. A and B: dehiscent; C and D: slightly dehiscent; E, F and G: indehiscent. (Adapted from Kulpa and Danert, 1962).

in color. The flowers open early in the day and by noon most petals have dropped. The five stamens are joined at their basal edge. The stamens and the five anthers vary in color with the same range of colors occurring as in the petals. The anthers can also be orange or yellow in color. The colorations of the different flower parts are inherited independently from each other. Self-pollination is common in cultivated flax and occurs simultaneously with the opening of the flowers.

The five carpels produce a more or less round capsule-type fruit, which is 6–9 mm wide (Figure 2.4). Each carpel forms two septa, which divide the capsule. Therefore,

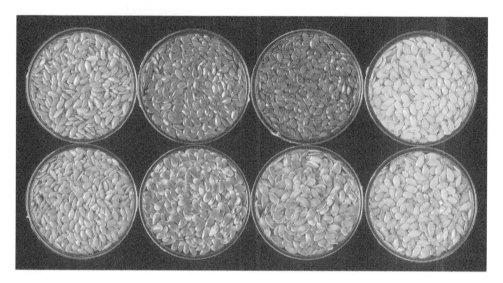

Figure 2.5 Flaxseeds of different coloration. (Photo: R. Underwood).

each capsule has ten lodicules and the maximum number of seeds possible within a capsule is also ten. Typical fiber flax does not form seeds in all lodicules, while large-seeded flax often reaches the maximum number of ten seeds. Irregular genotypes with more than five carpels can also be found. Depending on the genotype, the ripe capsules are either completely closed or open slightly along the septa. Some genotypes have capsules opening completely along the septas, which causes the seeds to shatter. Fiber flax has slightly opening capsules and in the typical oilseed flax the capsules do not open at all.

The flattened seeds are ovoid or oblong elliptic, rounded at the base, acute at the apex, and 3.3–5 mm long. Genotypic differences in seed size are also reflected by the variation in weight of 1,000 seeds, which ranges from 4–13 g. Seed color varies considerably and all variations between yellow, dark brown and olive can be found (Figure 2.5).

The stem

The stem is of special interest, as it contains fibers, which can be isolated and have several uses. This is one of the principal reasons for the initial cultivation of flax. The branching of the flax stem determines the quality of the fiber. A long, primary fiber can only be obtained from a part of the stem that has no side branches. Fiber flax is planted at high density, so each plant develops only one main shoot. The "technical stem length" is defined as the length from the stem base to the first secondary branches (Figure 2.6). Only this part of the stem produces the fiber used in the textile industry. This implies that these plants do not have many side branches and, therefore, will have low seed productivity.

Recently, interest in flax fiber for technical (i.e. nontextile) purposes has increased. This has resulted in technical stem length losing its importance as a selection criterion in breeding programs, as short fibers are also suitable for such uses. A

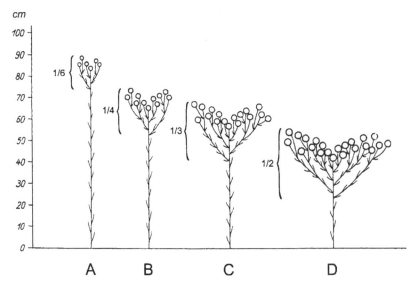

Figure 2.6 Technical stem length in flax. The part of the stem without side branches defines the technical stem length. The part of the stem with branches is indicated. A: typical fiber flax; B and C: intermediate flax; C: large-seeded flax. (Adapted from Kulpa and Danert, 1962).

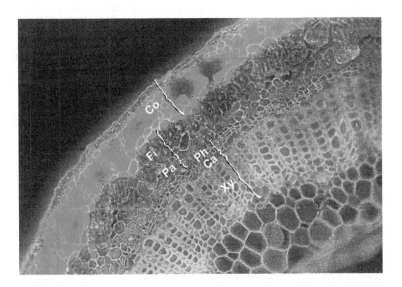

Figure 2.7 Cross-section of flax stem (cv. "Novotorzhskij"). Co: cortex; Fi: fiber fascicle; Pa: parenchyma; Ph: phloem; Ca: cambium; Xy: secondary xylem (Zhuzhenko *et al.*, All-Russian Flax Research Institute, Torzhok, Russian Federation, 1993).

cross-section of the flax stem shows the fibers are located in the exterior part of the stem, between the cortex and the phloem, i.e. in the pericambial part of the stem (Figures 2.7 and 2.8). The stem contains 20–51 fiber fascicles, and each fascicle is composed of 10–30 longitudinal cells (Hegi, 1925). The elementary cells are prolonged,

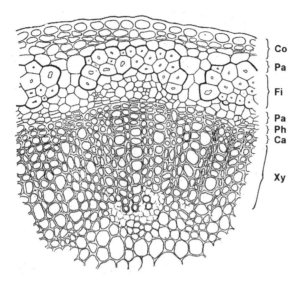

Figure 2.8 Cross-section of flax stem. Co: cortex; Pa: parenchyma; Fi: fiber fascicle;
Ph: phloem; Ca: cambium; Xy: secondary xylem (Elladi, 1940).

between 20 and 30 mm long, but can extend up to 120 mm. The longitudinal fascicles are anastomosed with each other. The fibers consist mostly of cellulose. Antagonism exists between the xylem-formation activity of the cambium and the formation of the fiber fascicles in the outer stem. Densely planted flax develops more fiber fascicles and has reduced xylem production of the cambium (Elladi, 1940). Therefore, thin stems of densely planted flax contain more fiber than thicker stems from wider plant spacing. Dense planting also suppresses the formation of secondary shoots at the stem base and branching in the apical parts of the stem, which makes the stem fibers more suited for classical fiber use.

The seed

The seed contains oil, which is used for nutritional and industrial purposes. Man has paid special attention to the characteristics of the flaxseed. Selection over thousands of years has resulted in several distinct genotypic differences regarding seed-related traits. Certain seed characteristics have significant impact on seed oil production. The anatomical components of flaxseeds are the testa, the endosperm and the embryo. Cross-sections of the seeds are given in Figures 2.9 and 2.10. The description presented here is based on the observations made by Rüdiger (1942, 1954). The outer surface of the testa is shiny and slightly wavy. A single layer of epidermal cells covers the seed. Below the epidermal cells are one to five layers of parenchyma cells. They are called ring-cells, because their shape is more or less round when viewed in a pericline (parallel to the seed surface) section. There is considerable intercellular space in this layer. The ring-cells of this parenchyma may contain dark, tannin-like substances and occasionally chlorophyll. If these substances are present, the ring-cells also contribute to the outer appearance of the seed color. The epidermis and the ring-cells originate from

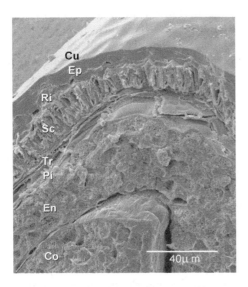

Figure 2.9 Cross-section of flaxseed. Cu: cuticula; Ep: epidermis; Ri: ring-cells; Sc: sclerenchyma; Tr: transversal-cells; Pi: pigment-cells; En: endosperm; Co: cotyledon. (Photo: E. Kokko).

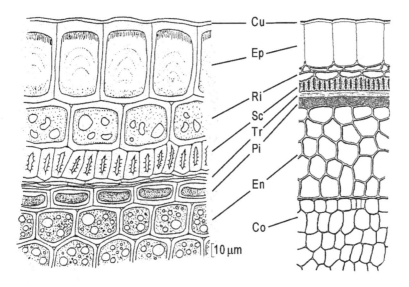

Figure 2.10 Cross-section of flaxseed. Cu: cuticula; Ep: epidermis; Ri: ring-cells; Sc: sclerenchyma; Tr: transversal-cells; Pi: pigment-cells; En: endosperm; Co: cotyledon (Gassner, 1951; Rüdiger, 1954).

the outer integument of the ovule. Below the ring-cells of the parenchyma there is a single layer of sclerenchyma fibers, which are oriented parallel to the longitudinal axis of the seed. The sclerenchyma cells form a unicellular layer, 16–25 μm thick. They are mostly colorless, but can sometimes be dark yellow. Within the wild *Linum* species, the sclerenchyma cells are thicker compared to cultivated flax. In cultivated flax there are genotypic differences in the thickness of this layer (Rüdiger, 1942). The next layer is composed of the so-called transversal-cells which have an irregular orientation and are more or less collapsed in mature seeds. These cells make up at least two cell layers. The innermost layer of the testa contains the pigment-cells which, in a pericline section, are square shaped (Figure 2.9) and influence the outer appearance of the seed color. These isodiametric or pericline-square shaped cells often contain tannic pigments of yellow-brown color. In yellow-seeded flax these pigment-cells are often absent or in some cases may be present, but do not contain any pigments. The sclerenchyma, transversal- and pigment-cells originate from the inner integument of the ovule. The next layer is the endosperm which contains oil and protein. The endosperm surrounds the embryo which has two large cotyledons. The cotyledons fill more than 2/3 of the inner seed volume, while the endosperm occupies 1/3 or less of the inner seed volume. The cotyledons are white or yellowish and also contain oil.

The epidermal cells produce mucilage, composed of polysaccharides, polypeptides and glycoproteins, which is deposited in the extracellular space just below the cell-wall facing the outside of the seed (Heinze and Amelunxen, 1984). During ripening, the dry epidermal cells shrink by folding along the anticline cell walls. Upon contact with water the cell walls stretch, the cuticula cracks, and the mucilage adsorbs water and leaves the epidermal cells (Rüdiger, 1942). One of the nutritional benefits of flaxseed consumption is attributed to this mucilage. The mucilage protects the epithelia of the digestive system. Cultivated flax produces more mucilage than the wild *Linum* species.

Flaxseed color is influenced by several anatomical seed parts. The presence or absence of the pigment-cell layer and whether these cells contain any pigments plays a significant role. All shades of brown seed color result from pigmentation of the pigment cells. Yellow, spotted or mottled seeds are the result of the irregular presence of pigments in the pigment cells. If there are no pigments present, then the color of the cotyledons becomes the main factor determining the outer seed color. Finally, the ring-cells' and sclerenchyma's coloration can also influence the seed color. Thus, a greenish to dark green seed color may result from the pigmentation of the outer parenchyma cell-layer, while the pigment layer may be colorless. Due to the numerous and complex factors influencing seed color, the genetics for this trait are also complex. Barnes *et al.* (1960) stated that at least three genes are involved in the determination of seed color and that two of these genes simultaneously influence petal color.

Brown seed color is and has always been the most common in flax. The Dutch flax breeder Tine Tammes (1927) was the first to initiate systematic studies on seed and flower color, in addition to other important traits. Her work provides the framework for understanding the genetics of these traits. The early collection missions conducted by N.I. Vavilov and his colleagues revealed that yellow seed color was selected in Central Asia, the Middle East, and China (Elladi, 1940). Rüdiger (1942) observed that the seed coat in yellow-seed flax was thinner than in the brown-seed flax. Recently, yellow-seed commercial varieties with low α-linolenic acid content were released in Canada and are marketed under the name "solin" (Kenaschuk and Rowland, 1993).

Phenotypic variation of cultivated flax

The phenotypic range of variation for several agrobotanic characters in cultivated flax is very high. Due to careful selection by man, many landraces with genetically fixed characteristics have evolved. This is because domestication has favored genotypes that

Table 2.1 Range of variation for some important characters. Observations of 73 accessions of flax, representing all 26 infraspecific groups proposed by Kulpa and Danert (1962)

Character	Mean (5 years)	Min.	Max.	CV
Weight of 1,000 seeds (g)	6.3	4.38	12.7	27.4
Width of petals (mm)	9.6	3.3	14.7	25.8
Width of capsule (mm)	6.9	5.4	8.2	8.6
Number of days from sowing to end of flowering	92.4	80.0	116.0	7.8
Height of plant (cm)	66.5	38.0	95.0	17.3

Note: Min. = Minimum; Max. = Maximum; CV = Coefficient of Variation in %.
Source: Diederichsen and Hammer, 1995.

Table 2.2 Range of variation for some qualitative characters in flax. Observations of 73 accessions of flax, representing all 26 infraspecific groups proposed by Kulpa and Danert (1962)

Character	Observed expressions
Petal color	White, light blue, blue, dark blue, pink, violet, red-violet
Petal shape	With/without longitudinal folding
Petal, folding of the margins	Margins folded inwards/margins plain
Capsule dehiscence	Completely dehiscent/completely indehiscent
Ciliation of the false septa	Present, not present
Seed color	Yellow, yellowish brown, olive, light brown, medium brown, dark brown, mantled brown/yellow

Source: Diederichsen and Hammer, 1995.

always show the desired characteristics, in spite of environmental differences. Therefore, seed or fiber yield has become more stable over time. Hence, an important characteristic, like weight of 1,000 seeds, is of higher heritability in cultivated flax than in its wild progenitor *L. angustifolium* Huds. (Diederichsen and Hammer, 1995). Observations of 73 accessions of cultivated flax, representing all infraspecific groups proposed by Kulpa and Danert (1962), give a good indication of the range of variation within the species (Tables 2.1 and 2.2). Several qualitative characters are useful for characterization because their expression is independent from environmental influences and some of them are of great economic relevance. Seed color is one example.

The range of variation in seed size is enormous. Large-seeded flax originates from the Mediterranean area, reflecting Vavilov's (1926) observation that many cultivated plant species from this area have infraspecific groups with large seeds. Large seed is correlated with large petals, and this character also shows a wide range of variation. The large-seeded flax from the Mediterranean area has a spectacular inflorescence with large petals and widely opened, bowl-shaped flowers. In other flax genotypes, the flowers can be tube shaped and do not open at all. The width of the capsule increases with increasing seed size.

The length of the vegetation period is highly influenced by day-length. In northern latitudes, fiber flax matures more rapidly than the branched, large-seeded oilseed types. Day-length sensitivity is an important factor in the breeding of flax cultivars for northern areas where growing seasons are of limited length. The vegetative period of flax from the Mediterranean area or India seems to be less influenced by day-length.

Careful observations of the morphological traits and their inheritance were conducted by the Dutch scientist and flax breeder Tammes at the beginning of the twentieth century (Tammes, 1927; Tammes, 1928; Tammes, 1930). In total, 44 genes and their functions have been identified in flax. Of these, 20 genes are known to affect the color of parts of the inflorescence, four affect seed color, five affect other morphological traits, two affect male sterility, two affect oil quality, and 11 affect disease resistance (Kutuzova, 1998). Several interactions between these genes are known. However, genetic mapping has not been conducted and it is not known how these genes are distributed on the chromosomes.

Taxonomy

The genus Linum L. in plant systematics and its geographical distribution

The genus *Linum* is the type genus for the flax family, Linaceae (DC.) Dumort. The flax family is geographically widespread with about 300 species worldwide (Hickey, 1988). Several of the species are shrubs and occur in tropical areas, while perennial and annual species are found in temperate areas of the world.

The flax family is positioned in the plant kingdom as follows: Division: Pteridophyta; Sub-Division: Angiospermae; Class: Dicotyledoneae; Sub-class: Rosidae; Order: Geraniales. Within the flax family, the genus *Linum* belongs to the tribe Linoideae H. Winkl. There are four other genera within this tribe: (1) *Reinwardtia* Dumort. occurring in India and South-East Asia; (2) *Tirpitzia* Hallier occurring in South China; (3) *Hesperolinon* (A. Gray) Small occurring in the western USA; and (4) *Radiola* (Dillen.) Roth. occurring in Europe and adventive in the USA.

Linum is the largest genus within the family. Three geographical centers of distribution exist in the genus *Linum*, each with large interspecific diversity: (1) the Mediterranean area; (2) southern North America and all of Mexico; and (3) South America (Figure 2.11). However, several other species also occur in the temperate areas of Europe, Asia and the Americas.

The genus *Linum* has received considerable attention from botanists during the last three centuries. The wide range of diversity within the genus continues to challenge its systematic treatment by botanists. Several proposals for dividing the genus into sections exist, and the status of many species remains to be clarified.

The first proposal for a natural system for arranging the species was published in 1837 by Reichenbach (Vul'f, 1940). A thorough overview of the genus was presented by Winkler (1931). Yuzepchuk (1949) delivered a more elaborated, systematic treatment and distinguished nine sections, some with further taxonomic subdivisions. However, the systematic treatment of the genus continues to be discussed. A recent investigation of pollen morphology could not confirm any of the previously proposed groupings (Grigoryevka, 1988).

Winkler's proposal for grouping still appears to be the most useful, because its structure is clear and based on traits which are suitable for taxonomic distinctions. Chemotaxonomic investigations of the fatty oil composition in the seeds support this morphological grouping (Rogers, 1972). The systematic hierarchy proposed by Winkler (1931) is presented here with some minor changes which incorporate results of recent publications, including the Flora Europaea (Ockendon and Walters, 1968). A recent revision of the genus *Linum* is not available, and there is need for better

Figure 2.11 Geographical distribution of the sections of the genus *Linum* L. (After Vul'f (1940) and recent floras).

clarification of the questions regarding its systematic treatment. The genus *Linum* can be subdivided into the following six sections:

1 Sect. Linum

Characteristics   Flowers usually large; fruiting pedicels elongate; petals free, blue, pink, or white; sepals without obvious longitudinal veins; stigmas longer than wide, clavate or linear. Leaves alternate, without stipular glands and glabrous. Perennials, biennials or annuals.

Important species and range of distribution   *Linum usitatissimum* L. belongs to this section. *L. usitatissimum* does not occur as a wild plant and its cultivation is common in temperate climates world-wide. A close relative and the supposed wild progenitor of cultivated flax is the species *L. angustifolium* Huds., which is a biennial or perennial plant of the Mediterranean and submediterranean area, Ireland and the south of Great Britain. The similar *Linum decumbens* Desf. occurs in Italy. Many other perennial species of this section can be found in the Mediterranean area, and their ranges of distribution sometimes extend into the northern temperate climates of Europe and Asia. The impressive flowers of several species such as *L. austriacum* L., *L. perenne* L. or *L. narbonense* L. have resulted in their being cultivated as ornamental plants and commonly found in European botanic gardens. The annual *L. grandiflorum* Desf., which originates from Algeria, has large red flowers and is also a common ornamental plant. Some perennial species appear to have dispersed early in evolutionary time from the common center of origin and can be found in remote areas like Siberia (*L. amurense* Alef., *L. baicalense* Juz.) or in North America (*L. lewisii* Pursh).

The range of variation of the phenotypic characters used to distinguish among the species often overlaps within the section Linum. To resolve this problem, Ockendon and Walters (1968) proposed lumping species into groups, e.g. the *Linum perenne* group, and declaring the exact determination of the species as preliminary until a revision of the genus was completed. Unfortunately, such a revision has not yet been conducted. Another approach was to define many species and to single out more infraspecific taxa to get a better structure as Yuzepchuk (1949) did in the Flora of the USSR. However, this more elaborate classification has remained preliminary and inconsistent in discriminating species. Molecular investigations of seven species of the section Linum including cultivated flax and its wild progenitor confirmed the established morphological system (Fu *et al.*, 2002b).

2 Sect. Dasylinum (Planch.) Juz.

Characteristics   Like Sect. Linum, but pedicels or leaves pubescent. Perennial species.

Important species and range of distribution   Some species are cultivated as ornamental plants (e.g., *L. hypericifolium* Salisb., *L. hirsutum* L., *L. viscosum* L.). Most occur in the Mediterranean area, although some are restricted to Asia Minor.

3 Sect. Linastrum (Planch.) Benth.

Characteristics   Like Sect. Linum, but flowers small and usually yellow.

Important species and range of distribution   This section has a wide range of distribution. Species occur in the Mediterranean area, Africa and South America.

4 Sect. Cathartolinum (Reichenb.) Griseb.

Characteristics   Like Sect. Linum, but stigmas capitate.

Important species and range of distribution   About 50 species in this section are found in North America including Mexico, while only one (*L. catharticum* L.) occurs in Europe. *L. catharticum* has been used as a purgative in folk medicine. Small (1907) structured this section into eight subsections and several groups, indicating the wide range of diversity found in this section.

5 Sect. Syllinum Griseb.

Characteristics   Like Sect. Linum, but petals connected before opening of the flowers. Petals yellow or white. Stem with wings decurrent from leaf bases. Leaves with glands at base.

Important species and range of distribution   The perennial species of this section are distributed in the eastern Mediterranean area. An important species is *L. flavum* L. It has bright yellow flowers and is widely cultivated as an ornamental plant. Also in this group the differentiation of several species is not solved and the *L. flavum* group defined by Ockendon and Walters (1968) contains several species.

6 Sect. Cliococca (Babingt.) Planch.

Characteristics   Petals shorter than sepals.

Important species and range of distribution   This section contains only one species (*L. selaginoides* Lam.) which is found in South America.

The sect. Eulinum Planch., which is often mentioned in the literature, is here divided into the two sections Linum and Dasylinum. The main difference between these two is the presence of hairs on several parts of the plants in the section Dasylinum.

The diversity of species in the genus *Linum* differs in various parts of the world. Table 2.3 presents an overview based on floristic inventories. The Mediterranean area and adjacent Asia Minor have a very rich diversity of species, while northern Europe has fewer endemic *Linum* species. In Scandinavia, for example, only two species (*L. catharticum* L. and *L. austriacum* L.) can be found naturally. Turkey, on the other hand, which belongs partly to the Mediterranean area and partly to Asia Minor, has a

Table 2.3 Distribution of species in the genus *Linum* L. for different areas of the world

Area	Number of species	Source
Northern America and Mexico	63	Small, 1907
Former Soviet Union	45	Yuzepchuk, 1949
Turkey	38	Davis *et al.*, 1967
Europe	36	Ockendon and Walters, 1968
France	24	Fournier, 1977
Italy	20	Pignatti, 1982
Bulgaria	19	Stojanov and Stefanov, 1948
Morocco	19	Jahandiez and Maire, 1932
Lebanon and Syria	17	Mouterde, 1986
Algeria	15	Battandier, 1888–1890
Iran	13	Rechinger, 1974
Germany	10	Schubert *et al.*, 1988
Portugal	10	Palhinha, 1939
USA, north-east	10	Gleason and Cronquist, 1998
Central Asia	9	Vvedenskij, 1983
Canada	8	Scoggan, 1978
Switzerland	8	Aeschimann and Burder, 1994
Afghanistan	7	Kitamura, 1960
Alava/Spain	7	Uribe-Echebarria and Alejandre, 1982
Mallorca/Spain	6	Barcelo, 1979
Australia	6	Hnatiuk, 1990
Libya	6	Jafri and El-Gadi, 1977
Pakistan, West	6	Stewart, 1972
Cyprus	5	Meikle, 1985
India	5	Harja, 1993
Egypt	4	Monastir and Hassib, 1956
Tropical East Africa	4	Smith, 1966
Belgium	3	Lawalrée, 1964
Iraq	3	Rechinger, 1964
Arabia	2	Blatter, 1978
Japan	2	Ohwi, 1965
Scandinavia	2	Gram and Jessen, 1958

rich diversity with 38 *Linum* species. In the Americas, the species in the Section Cathartolinum are dominant. Although a good indicator, the figures in Table 2.3 have to be considered carefully, since the taxonomic systems used differ considerably among authors.

Plant systematics attempts to establish a system reflecting the evolutionary relationships among plant species. Natural systems in plant taxonomy are supportive for plant breeding, as they facilitate the targeted search for desired traits of economic importance in closely related wild species of a crop. The challenge for plant breeding is to introduce the desired traits from wild species into cultivated species.

Traditionally, taxonomic systems are based on morphological characters. Since the morphological systems previously proposed for the genus *Linum* were considered imperfect, Chennaveeraiah and Joshi (1983) conducted cytological studies on 19 species. They proposed the phylogenetic relationships based on chromosome numbers and similarities as outlined in Figure 2.12. Within the genus *Linum*, the chromosome number varies between 2n = 16 to 2n = 60, with most of the species having either 2n = 18 or 2n = 30 chromosomes (Chennaveeraiah and Joshi, 1983). Gill (1987) presented chromosome numbers of 41 species of the genus ranging from 2n = 16 to 2n = 80. However, the phylogenetic relationships within the genus are not yet clarified and all proposed systems are artificial.

A thorough revision of this diverse genus is necessary using the full range of morphological, chemical, and molecular characters and techniques available.

Cross-hybridization between cultivated flax and wild species has been tried since the beginning of scientifically based plant breeding in flax. Interspecific hybridization between cultivated flax and *L. angustifolium* has been successfully conducted (Tammes, 1923). In addition, Gill and Yermanos (1967) report successful hybridization with *L. africanum*, *L. corymbiferum* and *L. decumbens*. Using embryo-rescue techniques, hybrids with *L. monogynum* have been produced (Nickel, 1993). None of these interspecific crosses has had any practical use in flax breeding. So far, only *L. grandiflorum* has been used in breeding for cultivated flax. By pollinating *L. usitatissimum* with *L. grandiflorum*, the induction of dihaploid embryo growth in cultivated flax is possible.

Nomenclature of cultivated flax

Flax cultivation has its origin in Asia Minor and was introduced to other agricultural areas during the Neolithic revolution (8000 B.C.). It belongs to the "founder crops" that initiated agriculture in the Old World (Zohary, 1999). As the plant was introduced to new areas, it underwent selection for adaptation to very different environments, resulting in a wide range of diversity. Simultaneously, selection for the use of the seed oil or the fiber resulted in very distinct plant types. Selection for use of the seed focused on the development of the generative parts of the plant, while selection for use of the fiber resulted in tall, rarely branched plants. The common names for cultivated flax reflect this divergence, as there exist, in several languages, different names relating to the specific uses of the plant (Table 2.4). The English name, "flax," refers to the species in general. However, in Europe "flax" is usually associated with the fiber plant, while the name "linseed" is applied to oilseed flax. In the Americas the name "flax" is usually associated with the oil plant. Linseed and fiber flax belong to the same botanical species and can easily be crossed with each other. Intermediate types that can be used for either oil or fiber also exist and, in fact, make up the vast majority of all flax genotypes.

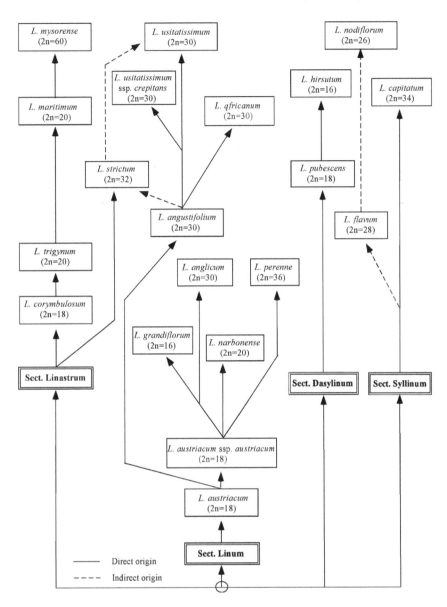

Figure 2.12 Evolutionary relationships among *Linum* L. species based on cytological data. (After Chennaveeraiah and Joshi, 1983).

A recent approach to use flax fiber not for long fibers, but for other technical purpose, has resulted in an increased interest in fiber quality parameters in all types of flax.

The accepted scientific name of the flax species cultivated for the use of its fiber or oil is *Linum usitatissimum* L. and was introduced by Carl von Linné in 1753. The classical Latin name of flax is *linum*, which is closely related to the classical Greek name *linon* (λινου). The translation of the Latin species epithet *usitatissimum* is "the most useful one," reflecting the several uses made of this plant. Linné (1753) deposited in

Table 2.4 Common names of flax in different languages

Language	General	Oil plant	Fiber plant
English	Flax	Linseed, oil-flax	Fiber-flax
German	Lein	Öllein	Flachs, Faserlein
Russian	Len'	Kudrjaš, len'-masličnyj	Dolgonec
Spanish	Lino	Lino oleaginoso	
French	Lin	Lin oléagineux	Lin à fibers
Danish	Hrr		
Turkish	Killu		
Hindi	Alsi, tisi		
Arab	Kattan		
Dutch	Vlaas		
Chinese	Hu ma, ya ma		

Table 2.5 The infraspecific groups of *Linum usitatissimum* L. according to Černomorskaja and Stankevič (1987)

Main characteristics	English	Russian	Formal name according to Černomorskaja and Stankevič (1987)	Placement in the system of Kulpa and Danert (1962)
One stem; tall	Fiber flax	Len-dolgunec	Ssp. *usitatissimum*	Convar. *elongatum*
One stem; medium height	Intermediate flax	Len-mežeumok	Ssp. *intermedium* Czernom.	Convar. *usitatissium*
Several stems; short; late mature	Crown flax	Len-kudrjaš	Ssp. *humile* (Mill.) Czernom.	Convars. *mediterraneum* and *usitatissimum*
Weight of 1,000 seeds larger than 7 g	Large-seeded flax	Len-krupnosemjannyj	Ssp. *latifolium* (L.) Stankev.	Convar. *mediterraneum*
Prostrate growth habit	Semiwinter flax	Len-poluzimnyj, Len-steljuščyj	Ssp. *bienne* (Mill.) Stankev.	Convar. *usitatissimum*

his herbarium a specimen of *L. usitatissimum* L., which is now recognized as the type species (Herb. Linn. No. 396–1). Since the stem of this plant is somewhat branched in the upper part, Kulpa and Danert (1962) stated the herbarium specimen is neither a typical fiber flax nor a large-seeded type, but an intermediate flax type. It shows more branching than fiber flax, but less than the large-seeded flax from the Mediterranean area. Linné's species epithet *usitatissimum* is the autonym and designated name for all infraspecific taxa (groups), which morphologically include the intermediate appearance of the type. Černomorskaja and Stankevič (1987) did not agree with this opinion and based their system on the assumption that Linné's herbarium specimen is a typical fiber flax. Therefore, the autonym *usitatissimum* used by Černomorskaja and Stankevič (1987) refers to a different infraspecific group of flax than the same name used by Kulpa and Danert (1962) (see Table 2.5). The rules outlined by the International Code for Botanical Nomenclature (Greuter, 1994) must be followed to avoid such confusion. Confusion or inconsistency in naming infraspecific taxa has resulted in the rejection of recognizing any formal grouping below the species level (Harlan and de Wet, 1971; Hetterscheid and Brandenburg, 1995). One has to be very careful when assigning formal botanical names to infraspecific groups. The usefulness of a formal approach to the naming of infraspecific groups in the study of plant genetic resources was stressed by Hanelt (1988).

Linné (1753) distinguished four different types of *Linum usitatissimum*, thus recognizing the diversity within the species. The English botanist Philip Miller (1768) restricted the name *L. usitatissimum* L. to the fiber flax and called the typical oil flax *L. humile* Mill. Miller's descriptions of the flax types are confusing and the name *L. humile* Mill. was later applied to cultivated flax with spontaneously opening capsules. Miller, however, does not mention the spontaneous opening of the capsules for the *L. humile* Mill. he describes. Common names exist for cultivated flax with dehiscent capsules, for example, the German name "Klenglein" translates in English as "Flax Sounding." This name is inspired by the noise the capsules of this particular group of flax make when they are mature and suddenly open with a cracking sound. In 1824, Boeninghausen considered this dehiscent cultivated flax as a distinct species and named it *L. crepitans* Boenning. (Kulpa and Danert, 1962). A further group distinguished by Miller (1768) were flax types with winter hardiness. He called them *L. bienne* Mill. Later it became obvious that the herbarium specimen Miller was referring to when he named *L. bienne* Mill. was not cultivated flax, but a wild species of flax, which was later described properly by Hudson (1788) and given the name *Linum angustifolium* Huds. Kulpa and Danert (1962) proposed that this name should be accepted for the species, due to the lack of clarity of Miller's description, which could claim priority. Even in some recent taxonomic publications the name *L. bienne* Mill. is considered to refer to cultivated flax (Černomorskaja and Stankevic, 1987). This name should be rejected as being doubtful.

A very elaborate classification for cultivated flax was published by Howard (1924). However, this grouping is restricted to Indian flax only and distinguishes 26 varieties. The system of Elladi (1940) is the most complicated and distinguishes 119 botanical varieties based on morphological characteristics and geographical origin of the different flax types. This system reflects the evolutionary development of cultivated flax during domestication. Elladi's classification has more theoretical than practical value since it is very complicated to apply, and recent crosses between the distinguished and ecomorphological varieties do not fit into the system. Elladi's system has never been applied to any collection of flax genetic resources and has been criticized for neglecting the rules of taxonomic nomenclature (Chrzanovskij *et al.*, 1979).

The classification presented by Dillman (1953) also proposes to distinguish morphological and geographical groups of flax. This system does not divide the genepool into formal botanical varieties, but makes distinctions between some morphological and geographical groups and assigns informal names to them. The major problem with this system is the inconsistency in applying the criteria for establishing the groups, thus limiting its usefulness. Dillman's (1953) classification is more artificial than Elladi's (1940). However, Dillman's system was partly applied to the world flax collection of the United States Department of Agriculture (USDA). The USDA world germplasm collection is grouped into six plant types: (1) fiber flax; (2) spring-type seed flax; (3) winter-type seed flax; (4) short, large-seeded, Indian seed flax; (5) Ethiopian forage-type flax; and (6) Mediterranean or Argentine seed flax.

The most recent proposal for a formal taxonomic grouping of flax came from Černomorskaja and Stankevič (1987). Their system further simplifies the groupings and distinguishes five subspecies. With slight modifications this system has been applied to the flax collection at the All Russian Flax Research Institute at Torzhok, Russia. It is a natural grouping consisting of five groups: (1) fiber flax; (2) intermediate flax (only one stem, shorter than fiber flax); (3) crown-flax (several stems, short); (4) large-seeded flax; and (5) prostrate flax (winter flax or semi-winter flax). Cultivated flax with spontaneously opening capsules is not recognized as a separate group, but belongs in the fiber flax group. This natural system of grouping does not distinguish flax with spontaneous opening capsules, but nevertheless has a distinct category for winter flax. It is difficult to distinguish winter flax from other flax types, since only certain environmental conditions allow for observation of this behavior. The system of Černomorskaja and Stankevič (1987) is often used in Russian literature, and the groups proposed by them and their scientific names are listed in Table 2.5.

The above examples illustrate how the phenotypically obvious genotypic diversity existing in flax has been a scientific challenge to taxonomists since they started to consider classification groupings. Several systematic treatments of cultivated flax already existed when Kulpa and Danert published their proposal in 1962. They carefully considered eight different classifications for cultivated flax proposed between 1866 and 1953. The system proposed by Kulpa and Danert (1962) distinguishes four convarieties and, in a second step, subdivides them into 28 varieties within the convarieties. Compared to Elladi's proposal (1940), which named 119 varieties, their system is simpler, but is still difficult to apply. However, the convarieties of Kulpa and Danert (1962) are very distinct and represent flax groups which have been selected for special purposes, thus forming natural groups. The 28 described varieties are based on morphological characters with the grouping being more artificial. Some flower traits used by these authors are not clear, while obvious characteristics like the folding of the petals are not mentioned at all. This classification with 28 varieties has been applied to the genebank collection at the IPK Gatersleben, Germany, and allows for targeted requests at the Gatersleben genebank. The accepted scientific name of cultivated flax and those of the four infraspecific groups according to Kulpa and Danert (1962) are listed with their synonyms below. The species grouping into four convarieties is natural and takes essential morphological traits into account. A further grouping into botanical varieties is possible. However, the varieties distinguished in the original publication are not quoted here; they are in need of clarification to be applicable and useful.

Accepted scientific name of cultivated flax and synonyms

The accepted scientific name and synonyms of cultivated flax and its infraspecific groups are presented according to Hammer (2001) and Kulpa and Danert (1962). Changes based on recent taxonomic studies were incorporated. The accepted name is printed in bold letters and followed by the synonyms.

Linum usitatissimum L., Sp. Pl. (1753) 277, ssp. **usitatissimum.** – *L. sativum* Hasselqu., Iter Palaest. (1757) 462; *L. arvense* Neck., Delic. Gallo-Belg. 1 (1768) 159; *L. utile* Salisb., Prodr. (1796) 177; *L. usitatissimum* β *humile* Pers., Syn. 1 (1805) 344; *L. usitatissimum* var. *vulgare* Boenningh., Prodr. Fl. Monast. Westph. (1824) 94; *L. usitatissimum* var. *indehiscens* Neilr., Fl. Nied.-Oest. (1856) 864; *L. usitatissimum* var. *humile* Alef., Landw. Fl. (1866) 103; *L. usitatissimum* var. *typicum* Pospich., Fl. Oest. Küstenl. (1898); *L. typicum* (Pospich.) Schilling in Herzog, Technol. Textilpfl. 5, 1 (1930) 64; *L. indehiscens* (Neilr.) Vav. et Ell., Kul't. Fl. SSSR 5, 1 (1940) 117; incl. *L. dehiscens* Vav. et Ell. ssp. *crepitans* (Dum.) Vav. et Ell., l. c., 110.

Since the beginning of the twentieth century, it has been known that cultivated flax can be crossed with the wild species pale flax (*L. angustifolium* Huds.) (Tammes, 1911; 1923). Cytogenetic studies reveal both species to have 2n = 30 chromosomes and to differ in caryotype by one translocation (Gill and Yermanos, 1967). The geographical range of distribution of *L. angustifolium* Huds. and studies on the distribution of weeds associated with flax confirm *L. angustifolium* as being the wild progenitor of cultivated flax (Hjelmquist, 1950; Helbaek, 1959). This evolutionary relationship was mentioned very early by Heer (1872), who conducted morphological observations of recent and historic findings on flax. Thellung (1912) unified cultivated and pale flax under the single species name, *L. usitatissimum* L. Recent molecular investigations confirmed the close relationship between these two taxa (Fu *et al.*, 2002b). Therefore, according to the rules of botanical nomenclature, cultivated flax and the wild progenitor each receive in this case the rank of a subspecies (Hammer, 2001). Cultivated flax is then named *L. usitatissimum* L. ssp. *usitatissimum* and the wild progenitor *L. usitatissimum* L. ssp. *angustifolium* (Huds.) Thell. The unification of these two flaxes into one species reflects the concept that all subordinated taxa belonging to one species can hybridize with each other. According to the genepool concept of Harlan and de Wet (1971), both subspecies belong to the primary genepool for breeding activities in cultivated flax. From the viewpoint of an evolutionary biologist, this taxonomic widening of the species *Linum* is consistent. However, for practical reasons it is not always necessary to quote the complete name and cultivated flax can be referred to as *L. usitatissimum* L. and pale flax as *L. angustifolium* Huds. Usually it is clear from the context which subspecies is discussed.

Key for determination of the convarieties of L. usitatissimum

A Capsules open septicidaly and loculicidally during ripening. Seeds are easily shattered. Later empty capsules are dropped.   1. convar. *crepitans*

A* Capsules not opening during ripening or only slightly septicidally separating. Seeds are not easily shattered. Capsules are not dropped.   B

B Plant height more than 70 cm and only the upper 1/3 or less of the entire stem length with side branches; if less than 70 cm then stem branches only in the upper 1/5 of the entire stem length.   2. convar. *elongatum*

B* Plant height usually less than 70 cm; more than 1/5 of the entire stem length with side branches.   C

C Weight of 1,000 seeds more than 9 g; plants usually without basal branches.   3. convar. *mediterraneum*

C* Weight of 1,000 seeds less than 9 g; plants often with basal branches.   4. convar. *usitatissimum*

1. convar. *crepitans* (Boeninningh.) Kulpa et Danert, Kulturpflanze, Beih. 3, (1962) 374. – *L. usitatissimum* var. *crepitans* Benningh., Prodr. Fl. Monast. Westph. (1824) 94; *L. crepitans* (Boenningh.) Dum., Fl. Belg. (1827) 111; *L. humile* Planch. in London Bot. 7 (1847) 165, p. p., et auct. vix Miller (1768); *L. usitatissimum* ζ *dehiscens* Neilr., Fl. Nied.-Oest. (1859) 864; *L. dehiscens* Vav. et Ell. in Kul't. Fl. SSSR 5, 1 (1940) 113; *L. usitatissimum* ssp. *usitatissimum* apud Czernom. et Stankev., Sel. i gen. techn. kul't. 113 (1987) 58, p. p.

The convar. *crepitans* has been used in Central and South-East Europe as a fiber plant. Seed shattering makes it difficult to harvest the seeds. This type of flax is no longer cultivated in agriculture and only the germplasm collection conducted early in the twentieth century helped maintain and preserve this type from extinction. Its range of variation is limited. With the exception of the dehiscence of the capsules the plants are phenotypically similar to those of the convar. *usitatissimum*.

2. convar. *elongatum* Vav. et Ell. in Kul't. Fl. SSSR 5, 1 (1940) 153, pro prole sub *L. indehiscens* Vav. et Ell. ssp. *eurasiaticum*. – *L. usitatissimum* L. ssp. *vulgare* (Boenningh.) Ell., Tr. prikl. bot. gen. i sel. 22, 2 (1929) 457, p. p.; *L. usitatissimum* ssp. *eurasiaticum* Vav. ex Ell., in Zhuk., Turquie Agric. (1933) 442, p. p.; *L. usitatissimum* ssp. *usitatissimum* apud Czernom. et Stankev., Sel. i gen. techn. kul't. 113 (1987) 58, p. p.

The convar. *elongatum* describes typical fiber flax. It has long stems which are only branched at the top of the stem. This type of flax has been of great importance in the temperate and northern areas of Europe and in particular eastern Europe. China is also a center of diversity for fiber flax. Fiber flax has a shorter vegetative period than the large-seeded flax. This group is identical with the fiber flax as defined by Dillman (1953).

3. convar. *mediterraneum* (Vav. ex Ell.) Kulpa et Danert, Kulturpflanze, Beih. 3, (1962) 376. – *L. usitatissimum* var. ζ L., Sp. Pl. (1753) 277; *L. usitatissimum* ssp. *mediterraneum* Vav. ex Ell., in Zhuk., Turquie Agric. (1933) 442; *L. usitatissimum* ssp. *bienne* (Mill.) Rothm., Agron. Lusitana 6 (1944) 259, p. p.; *L. usitatissimum* ssp. *latifolium* (L.) Stankev., Sel. i genet. techn. kul't. 113 (1987) 59.

The convar. *mediterraneum* represents the large-seeded flax with large flowers and capsules, and branched stems. It is only used for seed production. Flax of this type originates from the Mediterranean area and has a long vegetative period. This group is identical with the Mediterranean seed flax defined by Dillman (1953).

4. convar. *usitatissimum* – *L. humile* Mill. ssp. *transiens* (Ell.) Rothm., Agron. Lusitana 6 (1944) 257; *L. usitatissimum* ssp. *bienne* (Mill.) Rothm., Agron. Lusitana 6 (1944) 259, p. p; incl. *L. usitatissimum* ssp. *humile* (L.) Czernom., Sel. i genet. techn. kul't. 113 (1987) 58, ssp. *intermedium* Czernom., l. c., 60 et ssp. *bienne* (Mill.) Stankev., Sel. i genet. techn. kul't. 113 (1987), 61, p. p.

The convar. *usitatissimum* refers to the intermediate flax, or dual purpose flax. This is the most common type of flax in the world. It has adapted to a wide range of climates. Within this convariety further segregation into different morphotypes is possible. This group covers the spring-type seed flax, winter-type seed flax, and Indian and Ethiopian (Abyssinian) flax, as defined by Dillman (1953).

Accepted scientific name of pale flax, the wild progenitor of cultivated flax, and synonyms

Linum usitatissimum ssp. angustifolium (Huds.) Thell., Fl. Adv. Montp. (1912) 361. – *L. bienne* Mill., Gard. dict. ed. 8 (1768) n. 8, nomen dub. rejiciend.; *L. angustifolium* Huds., Fl. angl. ed. 2 (1788) 13;. *L. ambiguum* Jord., Cat. jard. Dijon (1848) 27; *L. hohenhackeri* Boiss., Diagn., ser. 2, 1 (1853) 97; *L. usitatissimum* ssp. *hispanicum* Thell., Feddes Rep. 11 (1912) 75; *L. usitatissimum* ssp. *bienne* (Mill.) Thell., l. c., 129; *L. dehiscens* Vav. et Ell. ssp. *angustifolium* (Huds.) Vav. et Ell., Kul't. Fl. SSSR 5, 1 (1940) 111.

Pale flax (Figure 2.13) has a biennial or perennial growth habit. The flowers are homostylous and self pollinated. The capsules open spontaneously and the seeds shatter. It occurs in the Mediterranean area and the Atlantic western Europe. Investigation

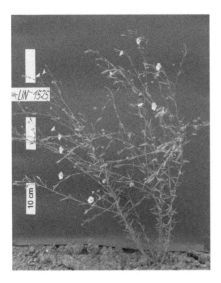

Figure 2.13 Pale flax, *Linum angustifolium* Huds., flowering plant. (Photo: A. Diederichsen).

of morphology showed that some characters of pale flax show a wider range of variation than in cultivated flax (Diederichsen and Hammer, 1995). It can be expected that molecular investigations will also reveal more genetic diversity in the wild progenitor pale flax than in cultivated flax.

Germplasm conservation

In situ *and* ex situ *conservation of flax germplasm*

When discussing the different approaches for conservation of flax germplasm a distinction has to be made between the wild species and cultivated flax. Several of the wild species occur in native habitats remote from civilization (Figure 2.11, Table 2.3). Activities to preserve such natural ecosystems will also contribute to preservation of the respective species at their natural sites. Some of the wild species gained importance as ornamental plants and their distribution by man has widened their geographical range of distribution, e.g. species of the section *Linum* in Australia or North America.

The lack of a recent taxonomic revision of the genus makes it difficult to judge information from floristic inventories as listed in Table 2.3. However, the *in situ* approach has to be favored for effective conservation of these species and *ex situ* collection can support this by allowing for detailed studies of the diversity of these species.

With cultivated flax there is no doubt that collection and *ex situ* conservation have preserved (infraspecific) diversity, which would have been lost without such activities. In spite of recent tendencies in industrialized countries to revive the cultivation of fiber flax this entire group is decreasing and has already disappeared from many areas (Diederichsen and Hammer, 1995). To find relics of fiber flax cultivation in *on farm* situations is very rare. Recently Laghetti *et al*. (1998) reported such an event from the island Ustica in Italy. Landraces, both of fiber flax and linseed, have vanished or are replaced by modern breeding lines. For cultivated flax, *in situ* conservation, which should better be called *on farm* conservation, is nonexistent in industrialized countries.

In developing countries, the diversity of flax is also declining due to replacement by uniform breeding lines. There are no effective programs in place to change this situation and genebanks have to coordinate their efforts to continue to preserve the diversity of cultivated flax.

The relative superiority of *in situ* conservation for wild flax and *ex situ* conservation for cultivated flax reflects the general picture for conservation of diversity of crops and their wild relatives as described by Hammer *et al.* (1999).

Major ex situ *collections of flax germplasm*

The total number of flax germplasm accessions existing in world genebanks is about 53,000. The eight largest flax collections hold more than 55 percent of the total number of accessions (Table 2.6). There are at least another 81 genebanks, institutions or breeding stations in the world, which maintain collections of flax germplasm. About 38 of these 81 collections hold more than 100 accessions each, while the other sites maintain smaller collections or only a few accessions of *Linum* germ-plasm. A directory of information on germplasm collections is provided by the Food and Agriculture Organization (FAO) (http://apps3fao.org/wiews/).

The total number of flax germplasm accessions is impressive, but the figure must be considered with caution. For example, a high degree of duplication exists between the two major collections held in the Russian Federation, the collection at the All-Russian Flax Research Institute (VNIIL) and the N.I. Vavilov Institute for Plant Industry (VIR). The collection at the VNIIL has been intensively integrated into the breeding program for fiber flax. It has been screened for several important traits and considerable characterization and evaluation data exist for their accessions. A catalog with basic passport data of the VNIIL accessions was published in 1993 (Zhuzhenko *et al.*, 1993). Evaluation data are partly accessible in published form (Marcenko, 1994; Rozhmina and Zhuzhenko, 1998). The VNIIL germplasm is derived from the VIR collection which is made up of many accessions dating back to the early period of collecting plant genetic resources at the beginning of the twentieth century. This collection is very valuable since much of the germplasm at the VIR has vanished from cultivation in agriculture. The passport data for the VIR collection are available on the Internet

Table 2.6 Genebank collections of *Linum* germplasm

Name of institution	Country	No. of accessions
All Russian Flax Research Institute, VNIIL, Torzhok	Russia	6,100
N.I. Vavilov Institute of Plant Industry, VIR, St. Petersburg	Russia	5,700
Research Institute for Technical Cultures, RITC	China	4,000
Breeding company DSV, Lippstadt	Germany	3,500
Ethiopian Genebank, Addis-Abbaba	Ethiopia	3,100
North Central Plant Introduction Center, Ames, Iowa	USA	2,800
Plant Gene Resources of Canada, PGRC, Saskatoon	Canada	2,800
Research Institute for Cereals and Industrial Crops, RICIC, Funduela	Romania	2,700
Other collections (81)		22,300
Total		53,000

Sources: FAO; Zhuzhenko and Rozhmina, 2000.

(http://www.vir.nw.ru), while characterization data are accessible by direct contact with the curator of the flax collection at the VIR.

In China, fiber flax production is an important industry and the large collection at the Research Institute for Technical Cultures (RITC) meets this need. The Chinese collection contains 2,000 accessions, which were received from the USDA (Pan, 1990), again reflecting a duplication between large *ex situ* collections of flax germplasm.

The German plant breeding company Deutsche Saatveredelung (DSV) is engaged in flax breeding and also has a large collection of flax germplasm. This collection consists mainly of breeding material and the diversity it contains is limited compared to that maintained in genebanks.

Ethiopia is one of the classical centers of flax diversity named by N.I. Vavilov (1926). Obviously a large collection of flax germplasm exists there, but unfortunately access to this collection is presently restricted.

The collections for flax germplasm of the United States National Plant Germplasm System (NPGS) in the USA and of Plant Gene Resources of Canada (PGRC) are to a large extent duplicates of one another. Passport and evaluation data for these collections are available via the Internet (USA: http://www.ars-grin.gov/; Canada: http://www.agr.gc.ca/pgrc-rpc).

The collection of flax germplasm at the German genebank, at the Institut für Pflanzengenetik und Kulturpflanzenforschung (IPK), Gatersleben, has passport data accessible via the Internet (http://fox-serv.ipk-gatersleben.de). This is the only collection of flax germplasm which has been subjected to a formal infraspecific classification.

In summarizing all information about existing *ex situ* collections of flax germplasm, it becomes obvious that considerable duplication exists among the different *ex situ* collections of the world as listed in Table 2.6. About 30 percent of the world flax germplasm accessions listed in Table 2.6 can be assumed to be unique. In particular, the more recently bred varieties will account for much of the duplication, since on the national and international level plant breeders tend to use similar germplasm. This, of course, implies duplication of rejuvenation and storage expenses at genebank sites. However, the continued preservation of germplasm in genebanks can be severely affected by political change, genebank priorities, and problems with rejuvenation. Therefore, a certain degree of duplication helps to ensure the preservation of the diversity. Duplication of germplasm for security reasons could be better organized than it is at present. Detailed observations on accessions, which are supposed to be duplicates, often show these accessions to be significantly different.

The wild species in the genus *Linum* are rarely found in genebank collections because they are perennials and are difficult to regenerate due to cross-pollination. The German genebank at Gatersleben has the largest collection of these species. Several botanical gardens, especially those in Europe, have seeds of wild species of the genus available for use.

Characterization and evaluation of flax germplasm in ex situ *collections*

All collections listed in Table 2.6 have been subjected to characterization for some traits. However, this information is often difficult to access by external users. A list of 63 descriptors for flax was established for genebank use for four major collections in Eastern Europe (Rykova *et al.*, 1989). This comprehensive list has been partly applied to the collection preserved at the Vavilov Institute. A new initiative to develop a descriptor list was started in 1993 by a European working group on flax and other fiber plants. This working group is composed of several institutions and plant breeding companies in 11 countries (Bulgaria, Czech Republic, France, Germany, Netherlands, Poland, Romania, Russian Federation, United Kingdom, Ukraine and USA). The group proposed a descriptor list for flax based on 24 traits including the most important oil- and fiber-quality traits (Pavelek, 1998). The list is being applied by some of the institutions participating in this network. A centralized database, which includes parts of the collections of the participating countries, is located at Šumperk, Czech Republic.

The United States Department of Agriculture (USDA) has put significant effort into its Genetic Resources Information Network (GRIN) and has characterization data on 18 characters of flax for more than 2,800 accessions. These data are available via the Internet (http://www.ars-grin.gov/). Plant Gene Resources of Canada (PGRC) has recently collected characterization and evaluation data (Diederichsen, 2001) on the 2,800 flax accessions within its collection, using the descriptor list given in Table 2.7. These data are also available on the Internet (http://www.agr.gc.ca/pgrc-rpc).

The characterization of flax accessions in genebanks is based on morphological traits which are minimally influenced by environmental conditions. Such characteristics ensure identification of a genebank accession, and can be obtained during field regeneration without the restriction of establishing field trials. Plant breeders, however, are primarily interested in quantitative traits such as yield, which are much more influenced by environment, and demand intense observation over several years to achieve exact results. Detailed information is often not available for genebank accessions. However, Atlin *et al.* (1992) showed that, even in breeding programs, a reduction of plot size in flax to single rows can still produce results which are useful in selection decisions.

Modern computer software allows for the handling of large amounts of data, and electronic communication via the Internet is possible worldwide. Genebanks should make more use of this technology to share information on their accessions. Cooperation among different sites maintaining flax germplasm collections needs to be intensified to avoid duplication of effort and to increase the use of the existing flax germplasm collections.

Regeneration and storage of flax germplasm

Cultivated flax is self-pollinating. However, during flowering, flax is occasionally visited by insects, which may result in cross-pollination. Robinson (1937) reported out-crossing rates up to 1.47 percent. More recent studies report out-crossing up to 5 percent (Williams, 1988) and isolation distances between different genotypes of flax have been demanded to avoid cross-pollination (Nickel, 1993). Cross-pollination by wind is rare because flax pollen grains are very heavy. Observations made during the regeneration of flax accessions preserved in genebanks showed that off-types appear regularly. Flax accessions were grown in the field without isolation from each other during a previous regeneration cycle. Plants which are off-types should be removed because they obviously indicate cross-pollination and introgression of genes from different populations. Isolation of genotypes during flowering by distance, bagging, or cages would minimize this problem.

Another problem experienced during regeneration of accessions is physical mixture between genotypes. This can occur during seeding, harvesting or seed cleaning.

Table 2.7 Descriptors for cultivated flax used by Plant Gene Resources of Canada (PGRC)

No.	Character	Scale
	Phenology traits	
1	Emergence date	Date
2	Start of flowering	Date
3	End of flowering	Date
4	Length of flowering	Number of days
5	Days until maturity	Number of days
6	Harvest date	Date
	Flower	
7	Sepal dotting	1=None, 2=Intermediate, 3=Many
8	Anther color	1=White, 3=Blue, 5=Pink, 7=Crème-colored (Yellow), 9=Orange, 11=Gray (turquoise)
9	Filament color	1=White, 3=Blue, 4=Dark blue, 5=Pink, 6=Violet
10	Style color	1=White, 3=Blue, 4=Dark blue, 5=Pink, 6=Violet, 7=Crème-colored
11	Petal color (basal)	1=White, 2=Light blue, 3=Blue, 4=Dark blue, 5=Pink, 6=Violet 10=Red-violet (lavender)
12	Petal width (W)	mm
13	Petal length (L)	mm
14	Petal ratio W/L	Ratio
15	Petal longitudinal folding	No=Absent, Yes=Present
16	Petal margin folding	No=Plain, Yes=Folded inwards
17	Petal overlap	1=Petals overlap more than 50% of length, 2=Petals overlap less than 50% of length
18	Flower shape	1=Tube, 2=Funnel, 3=Bowl
	Capsule	
19	Capsule width	mm
20	Capsule shape	1=Ovate, 2=Round, 3=Flattened
21	Capsule dehiscence	1=Dehiscent, 3=Medium opened, 5=Slightly opened, 7=Weak, 9=Indehiscent
22	Ciliation of septa	No=Hairs absent, Yes=Hairs present
	Plant, habit	
23	Branching	1=1/1, 2=1/2, 3=1/3, 4=1/4, 5=1/5, 6=1/6 (of total stem length branched)
24	Plant height	cm
25	Growth habit	1=Postrate, 2=Intermediate, 3=Erect
	Seed	
26	Seed color	1=Light brown (7.5YR5/6), 2=Medium brown (7.5YR4/6), 3=Dark brown (7.5YR3/2), 4=Yellow (2.5Y6/6), 5=Olive (5Y5/6), 6=Mottled brown/yellow; (Color codes according to Munsell (Anonymous, 1976))
27	Weight 1,000 seeds	grams
28	Yield from 3m row	grams

Mechanization of seeding and harvesting does not always allow the maintenance of high purity standards necessary for preserving the genetic integrity of genebank accessions. Existing characterization data have to be used as a reference for detecting deviations in the performance of an accession in different growing seasons.

Observations at the Canadian genebank have shown that some genotypes of yellow-seeded flax tend to lose viability during seed storage much faster than brown-seeded flax (Diederichsen, unpublished). Long-term storage can obviously result in genetic shift of genebank accessions when genotypes differ in their ability to maintain viability during long-term storage. This underlines the need to properly characterize accessions within genebanks, and to record any changes that may have occurred. Herbarium specimens collected during the first increase of an accession are very useful as reference samples for verifying the genetic identity of the accessions (Hammer, 1992).

Landraces of cultivated flax are composed of several genotypes, which can sometimes be distinguished phenotypically. In these cases, it is necessary to note the frequency of the different genotypes in the original landrace and, if possible, to keep these different strains distinct. This allows consistent characterization of the material and helps to avoid genetic change (shift/drift) in the collection. The qualitative characteristics given in the descriptor list (Table 2.7) are useful for making such decisions.

There are significant differences in the dehiscence of the ripe flax capsules. Accessions which are dehiscent must be observed during the ripening period so as not to lose the seeds due to shattering.

Linseed oil contains a high percentage of double and triple unsaturated C-C bounds, which are chemically more unstable than saturated C-C bounds. The seeds should be stored under cold and dry conditions to maintain viability. Seeds with a water content of 6 percent did not show any decrease in viability after having been stored for 15 years at a temperature of −15°C (Specht *et al.*, 1998). Seeds stored at ambient conditions should not be stored for more than five years to avoid loss of genotypes (Gladis, 1999; Rozhmina, VNIIL, Torzhok, personal communication).

Maintenance of germplasm of wild *Linum* species is much more challenging. Seed-set in the perennials is low during the first growing season and most of these species are highly out-crossing. The majority of the species in the genus *Linum* are self-incompatible due to the obligatory heterostylous flowering habit and cross-pollination by insects (Ockendon, 1968). Therefore, additional efforts such as isolation distance, cages or pollination management are required to maintain these species. This partially explains why germplasm of these species is not maintained at many sites. Taxonomic difficulties in correctly determining the species causes some additional confusion in the germplasm of the wild species available in *ex situ* collections. Cross-pollination among different species during seed increase, due to insufficient isolation among accessions in botanical gardens or genebanks, has resulted in interspecific hybrids. These hybrids are not suitable for taxonomic studies and may explain some of the present confusion in the taxonomy of the genus *Linum*.

Conclusions

A wide range of diversity in cultivated flax has been preserved in genebanks world-wide. This has been possible since collection activities for this species started early in the twentieth century. Without collecting and preserving flax germplasm in *ex situ* collections, most of the cultivated flax genetic diversity which has evolved under domestication since 8000 B.C. would have been lost to future utilization. About 53,000 accessions of flax are maintained in world genebanks and a high degree of duplication exists among the world flax collections. Duplication is beneficial as it secures the maintenance of germplasm. However, better coordination among international genebanks maintaining flax germplasm could enhance effective preservation of flax diversity.

Comparisons of passport, characterization and evaluation data among different collections would be an initial step. Characterization of germplasm using internationally standardized crop descriptors would also assist in the identification of particular infraspecific groups and unique germplasm. Molecular characterization needs to be encouraged as it will contribute to a better understanding of the diversity within cultivated flax. By combining different data sets (agrobotanic, evaluation and molecular data) it will become possible to better identify natural structures in the diversity. The classical approach was to describe infraspecific groups and assign names to them according to the rules of botanic nomenclature. Germplasm of these infraspecific groups represents cornerstones in the structure of the entire diversity of the species. The establishment of core collections, i.e. subsets of existing germplasm collections, is based on a similar principle and the use of computer programs will allow for the compilation, assimilation and analysis of even more information than was previously possible.

The history of the scientific discussions on cultivated flax's taxonomic grouping demonstrates how the general approach to the classification of the entire genepool of cultivated plant species has changed. After Linné (1753), scientists started to distinguish infraspecific groups and the increased knowledge led to very complicated systems. These systems were rarely of practical use but have significant scientific value. This is particularly valid for the first generation of taxonomists who elaborated the theories of N.I. Vavilov. Further considerations led to classification systems naming a few clearly distinguishable groups. Recently, some scientists have proposed to discard all formal infraspecific groupings in cultivated plants and to substitute the scientific species epithet with a new taxonomic category called "culton" (Harlan and de Wet, 1971; Hetterscheid and Brandenburg, 1995). The latter is obviously a reaction to the complicated infraspecific classifications that are confusing and do not support clear communication.

Plant breeders, genebank curators, crop researchers and botanists have different needs and, therefore, often divergent opinions about the usefulness of infraspecific groupings of crop plants. The interest of breeders is often limited to agronomic and quality traits and disease resistance. Plant breeding typically encompasses only a small segment of the entire range of available diversity. The entire genepool is of primary importance for further crop improvement by exploiting monogenic traits like disease resistance. Genetic resources are often neglected, once the desired germplasm has been found. Genebank curators need to cooperate and pool their efforts to provide an overview of the identified range of variation available and to communicate this information to clients of the genebanks. Crop researchers and botanists are interested in questions regarding the evolution of cultivated plants and their intraspecific and interspecific genetic relationships. A reasonable infraspecific grouping according to the formal rules supports this interest.

The systematics of the genus *Linum* is in need of clarification. Botanical, biochemical, cytological and molecular methods are available and need to be used in combination to better clarify the evolutionary or biogeographical relationships among the species within the genus. It will be necessary to acquire additional germplasm through exploration of the wild *Linum* species. Additional effort and resources for research on the wild species are urgently required and these efforts should be undertaken in combination with *in situ* and *ex situ* conservation of this germplasm. The investigation of the potential of the wild *Linum* species for the improvement of cultivated flax will be based on these studies. To achieve taxonomic consistence on the species and infraspecies level is essential for communication of all disciplines interested in the diversity of flax.

Acknowledgments

The authors thank Mr. Eric Kokko, Lethbridge Research Center, Agriculture and Agri-Food Canada (AAFC) for the electron-microscopy, Mrs. Gisele Mitrow and Dr. Paul Catling, Eastern Cereal and Oilseed Research Center AAFC, Ottawa, for assistance compiling taxonomic information, Mr. Ralph Underwood, Saskatoon Research Center AAFC, for preparation of the figures and the Saskatchewan Flax Development Commission for financial support.

References

Aeschimann, D., and Burder, H.M. (1994). *Flore de la Suisse et des territoires limitropes*. Edition Griffon, Neutchâtel, pp. 265–266.

Anonymous (1976). *The Munsell Book of Color*. Macbeth Division of Kollmorgen Instruments Corp., New Windsor.

Atlin, G.N., Kenaschuk, E.O., and Lockwood, D.J. (1992). Single-row plots for agronomic evaluation of flax (*Linum usitatissimum* L.) lines. *Can. J. Plant Sci.* **72**, 997–1000.

Barcelo, F.B. (1979). *Flora de Mallorca*. Editorial Moll, Palma de Mallorca, pp. 102–106.

Barnes, D.K., Culbertson, J.O., and Lambert, J.W. (1960). Inheritance of seed and flower colors in flax. *Agron. J.* **52**, 456–459.

Battandier, J.A. (1888–1890). *Flore de Algérie*. Typographie Adolphe Nourdan, Algiers, 174–176 p.

Blatter, E. (ed.) (1978). *Flora Arabica*. Bishen Singh Mahendrs Pal Singh, Dehra Dun, pp. 91.

Boenninghausen, C.F.M. (1824). *Prodromus florae monasteriensis Westphalorum*. Monasterii, Suntibus F. Regensberg, Münster. 332 p.

Candolle, A.L.P.P. (1883). *Origine des plantes cultivées*. G. Baillère, Paris.

Černomorskaja, N.M., and Stankevič, A.K. (1987). To the problem of intraspecific classification of common flax (*Linum usitatissimum* L). (Russ.). *Selekcija i genetika tech. kul't.* **113**, 53–63.

Chennaveeraiah, M.S., and Joshi, K.K. (1983). Karyotypes in cultivated and wild species of *Linum*. *Cytologia* **48**, 833–841.

Chržanovskij, V.G., Ponomarenko, S.F., and Dolguzasvili, V.Z. (1979). K voprosu proischož-denija i evolucii roda *Linum* L. sem. Linaceae (To the question of the origin and evolution of the genus *Linum* L. family Linaceae). *Izvestija Akademii Nauk SSSR, Moskva* **5**, 696–713.

Darwin, C. (1868). *The Variation of Animals and Plants under Domestication*. J. Murray, London.

Davis, P.H., Chamberlain, D.F., and Mathews, V.A. (1967). *Flora of Turkey*. University Press, Edinburgh, pp. 425–450.

Diederichsen, A. (2001). Comparison of genetic diversity of flax (*Linum usitatissimum* L.) between Canadian cultivars and a world collection. *Plant Breeding* **120**, 360–362.

Diederichsen, A., and Hammer, K. (1995). Variation of cultivated flax (*Linum usitatissimum* L. subsp. *usitatissimum*) and its wild progenitor pale flax (subsp. *angustifolium* (Huds.) Thell.). *Genetic Res. Crop Evol.* **42**, 263–272.

Dillman, A.C. (1936). Improvement in flax. *Yearbook of Agriculture*, USDA, pp. 745–784.

Dillman, A.C. (1953). *Classification of flax varieties, 1946. USDA, Technical Bulletin No. 1054*, USDA, Washington.

Dillman, A.C., and Brinsmade, J.C. (1938). Effect of spacing on the development of the flax plant. *J. Am. Soc. Agron.* **30**, 267–278.

Elladi, V.N. (1940). *Linum usitatissimum* (L.) Vav. consp. nov. – Len. (Russ.). In Vul'f, E.V., and Vavilov, N.I. (eds), *Kul'turnaja flora SSSR, fiber plants*, Sel'chozgiz, Moscow, Leningrad, Vol. 5, Part 1, pp. 109–207.

FAO Food and Agricultural Organization, Rome, Italy: World Information and Early Warning System (WIEWS). Internet: http://apps3.fao.org/wiews/.

Fournier, P. (1977). *Les Quatre Flores de la France.* Editions Lechevalier, Paris, 2nd ed. pp. 611–615.

Fu, Y.-B., Diederichsen, A., Richards, K.W., and Peterson, G. (2002a). Genetic diversity of flax *(Linum usitatissimu*m L.) cultivars and landraces as revealed by RAPDs. *Genetic Res. Crop Evol.* **49**, 167–174.

Fu, Y.-B., Peterson, G., Diederichsen, A., and Richards, K.W. (2002b). RAPD analysis of genetic relationships of seven flax species in the genus *Linum* L. *Genetic Res. Crop Evol.* **49**, 253–259.

Gassner, G. (ed.) (1951). *Mikroskopische Untersuchung pflanzlicher Nahrungs- und Genussmittel*, 2nd ed. Gustav Fischer, Jena.

Gill, K.S. (ed.) (1987). *Linseed.* Indian Council of Agricultural Research, New Delhi, India., pp. 342–355.

Gill, K.S., and Yermanos, D.M. (1967). Cytogenetic studies on the genus *Linum. Crop Sci.* **7**, 623–631.

Gladis, T. (1999). Kulturelle Vielfalt und Biodiversität. *VEN Samensurium* **10**, 22–36.

Gleason, H.A., and Cronquist, A. (eds). (1998). *Manual of Vascular Plants of Northeastern United States and Adjacent Canada.* New York Botanical Garden. New York, pp. 345–347.

Gram, K., and Jessen, K. (1958). *Vilde planter i Norden.* Gads, Copenhagen, pp. 836–837.

Greuter, W. (ed.). (1994). *International Code of Botanical Nomeclature (Tokyo Code).* Koeltz, Königstein.

Grigoryevka, V.V. (1988). The pollen grain morphology in the genus *Linum* (Linaceae) of the flora of the USSR. (Russ.). *Botaničeskij žurnal* **73**, 1409–1417.

Hammer, K. (1992). The 50th anniversary of the Gatersleben genebank. *Plant Genetic Resources Newsletter* **91/92**, 1–8.

Hammer, K. (2001). Linaceae. In Hanelt, P., and IPK (eds), *Mansfield's Encyclopedia of Agricultural and Horticultural Crops*, Vol. 2. *Angiospermae-Dicotyledones: Leguminosae-Balsaminaceae*, Springer, Berlin and New York, pp. 1106–1108.

Hammer, K., Diederichsen, A., and Spahillari, M. (1999). Basic studies toward strategies for conservation of plant genetic resources. In Serwinski, J., and Faberová, I. (eds), *Proceedings of the Technical Meeting on the Methodology of the FAO World Information and Early Warning System on Plant Genetic Resources, 21–23 June 1999*, Prague, Czech Republic, pp. 29–33.

Hanelt, P. (1988). Taxonomy as a tool for studying plant genetic resources. *Kulturpflanze* **36**, 169–187.

Harja, P.K. (1993). *Linum* L. In Sharma, B.D., and Sanjappa, M. (eds), *Flora of India*. Botanical Survey of India, Calcutta, pp. 557–580.

Harlan, J.R., and de Wet, J.M.J. (1971). Toward a rational classification of cultivated plants. *Taxon* **20**, 509–517.

Heer, O. (1872). Über den Flachs und die Flachskultur im Altertum. *Neujahrsblatt der Naturforschenden Gesellschaft Zürich* **74**, 1–26.

Hegi, G. (1925), *Illustrierte Flora von Mitteleuropa.* Lehmanns Verlag, Munich, Vol. 5, Part 1, pp. 3–38.

Heinze, U., and Amelunxen, F. (1984). Zur Schleimbildung in *Linum*-Samen – Elektronenmikroskpoische und chemische Analysen. *Ber. Deutsch. Bot. Ges.* **97**, 451–464.

Helbaek, H. (1959). Notes on the evolution and history of *Linum. Kuml*, pp. 103–129.

Hetterscheid, W.L.A., and Brandenburg, W.A. (1995). Culton versus taxon: Conceptual issues in cultivated plant systematics. *Taxon* **44**, 161–175.

Hickey, M. (1988). *100 Families of Flowering Plants*, 2nd ed. University Press. Cambridge.

Hjelmquist, H. (1950). The flax weeds and the origin of cultivated flax. *Botaniska Notiser*, 2, 257–298.

Hnatiuk, R.J. (1990). *Census of Australian Vascular Plants, Australian Flora and Fauna Series No. II*. AGPS Publication, Canberra, pp. 294–295.

Howard, G.L.C. (1924). *Studies in Indian Oil Seeds, No. 2 Linseed*. Indian Dept. Agriculture, pp. 135–183.

Hudson, W. (1788). *Flora Anglica, Flora of England*. 2nd ed. R. Foulder, London, p. 134.

Jafri, S.M.H., and El-Gadi, A. (1977). *Flora of Libya*. Al Faateh University, Faculty of Science, Tripoli, pp. 1–12.

Jahandiez, E., and Maire, R. (1932). *Catalogue des plantes du Maroc*. Imprimerie Minerva, Alger, pp. 449–452.

Kenaschuk, E.O., and Rowland, G.G. (1993). Flax. In Slinkard, A.E., and Knott, D.R. (eds), *Harvest of Gold: The History of Field Crop Breeding in Canada*, University of Saskatchewan, Saskatoon, pp. 173–176.

Kitamura, S. (ed.) (1960). *Flora of Afghanistan*. Kyoto University, Kyoto, pp. 248–249.

Kulpa, W., and Danert, S. (1962). Zur Systematik von *Linum usitatissimum* L. *Kulturpflanze* 3, 341–388.

Kutuzova, S.N. (1998). Genetika l'na. In Dragavcev, V.A., and Fadeeva, T.S. (eds), *Genetika kil'turnych rastenij (len', kartofel, morkov', zelennye kul'tury, gladiolus, jablona, ljucerna)*. VIR, St. Petersburg, pp. 6–52.

Laghetti, G., Hammer, K., Olita, G., and Perrino, P. (1998). Crop genetic resources from Ustica island (Italy): Collecting and safeguarding. *Plant Genetic Resources Newsletter* 118, 12–17.

Lawalrée, A. (1964). *Flore Générale de Belgique*. Jardin Botanique de l'État, Brussels, pp. 287–292.

Linné, C. von. (1753). *Species Plantarum*. 1st ed. Impensis Laurentii Salvii. Holmiae. 287–292 p.

Marcenko, A.N. (ed.) (1994). Selekcija, semenovodstvo, vozdel'yvanije i pervičnaja obrabotka l'na-dolgunca. *Sbornik načnych trudov VNIIL*, **28–29**, 378.

Meikle, R.D. (1985). *Flora of Cyprus*. The Herbarium Royal Botanical Gardens, Kew, London. pp. 1201–1205.

Miller, Ph. (1768). *The Gardener's Dictionary*. 8th ed. John and James Revington, London.

Monastir, A.H., and Hassib, M. (eds) (1956). *Illustrated Flora of Egypt*, Imprimerie Misr, Aiin Sham, pp. 264–265.

Mouterde, P. (1986). *Nouvelle Flore du Liban et de la Syrie*. Dar El Mareq Éditeurs, Beirut, pp. 448–456.

Nickel, M. (1993). *Untersuchungen zur Erweiterung der genetischen Variation des Fettsäuremusters beim Lein* (Linum usitatissimum L.) *mit Hilfe von Artkreuzungen und in vitro Techniken. – Dissertation*. Justus-Liebig Universität, Gießen.

Ockendon, D.J. (1968). Biosystematic studies in the *Linum perenne* group. *New Phytol.* 67, 787–813.

Ockendon, D.J., and Walters, S.M. (1968). *Linum* L. In Tutin, T.G., and Heywood, V.H. (eds), *Flora Europaea*, University Press, Cambridge, pp. 206–211.

Ohwi, J. (1965). *Flora of Japan*. Smithsonian Institution, Washington DC, p. 581.

Palhinha, R.T. (1939). *Flora de Portugal*. 2nd ed. Bertrand (Irmàos) Ltd. Lisbon, pp. 448–450.

Pan, Q. (1990). Flax production, utilization and research in China. *Proc. of the Flax Institute of the United States*. Flax Institute of the United States, Fargo, ND pp. 59–63.

Pavelek, M. (1998). Analysis of current state of international flax database. *Natural Fibers, Special Edition*, Institute of Natural Fibers, Poznan, Poland, pp. 36–44.

Pignatti, S. (ed.) (1982). *Flora d'Italia*, Edagricole, Bologna, Vol. 2, pp. 20–26.

Rechinger, K.H. (1964). *Flora of Lowland Iraq*. Hafner Publishing, New York, pp. 399–400.

Rechinger, K.H. (1974). *Flora des iranischen Hochlandes und der umrahmenden Gebirge*. Akademische Druck- und Verlagsanstalt, Graz. No. 106/20, 4.74, pp. 1–18.

Robinson, B.B. (1937). Natural cross-pollination studies in fiber flax. *J. Am. Soc. Agron.* **29**, 644–649.

Rogers, C.M. (1972). The taxonomic significance of the fatty acid content of seeds of *Linum. Brittonia* **24**, 415–419.

Rozhmina, T.A., and Zhuzhenko, A.A. (1998). Study of national Russian flax collection of VNIIL, *Natural Fibers, Special Edition*, Institute of Natural Fibers, Poznan, Poland, pp. 50–55.

Rüdiger, W. (1942). Beiträge zur Kenntnis der Anatomie des Leinsamens mit besonderer Berücksichtigung der Samenfarbe. *Bastfaser* **2**, 3–9.

Rüdiger, W. (1954). Anatomische Untersuchungen über die Färbung der Leisamen. *Wissenschaftliche Zeitschrift der Martin-Luther Universität Halle-Wittenberg* **4**, 153–160.

Rykova, R., Kutuzova, S., Kornejčyk, V., Rosenberg, L., Kovalinska, Z., and Bondira, N. (1989). *The International Comecon List of Descriptors for the Species Linum usitatissimum L.* VIR. Leningrad.

Schubert, R., Werner, K., and Meusel, H. (eds) (1988). *Exkursionsflora für die Gebiete der DDR und der BRD, Band 2, Gefäßpflanzen*, Volk und Wissen, Berlin, pp. 318–319.

Scoggan, H.J. (ed.) (1978). *The Flora of Canada, Part 3, Dicotyledonae (Saururaceae to Violaceae)*. National Museums of Canada, Ottawa, pp. 1038–1040.

Small, E. (1993). The economic value of plant systematics in Canadian agriculture. *Can. J. Bot.* **71**, 1537–1551.

Small, J.K. (1907). Family 4. Linaceae, *North American Flora*. New York Botanical Garden, New York, pp. 67–87.

Smith, D.L. (1966). Linaceae, *Flora Tropical East Africa*, Crown Agents for Overseas Governments and Administrations, London, pp. 7–10.

Specht, C.-E., Freytag, U., Hammer, K., and Börner, A. (1998). Survey of seed germinability after long-term storage in the Gatersleben genebank (part 2). *Plant Genetic Resources Newsletter* **115**, 39–43.

Stewart, R.R. (1972). *Flora of West Pakistan*. Gordon College, Rawalpindi, pp. 432–433.

Stojanov, N., and Stefanov, B. (1948). *Flora of Bulgaria*. Unoversityeskaja Pečatnica, Sofia. No. 360, pp. 725–730.

Tammes, T. (1911). Das verhalten fluktuierend variierender Merkmale bei der Bastardisierung. *Recueil des travaux botaniques Neerlands* **8**, 201–288.

Tammes, T. (1923). Das genotypische Verhältnis zwischen wildem *Linum angustifolium* und dem Kulturlein, *Linum usitatissimum. Genetica* **5**, 61–76.

Tammes, T. (1927). Genetische Studien über die Samenfarbe bei *Linum usitatissimum. Hereditas* **9**, 11–16.

Tammes, T. (1928). The genetics of the genus *Linum. Bibliographica Genetica* **4**, 1–36.

Tammes, T. (1930). Die Genetik des Leins. *Züchter* **2**, 245–256.

Thellung, A. (1912). Neues aus der Adventivflora von Montepellier. *Feddes Repertorium* **11**, 69–80.

Uribe-Echebarria, P.M., and Alejandre, A.J. (1982). *Approximacion al catálogo floristico de Álava*. Vitoria, pp. 67–68.

Vavilov, N.I. (1926). Studies on the origin of cultivated plants. *Tr. po prikl. bot., gen. i sel.* **16**, 3–248.

Vul'f, E.V. (1940). Linaceae (DC.) Dumort., Linovye. In Vul'f, E.V., and Vavilov, N.I. (eds), *Kul'turnaja flora SSSR, fiber plants*, Sel'chozgiz, Moscow, Leningrad, Vol. 5, Part 1: 97–108.

Vvedenskij, A.N. (ed.) (1983). *Opredelitel' rastenij srednej Azii, kriticeskij konspekt flory.* VAN, Tashkent, pp. 19–23.

Williams, I.H. (1988). The pollination of linseed and flax. *Bee World* **69**, 145–152.

Winkler, H. (1931). Linaceae, Trib. I. 3. Linoideae-Eulineae. In Engler, A. (ed.), *Die natürlichen Pflanzenfamilien nebst ihren Gattungen und wichtigeren Arten, insbesondere den Nutzpflanzen*, 2nd ed. W. Engelmann, Leipzig, Vol. 19a, pp. 111–120.

Yuzepchuk, S.V. (1949). Genus 836. *Linum*. In Shishkin, B.K., and Ebrov, E.G. (eds), *Flora SSSR*, Izdatel'stvo Akademii Nauk, Moscow, Leningrad, pp. 86–146.

Zhuzhenko, A.A., and Rozhmina, T.A. (2000). *Mobilization of Flax Genetic Resources (Russ.)*. RASCHN, Starica.

Zhuzhenko, A.A., Ushapovsky, I.V., Kurchakova, L.N., Rozhmina, T.A., and Kolosova, T.M. (1993). *National Collection of Russian Flax*. VNIIL, Torzhok.

Zohary, D. (1999). Monophyletic vs. polyphyletic origin of the crops on which agriculture was founded in the Near East. *Genetic Res. Crop Evol.* 46, 133–142.

3 Chemical studies on the constituents of *Linum* spp.

Neil D. Westcott and Alister D. Muir

Introduction

The genus *Linum* consists of about 200 species of annual or perennial herbs or sub-shrubs. Most of our knowledge of the chemical constituents of members of the genus is limited to far fewer of the species. Most of the knowledge is, not surprisingly, based on studies of *L. usitatissimum*, the common oilseed and fiber flax species. Traditionally the utility of flax was driven by the value of its oil and the durable linen fabrics made from the stem fibers. The presence of gums in flaxseed has long been known and these gums are now being slowly developed for food and non-food uses. More recently the principal dibenzylbutyl-type lignan found in the seed of both oilseed and fiber flax has attracted significant attention as a potential pharmaceutical or nutraceutical therapy for some forms of cancer, diabetes, cardiovascular disease and lupus nephritis (see Chapters 8, 9 and 10). Other species of the genera produce podophyllotoxin-type lignans that are of interest because of their antiviral and cytotoxic activities (see Chapter 12).

LIPIDS

L. usitatissimum is the best known species of *Linum* used as an oil source. The principal distinguishing Characteristic of this oil is its high concentration of α-linolenic acid 1 (ALA, 18:3n-3). See Chapter 7 for a detailed review of the biological activity of ALA. Typically high quality flaxseed oils grown in Canada will have about 55 percent of the total oil as ALA (DeClercq and Daun, 2002). The fatty acid profiles for a number of *Linum* species have been reported. Selected data from the New Crops Database that was prepared by the United States Department of Agriculture are presented in Table 3.1 (Abbott, 1999). The predominant chain length is 18 carbons with either linolenic or linoleic 2 (18:2n-6) acid being the principal fatty acid. In one study it was determined that plants of the section Linum, of the section Dasylinum and Old World species of the section Linastrum generally have high levels of linolenic acid and lower levels of linoleic acid, while plants of section Syllinum, of section Cathartolinum and New World species of the section *Linastrum* have the reverse order (Rogers, 1972). Data on additional Linum species were reported from Australia (Green, 1984). In this paper the occurrence of ricinoleic acid 3 (12-hydroxy-9-octadecenoic acid) in species in the Syllinum taxonomic section was confirmed. The reports on the concentration of any one fatty acid need to be tempered with the observation that as temperature increases the concentration of both linoleic and linolenic acids decreases with an increase in oleic acid concentration (Canvin, 1965). Mutation breeding has also produced a form of

Table 3.1 Percentage of methyl esters of the fatty acids in *Linum* species

Species	NCSS*	14:0	16:0	16:1	18:0	18:1	18:2	18:3	20:0	20:1	22:0	22:1	24:0
L. austriacum	40,919	0.0	6.3	0.5	4.9	14.8	54.1	18.2		0.0			
L. austriacum	46,139	0.0	3.7	0.1	2.0	19.3	17.2	55.7	0.9	0.3	0.2		0.0
L. austriacum	47,049		3.9	0.0	2.4	17.9	22.9	52.6	0.3				
L. glabrescens	47,033	0.0	3.5	0.1	1.4	14.0	22.6	56.2	0.2	0.3			
L. lewisii	61,018	0.0	4.5	0.1	2.8	21.2	24.2	43.4	1.2	0.6	0.3	1.0	0.1
L. littorale	47,163	0.1	7.7	0.3	3.4	9.8	73.4	2.3	0.4	0.1	0.4		
L. mucronatum	46,176	0.1	6.0	0.2	3.6	23.9	48.1	2.0	0.1	0.4	0.1	0.1	0.0
L. narbonense	43,683	0.0	3.4	0.1	1.8	10.6	30.0	53.6	0.0	0.3			
L. perenne	40,240	0.3	3.4	0.1	1.7	16.9	22.0	53.9	1.2	0.1			
L. suffruticosum	46,577		4.8	0.0	3.6	9.0	80.8	1.8					
L. usitatissimum	45,636		6.0		4.4	24.2	15.3	50.1					

Note
* NCSS: New Crops Seed Sample number as assigned in New Crops Database (www.ncaur.usda.gov/nc).

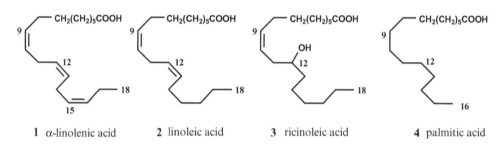

1 α-linolenic acid   **2** linoleic acid   **3** ricinoleic acid   **4** palmitic acid

Figure 3.1 Fatty acids from *Linum* spp.

L. usitatissimum, referred to as solin, whose oil is low in ALA and is greater than 70 percent linoleic acid (Dribnenki *et al.*, 1996). Additional mutation breeding has resulted in a *L. usitatissimum* that has nearly eliminated linolenic acid and quadrupled the level of palmitic acid 4 (16:0) (Saeidi and Rowland, 1999).

Protein

Part of the utility of *L. usitatissimum* is the possibility of using the protein fractions remaining after oil extraction. Canadian grown flaxseed has about 43 percent oil and 23 percent protein on a whole seed basis (DeClercq and Daun, 2002). As with other oilseeds the storage proteins can be classified as albumin and globulin. Globulins are soluble in dilute salt solutions but not in water. Albumins are soluble in water. Globulins, with a sedimentation coefficient of 12, account for about two-thirds of the total protein. Molecular weight estimates varied between 250 and 300 kDa for this fraction. The albumin fraction, sedimentation coefficient of 1.6S, has a molecular weight of 15–17 kDa. Preparation of a protein fraction by sucrose gradient centrifugation produces an albumin fraction with a sedimentation value of 2S. The 12S protein is rich in arginine, aspartic and glutamic acid and can be viewed as a nitrogen reserve. The 1.6S protein by contrast is rich in cysteine and methionine and may be viewed as a sulfur source. The amino acid composition of flaxseed meal is presented in Table 3.2 (Bhatty and Cherdkiatgumachi, 1990).

Table 3.2 Amino acid composition of flaxseed meal (Bhatty and Cherdkiatgumachi, 1990)

Amino acids (abbreviations)	Commercial flaxseed meal
Alanine (Ala)	5.5
Arginine (Arg)	11.1
Aspartic acid (Asn)	12.4
Cysteine (C-C)	4.3
Glutamic acid (Glu)	26.4
Glycine (Gly)	7.1
Histidine (His)	3.1
Isoleucine (Ile)	5.0
Leucine (Leu)	7.1
Lysine (Lys)	4.3
Methionine (Met)	2.5
Phenylalanine (Phe)	5.3
Proline (Pro)	5.5
Serine (Ser)	5.9
Threonine (Thr)	5.1
Tryptophan (Trp)	1.7
Tyrosine (Tyr)	3.1
Valine (Val)	5.6

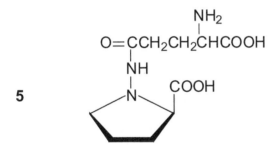

Figure 3.2 Structure of linatine.

Enzyme inhibition of some oilseeds has been reported (Bhatty, 1993). Laboratory prepared flaxseed meals had a trypsin inhibition activity (TIA) of between 42 and 51 units. Three of the four samples that were either toasted or heated had 20 units of TIA. By contrast raw canola or soy meal had 100 and 1650 TIA units, respectively. The low values in the flaxseed meal suggest that trypsin inhibition is not a serious issue.

A glutamyl derivative of D-proline, linatine 5, is a vitamin B6 antagonist found in flaxseed meal (Klosterman *et al.*, 1967). Flaxseed meal did not produce symptoms of vitamin B6 deficiency in rats but did in poultry and swine. Symptoms could be alleviated by administration of pyridoxine orally or subcutaneously (Kratzer and Williams, 1948; Klosterman *et al.*, 1967).

A series of cyclic polypeptides have been isolated from *L. usitatissimum* containing between eight and ten amino acids (Kaufmann and Tobschirbel, 1959; Morita *et al.*, 1997; Matsumoto *et al.*, 2001). They are commonly referred to as cyclolinopeptides and their composition is shown in Table 3.3 as structures 6 to 14. At least three of the peptides, CLA, CLB and CLE, have shown some immunosuppressive activity.

Table 3.3 Amino acid sequence of the cyclolinopeptides isolated from flaxseed meal

Cyclolinopeptide	Structure
CLA	Cyclo(-Pro-Pro-Phe-Phe-Leu-Ile-Ile-Leu-Val)
CLB	Cyclo(-Pro-Pro-Phe-Phe-Val-Ile-Met-Leu-Ile-)
CLC	Cyclo(-Pro-Pro-Phe-Phe-Val-Ile-Mso-Leu-Ile-)
CLD	Cyclo(-Pro-Phe-Phe-Trp-Ile-Mso-Lue-Luc-)
CLE	Cyclo(-Pro-Lue-Phe-Ile-Mso-Lue-Val-Phe)
CLF	Cyclo(Pro-Phe-Trp-Val-Mso-Leu-Mso)
CLG	Cyclo(Pro-Phe-Trp-Ile-Mso-Leu-Mso)
CLH	Cyclo(Pro-Phe-Trp-Ile-Mso-Leu-Met)
CLI	Cyclo(Pro-Phe-Trp-Met-Mso-Leu-Mso)

Note: Amino acid abbreviation as in Table 3.2 where Mso is methionine sulfoxide.

Polysaccharides

One long-recognized characteristic of *L. usitatissimum* that sets it apart from other oilseeds is the high content of polysaccharidic mucilage. The mucilage content of flax ranges from 6 to 8 percent dry weight (DW) (Mazza and Biliaderis, 1979). It is recognized that genotype influences the rheological and chemical properties of the polysaccharides from *L. usitatissimum* (Wannerberger *et al.*, 1991; Oomah *et al.*, 1995; Cui *et al.*, 1996).

The mucilage consists of both an acidic and a neutral polysaccharide in a ratio of 2:1 (Erskine and Jones, 1957; Fedeniuk and Biliaderis, 1994). The acidic polysaccharide has L-rhamnose, L-fucose, L-galactose and D-galacturonic acid in a molar ratio of 2.6:1:1.4:1.7. The neutral polysaccharide is composed of L-arabinose, D-xylose, and D-galactose in a molar ratio of 3.5:6.2:1. It is composed of branched arabinoxylans with a preponderance of terminal arabinopyranosyl units. All galactouronosyl units are located in the main chain. Most fucosyl and about half of the galactosyl units are present in non-reducing terminal groups (Muralikrishna *et al.*, 1987).

Phytochemicals

Cyanogenic glycosides

Cyanogenic glucosides occur in all organs of the flax plant. The monoglucosides, linamarin **15** and lotaustralin **16**, were found in germinating seeds, leaves, flowers and developing embryos. The diglucosides, linustatin **17** and neolinustatin **18**, were also found in developing embryos and are the main cyanogenic glucosides in mature seed (gentiobiose = 6-O-β-glucosyl-D-glucone). Roots and stems contained relatively low concentrations of cyanogenic glucosides (Niedzwiedz Siegien, 1998).

In a study of Canadian oilseed types it was determined that there was considerable variation in the content of cyanogenic glucosides. The principal cyanogenic glycoside was linustatin (213 to 352 mg/100 g of seed), followed by neolinustatin (91 to 203 mg/100 g). Linamarin was only detected at low levels (< 32 mg/100 g). Most of the variation was seen in different cultivars but there was also variation related to location and year (Oomah *et al.*, 1992). Cyanogenic glycosides have been reported to be present in several *Linum* species (*L. arboreum, L. campanulatum, L. flavum, L. gallicum,*

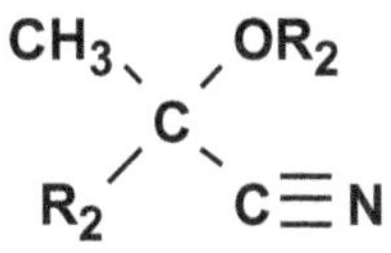

	Name	R_1	R_2
15	Linamarin	CH_3	β-glucose
16	Linustatin	CH_3	β-gentiobiose
17	Lotasutralin	C_2H_5	β-glucose
18	Neolinustatin	C_2H_5	β-gentiobiose

Figure 3.3 Cyanogenic glycosides from *Linum* spp.

L. maritimum, L. alpinum, L. catharticum, L. grandiflorum, L. narbonse, L. perenne, L. suffruticosum, L. kingii, L. lewisii and L. marginale) (Hegnauer, 1966; Hegnauer, 1989).

Lignans

Lignans are a class of phytochemicals that are formed from the coupling of two cinnamyl alcohols (Xia *et al.*, 2000). Lignans should not be confused with lignins that are more complex molecules containing several cinnamyl fragments. There are hundreds of lignans reported which differ in their degree of oxidation, substitution patterns and degree of coupling (Ayers and Loike, 1990). In the genus *Linum* the principal types of lignans belong to the dibenzyl butane group (e.g. secoisolariciresinol diglucoside, (SDG), **23**), aryltetralin group (e.g. 5-methoxypodophyllotoxin, **29**) and arylnaphthalene group (e.g. justicidin B, **42**) (Bakke and Klosterman, 1956; Broomhead and Dewick, 1990; Mohagheghzadeh *et al.*, 2002).

Coniferyl alcohol **19** or coniferin **20**, the glucoside of coniferyl alcohol, have been reported from several species (Berlin *et al.*, 1986; Broomhead and Dewick, 1990; van Uden *et al.*, 1991; van Uden *et al.*, 1994a; van Uden *et al.*, 1994b; Oostdam and van der Plas, 1996; Smollny *et al.*, 1998; Xia *et al.*, 2000). There is evidence that dibenzyl butane and aryltetralin lignans are biosynthesized from a dirigent protein-mediated coupling of two E-coniferyl alcohols to produce (+)-pinoresinol **21**, as presented in Figure 3.4 (Xia *et al.*, 2000). Pinoresinol is enatiospecifically converted into (+)-lariciresinol (**22**) and then to (−)-secoisolariciresinol (**23**). In *L. usitatissimum* this is di-glucosylated to produce SDG (**24**). Low levels of (+)-pinoresinol, (+)-lariciresinol and (−)-matairesinol (**25**) have been reported in flaxseed extracts (Mazur *et al.*, 1996; Meagher *et al.*, 1999; Liggins *et al.*, 2000).

Early investigators recognized that SDG was part of a larger complex in *L. usitatissimum* (Bakke and Klosterman, 1956). These investigators determined that it was necessary to treat the naturally occurring complex with alkali to liberate free SDG. As co-products they isolated 3-hydroxy-3methylglutaric acid **31** (HMGA), and two cinnamic acids, gluco-coumaric **32** and gluco-caffeic **33** acids, all as their methyl esters (Klosterman and Smith, 1954; Klosterman *et al.*, 1955; Klosterman and Muggli, 1959).

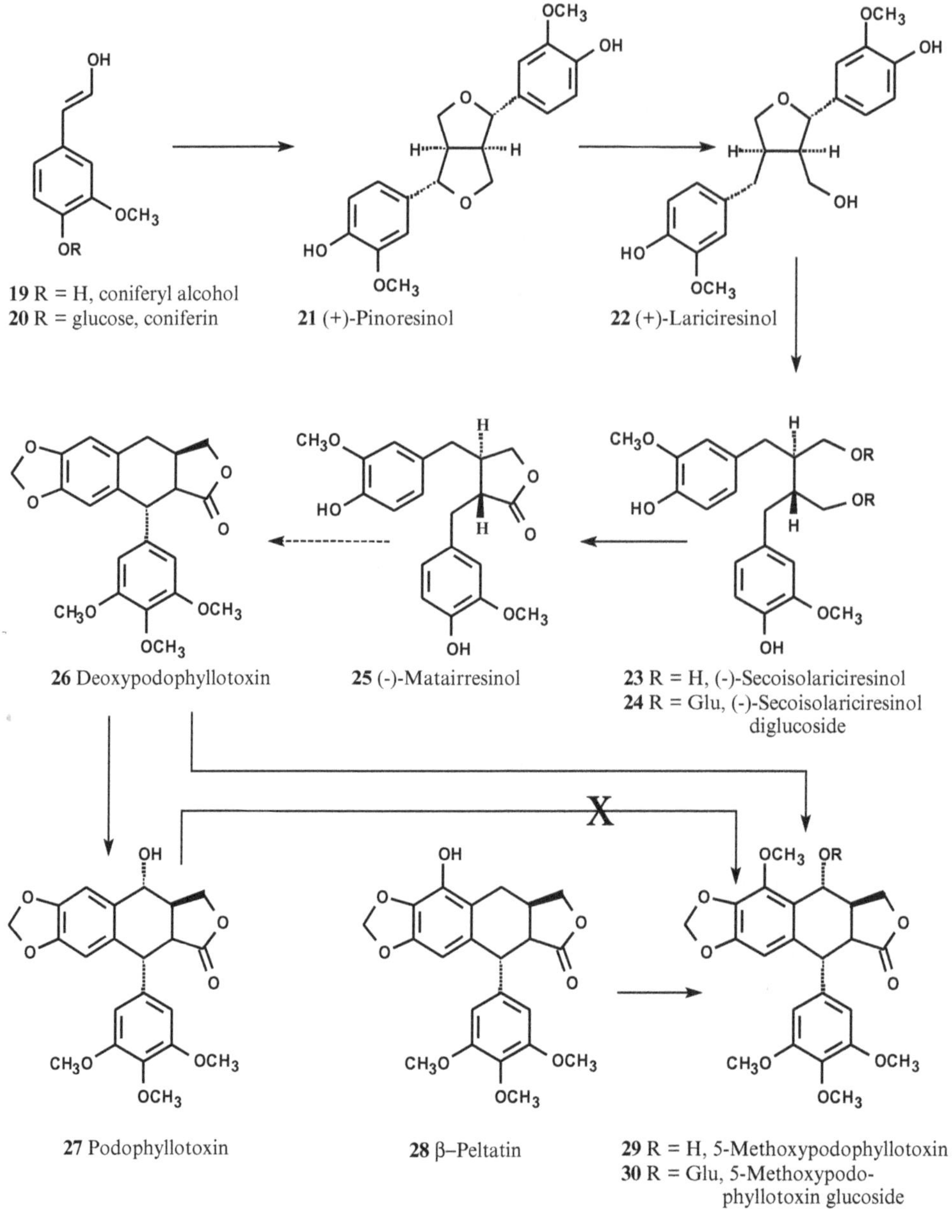

19 R = H, coniferyl alcohol
20 R = glucose, coniferin

21 (+)-Pinoresinol

22 (+)-Lariciresinol

26 Deoxypodophyllotoxin

25 (-)-Matairresinol

23 R = H, (-)-Secoisolariciresinol
24 R = Glu, (-)-Secoisolariciresinol
        diglucoside

X

27 Podophyllotoxin

28 β–Peltatin

29 R = H, 5-Methoxypodophyllotoxin
30 R = Glu, 5-Methoxypodo-
        phyllotoxin glucoside

Figure 3.4 Biosynthetic pathways leading to the synthesis of (−) secoisolariciresinol and 5-methoxyphodophyllotoxin beginning from coniferyl alcohol.

Subsequently it was reported that the complex also contained 3-methoxy-5-gluco-cinnamic acid **35** (Luyengi *et al.*, 1993). The assignment has been challenged by two independent groups (Hall *et al.*, 2000; Westcott *et al.*, 2000; Johnsson *et al.*, 2002). These groups have unequivocally shown that this is really 3-methoxy-4-gluco-cinnamic acid **34**. It is now known that that the HMGA and SDG form an ester-linked

31

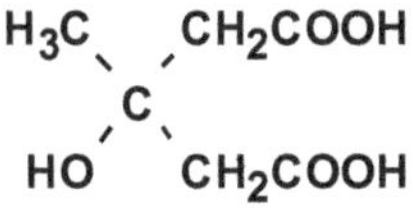
H3C
CH2COOH
C
HO
CH2COOH

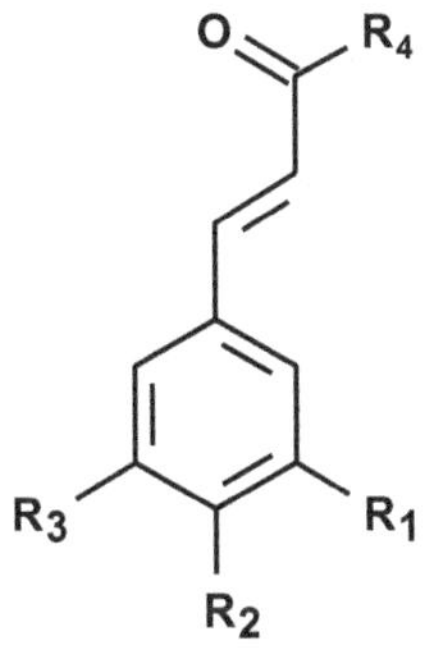
O
R4
R3
R1
R2

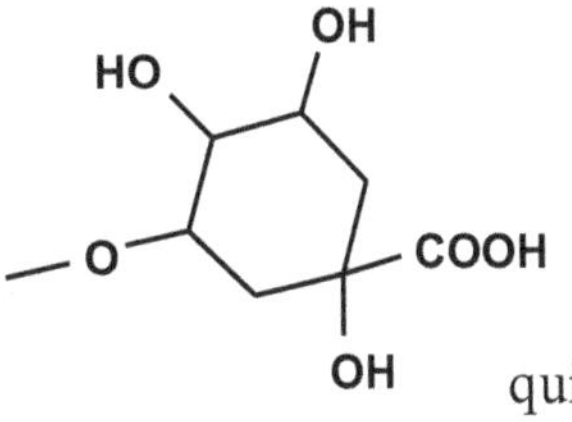
OH
HO
O
COOH
OH

quinoyl

	Name	R_1	R_2	R_3	R_4
32	4-Glucosyl-cinnamic acid	H	O-glu	H	OH
33	3-Hydroxy-4-glucosyl-cinnamic acid	OH	O-glu	H	OH
34	3-Methoxy-4-glucosyl-cinnamic acid	OCH_3	O-glu	H	OH
35	3-Methoxy-5-glucosyl-cinnamic acid	OCH_3	H	O-glu	OH
44	*p*-Coumaric acid	H	OH	H	OH
45	Caffeic acid	OH	OH	H	OH
46	Ferulic acid	OCH_3	OH	H	OH
47	Sinapic acid	OCH_3	OH	OCH_3	OH
48	*p*-Coumaryl-quinic acid	H	OH	H	quinoyl
49	*p*-Coumaryl glucose	H	OH	H	O-glu
50	Chlorogenic acid	OH	OH	H	quinoyl
51	Caffeoyl glucose	OH	OH	H	O-glu
52	Ferulol glucosey	OCH_3	OH	H	O-glu

glu = glucose

Figure 3.5 Cinnamic acid derivatives from *Linum* spp.

oligomer (Figure 3.6) (Ford *et al.*, 2001; Kamal-Eldin *et al.*, 2001). It is not known where the cinnamic acids fit into the oligomer described or if they are part of another ester based complex.

An examination of several cultivars of oilseed-type flax found that the concentration of SDG (24) could vary between 9 and 30 mg of SDG per g of defatted meal (Westcott

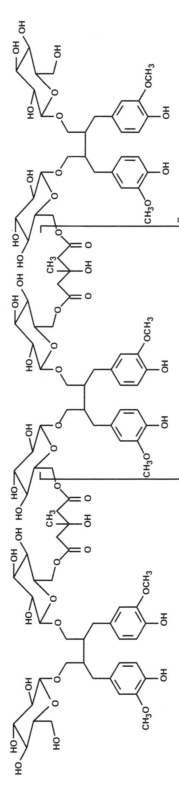

Figure 3.6 Postulated structure for the lignan oligomer containing secoisolariciresinol diglucoside and 3-hydroxy-3-methyl-gluctaric acid units.

Table 3.4 Predominant enatiomer of SDG in *Linum* species: + for (+) SDG; − for (−) SDG: and +/− for racemic mixture

Species	(+) SDG	(−) SDG	(+/−) SDG
L. angustifolium (= L. bienne)	+		
L. altaicum		−	
L. austriacum		−	
L. bienne	+	−	
L. campanulatum			+/−
L. capitatum	+		
L. decumbens			+/−
L. elegans Spruner			+/−
L. flavum			+/−
L. grandiflorum			+/−
L. hirsutum	+		
L. komarovii		−	
L. leonii		−	
L. lewisii (= L. perenne)		−	
L. narbonense		−	
L. pallescens		−	
L. perenne		−	
L. setaceum	+		
L. stelleroides		−	
L. strictum			+/−
L. tauricicum			+/−
L. tenuifolium			+/−
L. trigynum			
L. usitatissimum	+		

and Muir, 1996). Location and year had less influence on SDG concentration in the seed. In *L. usitatissimum* there is also a low concentration of the (+) enatiomer (Muir and Westcott, 1998). In a recent survey of *Linum* accessions it was determined that many *Linum* species produced a predominance of one of the two enatiomers of SDG in the seed but some species had both enatiomers (Table 3.4) (Westcott *et al.*, 2002).

The bio-coversion of (−)-secoisolariciresinol **23** into 5-methoxypodophyllotoxin **29** is not a trivial matter (refer to Figure 3.4). In *L. flavum* species, (−)-secoisolariciresinol is converted to (−)-matairesinol **25** and finally into 5-methoxypodophyllotoxin **29** (Xia *et al.*, 2000). Additional biosynthetic relationships from (−)-matairesinol to 5-methoxypodophyllotoxin have been elucidated using cell cultures of *L. flavum*. Suspension cultures of *L. flavum* converted podophyllotoxin **27** into its glucoside and not into 5-methoxypodophyllotoxin **29** (van Uden *et al.*, 1993). Feeding of deoxypodophyllotoxin **26** resulted in an accumulation of 5-methoxypodophyllotoxin **29** and its glucoside **30** (van Uden *et al.*, 1997). β-peltatin **28** was converted into 5-methoxypodophyllotoxin glucoside **30** by a *L. flavum* cell culture (van Uden *et al.*, 1997). Thus, it appears that 5-methoxypodophyllotoxin is biosynthesized via β-peltatin rather than from podophyllotoxin.

The investigation of aryltetralin-type lignans (podophyllotoxin and peltatin derivatives) has occurred in many *Linum* species (*L. arboreum, L. campanulatum, L. cariense, L. elegans, L. mucronatum, L. pampylicum, L. tauricum, L. thracicum, L. austricum, L. lewisii, L. monogynum, L. sibiricum, L. hirustum, L. vicosum, L. corymbulosum, L. boissieri, L. flavum,*

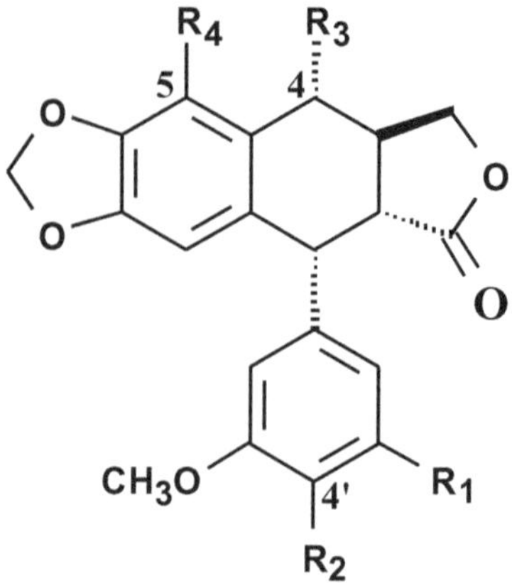

	Name		R_2	R_3	R_4
Podopyllotoxins					
26	3'-Deoxypodophyllotoxin	OH	OCH_3	H	H
27	Podophyllotoxin	OCH_3	OCH_3	OH	H
29	5-Methoxypodophyllotoxin	OCH_3	OCH_3	OH	OCH_3
30	4-O-β-D-Glucosyl-5-methoxypodophyllotoxin	OCH_3	OCH_3	O-glu	OCH_3
37	4-Acetyl-5-methoxypodophyllotoxin	OCH_3	OCH_3	O-Ac	OCH_3
40	4-O-β-D-Glucosyl-podophyllotoxin	OCH_3	OCH_3	O-glu	H
41	5'-Demethoxy-5-methoxypodophyllotoxin	H	OCH_3	OH	OCH_3
Peltatins					
28	β-Peltatin	OCH_3	OCH_3	H	OH
36	α-Peltatin	OCH_3	OH	H	OH
38	5-O-β-D-Glucosyl-α-Peltatin	OCH_3	OH	H	O-glu
39	5-O-β-D-Glucosyl-β-Peltatin	OCH_3	OCH_3	H	O-glu

Ac = acetate

Figure 3.7 Aryltetralin lignans from *Linum* spp.

L. capitatum, L. album, L. nodiflorum, L. carthharticum, and L. austriacum) (Weiss *et al.*, 1975; Berlin *et al.*, 1988; Broomhead and Dewick, 1988; Broomhead and Dewick, 1990; Wichers *et al.*, 1991; Konuklugil, 1996; Konuklugil, 1997a; Konuklugil, 1997b; Konuklugil, 1998; Smollny *et al.*, 1998; Konuklugil *et al.*, 1999; Mohagheghzadeh *et al.*, 2002). In a survey of 26 *Linum* species, sub-species or accessions from four sections of this genus it was determined that podophyllotoxin **27**, 5-methoxypodophyllotoxin **29**, α-peltatin **36** and β-peltatin **28** were detectable, some in concentrations as high as 24 mg/g (Konuklugil, 1996). All of the species studied contain aryltetralin-type lignans and it has been suggested that the genus *Linum* is characterized by them (Konuklugil, 1998).

In a study of roots, stems/leaves and roots of *L. flavum*, *L. flavum compactum* and *L. capitatum* aryltetralin lignans were detected (Broomhead and Dewick, 1990). In both *L. flavum* and *L. flavum compactum*, 5-methoxyodophyllotoxin **29**, its glucoside **30** and acetate **37** were the main constituents in roots and stems/leaves. The roots contain up to 3.5 percent DW of these compounds. *L. capitatum* contained both 5-methoxyo-dophyllotoxin **29** and its glucoside **30** in the roots, stems and leaves produced α- and β-peltatins (**36** and **28**, respectively) and their glucosides (**38** and **39**, respectively). Alcoholic extracts of the leaves and roots of *L. catharticum* contained podophyllotoxin **27** and its glucoside **40**, 5-methoxypodophyllotoxin **29** and its acetate **37** and gluco-side **30** plus α-pelatin **28** (Konuklugil, 1998).

While initial studies have examined intact plants, much of the more recent work has concentrated on the use of cell cultures (Petersen and Alfermann, 2001). In *L. album* suspension cultures glycosides of podophyllotoxin **40** and 5-methoxypodophyllotoxin **30** were the main products (Smollny *et al.*, 1998). Lesser amounts of deoxypodophyl-lotoxin **26**, 5′-demethoxy-5-methoxypodophyllotoxin **41**, lariciresinol **22**, pinoresinol **21**, matairesinol **25**, α- and β-peltatin (**36** and **28**, respectively) were also detected. Cell suspension culture of *L. nodiflorum* produces 5-methoxypodophyllotoxin **30**, podophyllotoxin **27** and deoxypodophyllotoxin **26** (Konuklugil *et al.*, 1999). As early as 1988 it was demonstrated that the cytotoxic activity of root cultures of *L. flavum* was due to the accumulation of 5-methoxypodophyllotoxin-4-β-D-glucoside **30** (Berlin *et al.*, 1988). The aglycon **29** exhibited a 250–500 times higher cytotoxic activity. On a dry weight basis root cultures contained between 0.7 and 1.3 percent podophyllo-toxins (Berlin *et al.*, 1986).

Callus, suspension and normal and hairy root cultures of *L. austriacum* produce the arylnaphthalene lignans justicidin B **42** and isojusticidin B **43** (Mohagheghzadeh *et al.*, 2002).

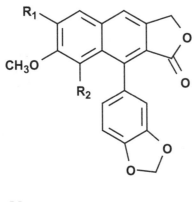

	Name	R$_1$	R$_2$
42	Justicidin B	OCH$_3$	H
43	Iso-Justicidin B	H	OCH$_3$

Figure 3.8 Arylnaphthalene lignans from *Linum* spp.

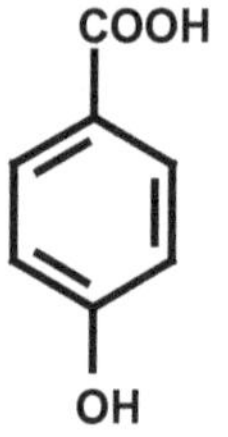

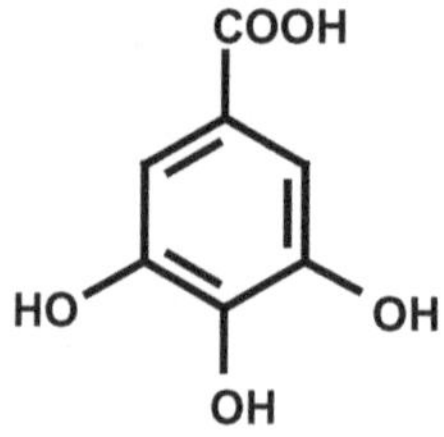

Figure 3.9 Structures of 4-hydroxybenzoic acid **53**, gallic acid **54** and phytic acid **55**.

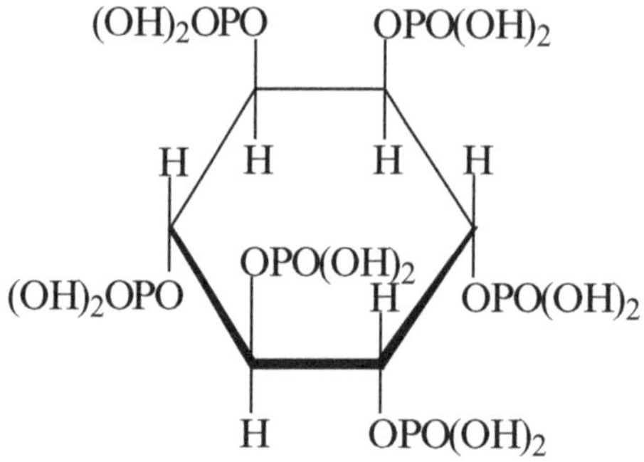

Phenolic and other acids

The occurrence of the glucosylated coumaric, caffeic and ferulic acids in the SDG complex was discussed earlier. In addition, the presence of coumaric **44**, caffeic **45**, ferulic **46** and sinapic acids **47** was reported to occur in the dehulled, defatted flaxseed (Dabrowski and Sosulski, 1984). In cotyledons and young shoots of *L. ustitassimum* a number of cinnamic acid derivatives were identified (Ibrahim and Shaw, 1970). These include *p*-coumaryl quinic acid **48**, *p*-coumaryl glucose **49**, chlorogenic (3-O-caffeoyl quinic) acid **50**, glucosyl caffeic acid **35**, caffeoyl glucose **43**, glucosyl ferulic acid **51**, feruloyl glucose **52** and a unspecified glycoside and an ester of sinapic acid. Chlorogenic **50**, 4-hydroxybenzoic **53**, and gallic **54** acids were also reported (Harris and Haggerty, 1993). Flaxseed meal contains 2 to 3 percent of phytic acid (inositol hexaphosphate) **55**, depending on cultivar and environment (Oomah *et al.*, 1996).

Isoprenoids

In the non-saponifiable fraction of linseed oil a number of oil-soluble materials have been identified (refer to Figure 3.10). The phytochemicals derived from triterpenes have been identified as cholesterol **56**, campesterol **57**, stigmasterol **58**, Δ5-avenasterol **59**, cycloartenol **60** and 24-methylenecycloartenol **61** (Middleditch and Knights, 1972).

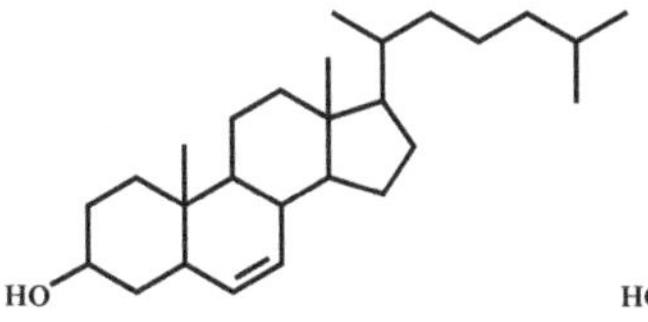

56 Cholesterol 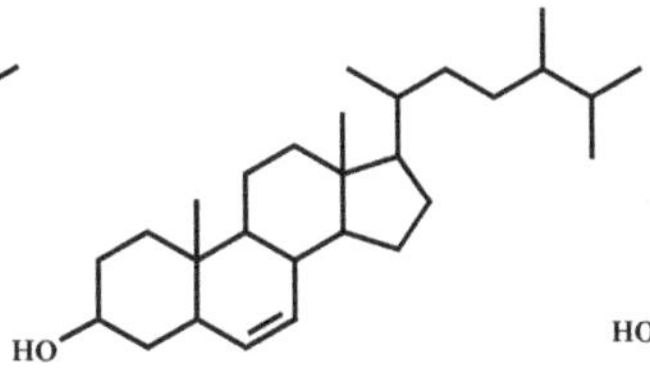   **57** Campesterol 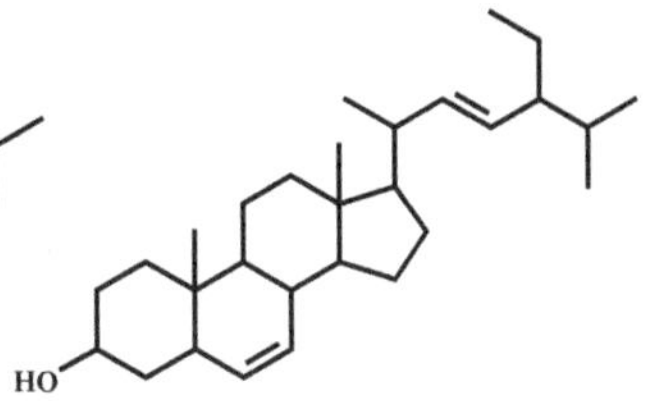   **58** Stigmasterol

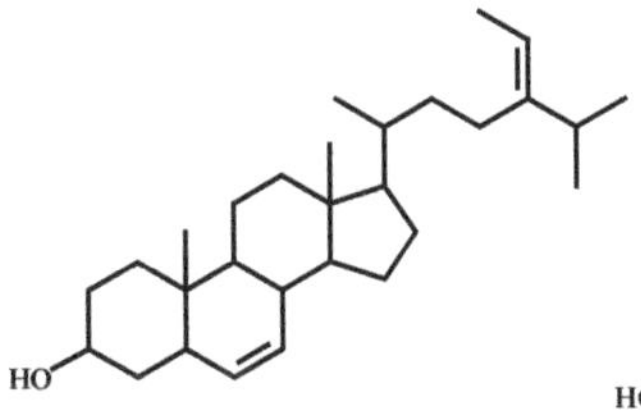

59 Δ5-Avenasterol 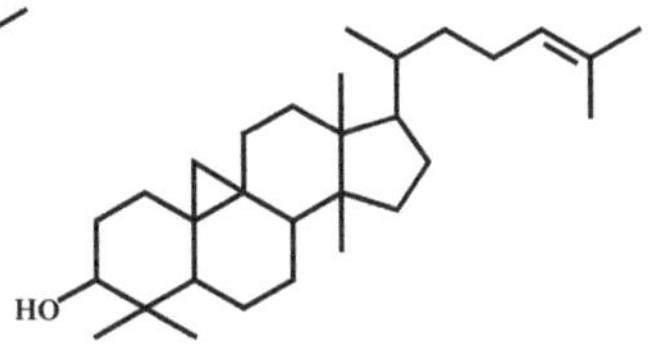   **60** Cycloartenol 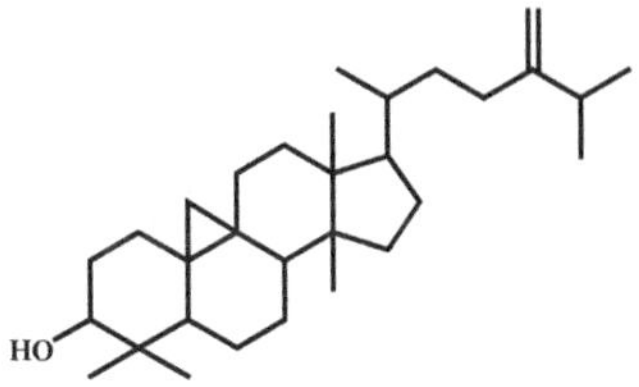   **61** 24-Methylenecycloartenol

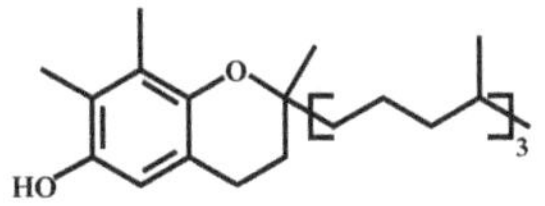

62 gamma-Tocopherol 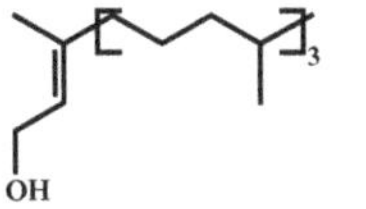   **63** Phytol 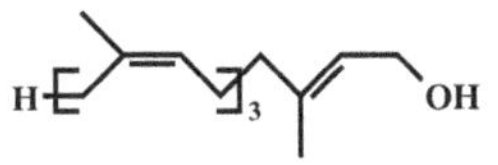   **64** Geranyl geraniol

Figure 3.10 Isoprenoids from *Linum* spp.

In the phospholipid fraction some of the sterols exist as glycosides (Aylward and Nichols, 1962). Tocopherols, natural fat-soluble anti-oxidants containing 40 carbons, are present in the oil at a concentration of about 10 mg/100 g seed. The concentration varies with cultivar and growing conditions. Gamma-tocopherol **62** accounts for about 80 percent of the total tocopherol in the oil (Oomah *et al.*, 1997). Two diterpenes, phytol **63** and geranyl geraniol **64**, were isolated from a linseed oil distillate fraction (Fedeli *et al.*, 1966).

Flavonoids

Interest in flavonoids is being stimulated by the knowledge that some have anti-oxidant properties or inhibitory activity against carcinogen-induced tumors. There are several classes of flavonoids that differ in the degree of oxidation. Within any class the substitution pattern will also vary.

An examination of color mutants of *L. usitatissimum* demonstrated the occurrence of several anthocyanin pigments. The anthocyanins included the 3-glucoylrutinosides of pelargonidin, cyanidin, and delphinidin (**65** to **67**, respectively); the 3-triglucosides of delphinidin and cyanidin (**68** to **70**, respectively); a 3-diglucoside of delphinidin **71**; 3-rutinosides of pelargonidin, cyanidin, and delphinidin (**72** to **74**, respectively); and 3-glucosides of pelargonidin, cyanidin, and delphinidin (**75** to **77**, respectively)

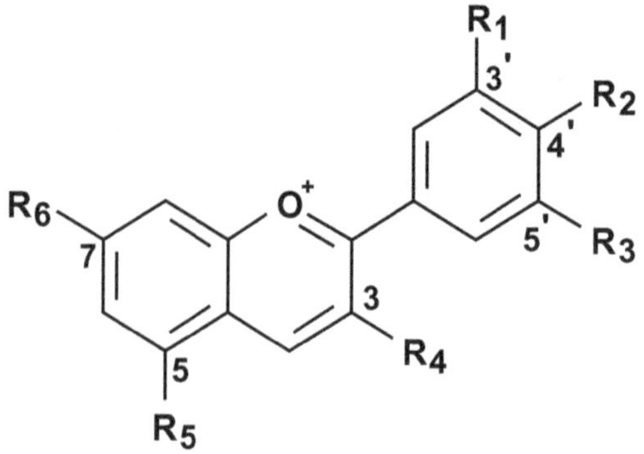

	Name	R₁	R₂	R₃	R₄	R₅	R₆
		R_1	R_2	R_3	R_4	R_5	R_6
65	3-Glucorutinosyl pelargonidin	H	OH	H	O-glu-rut	OH	OH
66	3-Glucorutinosyl cyanidin	OH	OH	H	O-glu-rut	OH	OH
67	3-Glucorutinosyl delphinidin	OH	OH	OH	O-glu-rut	OH	OH
68	3-tri-Glucosyl pelargonidin	H	OH	H	O-tri-glu	OH	OH
69	3-tri-Glucosyl cyanidin	OH	OH	H	O-tri-glu	OH	OH
70	3-tri-Glucosyl delphinidin	OH	OH	OH	O-tri-glu	OH	OH
71	3-di-Glucosyl delphinidin	OH	OH	OH	O-di-glu	OH	OH
72	3-Rutinosyl pelargonidin	H	OH	H	O-rut	OH	OH
73	3-Rutinosyl cyanidin	OH	OH	H	O-rut	OH	OH
74	3-Rutinosyl delphinidin	OH	OH	OH	O-rut	OH	OH
75	3-Glucosyl pelargonidin	H	OH	H	O-glu	OH	OH
76	3-Glucosyl cyanidin	OH	OH	H	O-glu	OH	OH
77	3-Glucosyl delphinidin	OH	OH	OH	O-glu	OH	OH
78	3-Xylorutinosyl delphinidin	OH	OH	OH	O-xyl-rut	OH	OH

rut = rutinose, xyl = xylose

Figure 3.11 Anthocyanins from *Linum* spp.

(Dubois and Harborne, 1975). An anthocyanidin triglycoside was isolated from the scarlet flowers of *L. grandiflorum* cv scarlet flax as a major anthocyanin. The structure of the main pigment was determined to be delphinidin 3-O-3-xylosylrutinoside 78 (Toki *et al.*, 1995). The 3-rutinosides of delphinidin and cyanidin were also identified.

The flavonols, 3,7-O-dimethoxy-herbacetin 79, 3,7-O-diglucopyranosyl-kaempferol 80, and 3,8-O-diglucopyranosyl-herbacetin 81 were isolated from common flaxseed meal (Qiu *et al.*, 1999). *L. usitatissimum* contains four mono C-glycoside flavones: orientin 82, iso-orientin 83, vitexin 84, iso-vitexin 85; and four di-C-glycosides: vicenin-1 86, vicenin-2 87, lucenin-1 88 and 7-rhanmosyl-lucenin-2 89 (Dubois and Mabry, 1971).

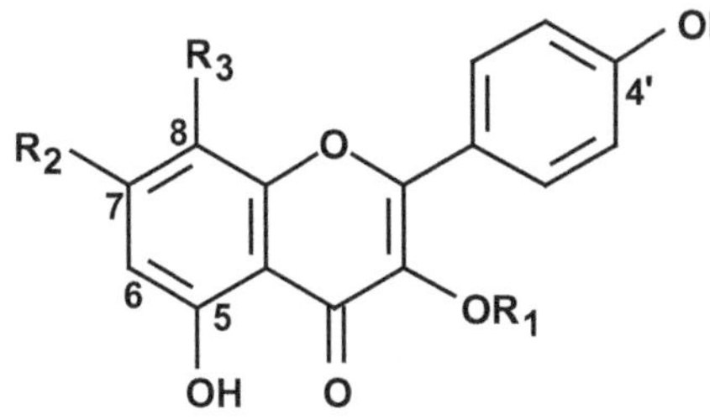

	Name	R$_1$	R$_2$	R$_3$
79	3,7-Dimethoxy-herbacetin	CH$_3$	OCH$_3$	OH
80	3,7-Diglucopyranosyl-kaempferol	glu	O-glu	H
81	3,8-O-Diglucopyranosyl-herbacetin	glu	H	O-glu

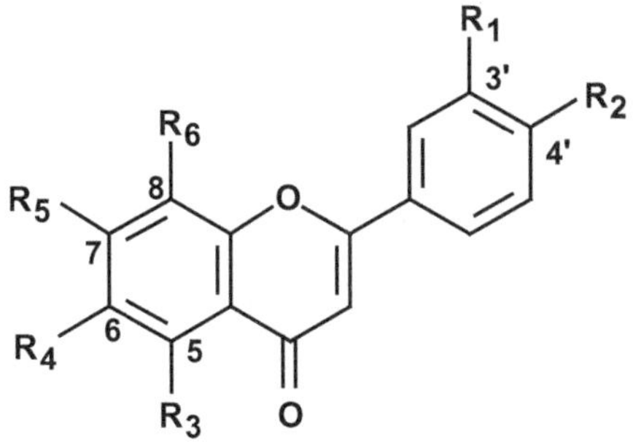

	Name	R$_1$	R$_2$	R$_3$	R$_4$	R$_5$	R$_6$
82	Orientin	OH	OH	OH	H	OH	C-glu
83	Iso-orientin	OH	OH	OH	C-glu	OH	H
84	Vitexin	H	OH	OH	H	OH	C-glu
85	Iso-vitexin	H	OH	OH	C-glu	OH	H
86	Vicenin-1	H	OH	OH	C-xyl	OH	C-glu
87	Vicenin-2	H	OH	OH	C-glu	OH	C-glu
88	Lucenin-1	OH	OH	OH	C-xyl	OH	C-glu
89	7-Rhamnosyl-lucenin-2	OH	OH	OH	C-glu	O-rham	C-glu
90	3',4'-Dimethoxy-7-rhamnosyl-luteolin	OCH$_3$	OCH$_3$	OH	H	O-rham	H
91	Linoside A	OH	OH	OH	H	OH	C-glu-2"rham-6"-Ac
92	Linoside B	OH	OH	OH	H	OH	C-glu-2"rham

rham = rhamnose

Figure 3.12 Flavonoids from *Linum* spp.

Iso-orientin has also been reported from *L. capitatum* (Stosic *et al.*, 1989). *Linum maritimum* contains 3', 4'-dimethoxy-7-rhamnosyl-luteolin **90** and two flavone-C-glycosides, linoside A **91** and linoside B **92** which are rhamnosyl derivatives of iso-orientin (Volk and Sinn, 1968; Wagner *et al.*, 1972).

References

Abbott, T.P. (1999). New Crops Database, United States Department of Agriculture, www.ncaur.usda.gov/nc

Ayers, D.C., and Loike, J.D. (1990). *Lignans: Chemical, Biological and Clinical Properties.* Cambridge University Press, Cambridge.

Aylward, F., and Nichols, B.W. (1962). Plant lipids. II. Free and combined sterols (sterol esters and glycosides in commercial oilseed phospholipids. *J. Sci. Food Agric.* 13, 86–91.

Bakke, J.E., and Klosterman, H.J. (1956). A new diglucoside from flaxseed. *Proc. N. Dakota Acad. Sci.* 10, 18–22.

Berlin, J., Bedorf, N., Mollenschott, C., Wray, V., Sasse, F., and Höfle, G. (1988). On the podophyllotoxins of root cultures of *Linum flavum. Planta Med.* 54, 204–206.

Berlin, J., Wray, V., Mollenschott, C., and Sasse, F. (1986). Formation of β-peltatin-A-methyl ether and coniferin by root cultures of *Linum flavum. J. Natural Prod.* 49, 435–439.

Bhatty, R.S. (1993). Further compositional analyses of flax: Mucilage, trypsin inhibitors and hydrocyanic acid. *J. Am. Oil Chem. Soc.* 70, 899–904.

Bhatty, R.S., and Cherdkiatgumachi, P. (1990). Compositional analysis of laboratory-prepared and commercial samples of linseed meal and of hull isolated from flax. *J. Am. Oil Chem. Soc.* 67, 79–84.

Broomhead, A.J., and Dewick, P.M. (1988). *Linum* species as a source of anticancer lignans. *J. Pharm. Pharmacol.* 40 Suppl, 55p.

Broomhead, A.J., and Dewick, P.M. (1990). Aryltetralin lignans from *Linum flavum* and *Linum capitatum. Phytochemistry* 29, 3839–3844.

Canvin, D.T. (1965). The effect of temperature on the oil content and fatty acid composition of the oils from several oilseed crops. *Can. J. Bot.* 43, 63–69.

Cui, W., Kenaschuk, E., and Mazza, G. (1996). Influence of genotype on chemical composition and rheological properties of flaxseed gums. *Food Hydrocolloids* 10, 221–227.

Dabrowski, K.J., and Sosulski, F.W. (1984). Composition of free and hydrolizable phenolic acids in defatted flours of ten oilseeds. *J. Agric. Food Chem.* 43, 2016–2019.

DeClercq, D.R., and Daun, J.K. (2002). Quality of Western Canadian Flaxseed, Canadian Grain Commission, http://www.cgc.ca/quality/qualmenu-e.htm#Flaxseed

Dribnenki, J.C.P., Green, A.G., and Atlin, G.N. (1996). Linola 989 low linolenic flax. *Can. J. Plant Sci.* 76, 329–331.

Dubois, J.A., and Harborne, J.B. (1975). Anthocyanin inheritance in petals of flax, *Linum usitatissimum. Phytochemistry* 14, 2491–2494.

Dubois, J.A., and Mabry, T.J. (1971). The C-glycosylflavonoids of flax, *Linum usitatissimum. Phytochemistry* 10, 2839–2840.

Erskine, A.J., and Jones, J.K.N. (1957). The structure of linseed mucilage. Part I. *Can. J. Chem.* 35, 1174–1182.

Fedeli, E., Capella, P., Cirimele, M., and Jacini, G. (1966). Isolation of geranylgeraniol from the unsaponifiable fraction of linseed oil. *J. Lipid Red.* 7, 437–441.

Fedeniuk, R.W., and Biliaderis, C.G. (1994). Composition and physicochemical properties of linseed (*Linum usitatissimum* L.) mucilage. *J. Agric. Food Chem.* 42, 240–247.

Ford, J.D., Huang, K.S., Wang, H.B., Davin, L.B., and Lewis, N.G. (2001). Biosynthetic pathway to the cancer chemopreventive secoisolariciresinol diglucoside-hydroxymethyl glutaryl ester-linked lignan oligomers in flax (*Linum usitatissimum*) seed. *J. Natural Prod.* 64, 1388–1397.

Green, A.G. (1984). The occurrence of ricinoleic acid in *Linum* seed oils. *J. Am. Oil Chem. Soc.* 61, 939–940.

Hall, T.W., Westcott, N.D., and Muir, A.D. (2000). Phenylpropanoid glucoside from flaxseed. *83rd CSC Conference, May 27–31, 2000. Calgary.*

Harris, R.K., and Haggerty, W.J. (1993). Assays for potentially anticarcinogenic phytochemicals in flaxseed. *Cereal Foods World* 38, 147–151.

Hegnauer, R. (1966). Dicotyledoneae: Daphniphyllaceae-Lythraceae. In *Chemotaxonomie der pflanzen; eine übersicht über die verbreitung und die systematische bedeutung der pflanzenstoffe, Band 4*, Burkhauser, Based, pp. 393–401.

Hegnauer, R. (1989). Nachträge zu Band 3 und Band 4: (Acanthaceae bis Lythraceae). In *Chemotaxonomie der pflanzen; eine übersicht über die verbreitung und die Systematische bedeutung der pflanzenstoffe. Band 8*, Birkhäuser, Basel, pp. 669–672.

Ibrahim, R.K., and Shaw, M. (1970). Phenolic constituents of the oil flax (*Linum usitatissimum*). *Phytochemistry* 9, 1855–1858.

Johnsson, P., Peerlkamp, N., Kamal-Eldin, A., Andersson, R.E., Andersson, R., Lundgren, L.N., and Aman, P. (2002). Polymeric fractions containing phenol glucosides in flaxseed. *Food Chem.* 76, 207–212.

Kamal-Eldin, A., Peerlkamp, N., Johnsson, P., Andersson, R., Andersson, R.F., Lundgren, L.N., and Aman, P. (2001). An oligomer from flaxseed composed of secoisolariciresinol diglucoside and 3-hydroxy-3-methyl glutaric acid residues. *Phytochemistry*, 58, 587–590.

Kaufmann, H.P., and Tobschirbel, A. (1959). Über ein Ooligopeptid aus Leinsamen. *Chem. Ber.* 92, 2805–2809.

Klosterman, H.J., Lamoureux, G.L., and Parsons, J.L. (1967). Isolation, characterization, and synthesis of linatine, a vitamin B_6 antagonist from flaxseed (*Linum usitatissimum*). *Biochem.* 6, 170–177.

Klosterman, H.J., and Muggli, R.Z. (1959). The glucosides of flaxseed. II. Linocaffein. *J. Am. Chem. Soc.* 81, 2188–2191.

Klosterman, H.J., and Smith, F. (1954). The isolation of β-hydroxy-β-methylglutaric acid from the seed of flax (*Linum usitatissimum*). *J. Am. Chem. Soc.* 76, 1229–1230.

Klosterman, H.J., Smith, F., and Clagett, C.O. (1955). The constitution of Linocinnamarin. *J. Am. Chem. Soc.* 77, 420–421.

Konuklugil, B. (1996). Aryltetralin lignans from genus *Linum. Fitoterapia* 67, 379–381.

Konuklugil, B. (1997a). Lignans from *L. flavum* var. *compactum* (Linaceae). *Biochem. Syst. Ecol.* 25, 75.

Konuklugil, B. (1997b). Lignans from *Linum boissieri. Fitoterapia* 68, 183.

Konuklugil, B. (1998). Arytetralin lignans from *Linum catharticum* L. *Biochemical Systematics and Ecology* 26, 795–796.

Konuklugil, B., Schmidt, T.J., and Alfermann, A.W. (1999). Accumulation of aryltetralin lactone lignans in cell suspension cultures of *Linum nodiflorum. Planta Med.* 65, 587–588.

Kratzer, F.H., and Williams, D.E. (1948). The improvement of linseed oil meal for chick feeding by the addition of synthetic vitamins. *Poultry Sci.* 27, 236–238.

Liggins, J., Grimwood, R., and Bingham, S.A. (2000). Extraction and quantification of lignan phytoestrogens in food and human samples. *Anal. Biochem.* 287, 102–109.

Luyengi, L., Pezzuto, J.M., Waller, D.P., Beecher, C.W.W., Fong, H.H.S., Che, C.T., and Bowen, P.E. (1993). Linusitamarin, a new phenylpropanoid glucoside from *Linum usitatissimum. J. Natural Prod.* 56, 2012–2015.

Matsumoto, T., Shishido, A., Morita, H., Itokawa, H., and Takeya, K. (2001). Cyclolinopeptides F-I, cyclic peptides from linseed. *Phytochemistry* 57, 251–260.

Mazur, W., Fotsis, T., Wähälä, K., Ojala, S., Salakka, A., and Adlercreutz, H. (1996). Isotope dilution gas chromatographic-mass spectrometric method for the determination of isoflavonoids, coumestrol, and lignans in food samples. *Anal. Biochem.* 233, 169–180.

Mazza, G., and Biliaderis, C.G. (1979). Functional properties of flaxseed mucilage. *J. Food Sci.* 54, 1392–1305.

Meagher, L.P., Beecher, G.R., Flanagan, V.P., and Li, B.W. (1999). Isolation and characterization of the lignans, isolariciresinol and pinoresinol, in flaxseed meal. *J. Agric. Food Chem.* 47, 3173–3180.

Middleditch, B.S., and Knights, B.A. (1972). Sterols of *Linum usitatissimum* seed. *Phytochemistry* 11, 1183–1184.

Mohagheghzadeh, A., Schmidt, T.J., and Alfermann, A.W. (2002). Arylnaphthalene lignans from *in vitro* cultures of *Linum austriacum*. *J. Natural Prod.* 65, 69–71.

Morita, H., Shishido, A., Matsumoto, T., Takeya, K., Itokawa, H., Hirano, T., and Oka, K. (1997). A new immunosuppressive cyclic nanapeptide, cyclolinopeptide B from *Linum usitatissimum*. *Bioorg. Med. Chem. Lett.* 7, 1269–1272.

Muir, A.D., and Westcott, N.D. (1998). Flaxseed lignans, sterochemistry, quantitation and biological activity (Paper 14). *Phytochemicals in human health protection, nutrition and plant defence. Pullman, WA.*

Muralikrishna, G., Salimath, P.V., and Tharanathan, R.N. (1987). Structural features of an arabinoxylan and a rhamnogalacturonan derived from linseed mucilage. *Carbohydrate Research* 161, 265–271.

Niedzwiedz Siegien, I. (1998). Cyanogenic glucosides in *Linum usitatissimum*. *Phytochemistry* 49, 59–63.

Oomah, B.D., Kenaschuk, E.O., Cui, W., and Mazza, G. (1995). Variation in the composition of water-soluble polysaccharides in flaxseed. *J. Agric. Food Chem.* 43, 1484–1488.

Oomah, B.D., Kenaschuk, E.O., and Mazza, G. (1996). Phytic acid content of flaxseed as influenced by cultivar, growing season, and location. *J. Agric. Food Chem.* 44, 2663–2666.

Oomah, B.D., Kenaschuk, E.O., and Mazza, G. (1997). Tocopherols in flaxseed. *J. Agric. Food Chem.* 45, 2076–2080.

Oomah, B.D., Mazza, G., and Kenaschuk, E.O. (1992). Cyanogenic compounds in flaxseed. *J. Agric. Food Chem.* 40, 1346–1348.

Oostdam, A., and van der Plas, L.H.W. (1996). A cell suspension of *Linum flavum* (L.) in phosphate limited continuous culture. *Plant Cell Rep.* 16, 188–191.

Petersen, M., and Alfermann, A.W. (2001). The production of cytotoxic lignans by plant cell cultures. *Appl. Microbiol. Biotechnol.* 55, 135–142.

Qiu, S.-X., Lu, Z.-Z., Luyengi, L., Lee, S.-K., Pezzuto, J.M., Farnsworth, N.R., Thompson, L.U., and Fong, H.H.S. (1999). Isolation and characterization of flaxseed (*Linum usitatissimum*) constituents. *Pharma. Biol.* 37, 1–7.

Rogers, C.M. (1972). The taxonomic significance of the fatty acid content of seeds of *Linum*. *Brittonia* 24, 415–419.

Saeidi, G., and Rowland, G.G. (1999). Seed color and linolenic acid effects on agronomic traits in flax. *Can. J. Plant Sci.* 79, 521–526.

Smollny, T., Wichers, H., Kalenberg, S., Shahsavari, A., Petersen, M., and Alfermann, A.W. (1998). Accumulation of podophyllotoxin and related lignans in cell suspension cultures of *Linum album*. *Phytochemistry* 48, 975–979.

Stosic, D., Gorunovic, M., Skaltsounis, A., Tillequin, F., and Koch, M. (1989). Flavonoids of the leafs from *Linum capitatum* Kit. *Acta Pharm. Jugosl.* 39, 215–218.

Toki, K., Saito, N., Harada, K., Shigihara, A., and Honda, T. (1995). Delphinidin 3-xylosylrutinoside in petals of *Linum grandiflorum*. *Phytochemistry* 39, 243–245.

van Uden, W., Bos, J.A., Boeke, G.M., Woerdenbag, H.J., and Pras, N. (1997). The large-scale isolation of deoxypodophyllotoxin from rhizomes of *Anthriscus sylvestris* followed by its bioconversion into 5-methoxypodophyllotoxin beta-D-glucoside by cell cultures of *Linum flavum*. *J. Natural Prod.* 60, 401–403.

van Uden, W., Holidi Oeij, K., Woerdenbag, H.J., and Pras, N. (1993). Glucosylation of cyclodextrin-complexed podophyllotoxin by cell cultures of *Linum flavum*. *Plant Cell, Tissue Organ Cult.* 34, 169–175.

van Uden, W., Pras, N., Batterman, S., Visser, J.F., and Malingre, T.M. (1991). The accumulation and isolation of coniferin from a high-producing cell suspension of *Linum flavum* L. *Planta* 183, 25–30.

van Uden, W., Pras, N., and Woerdenbag, H.J. (1994a). *Linum* species (flax): *in vivo* and *in vitro* accumulation of lignans and other metabolites. *Biotechnol. Agric. For.* 26, 219–244.

van Uden, W., Woerdenbag, H.J., and Pras, N. (1994b). Cyclodextrins as a useful tool for bioconversions in plant cell biotechnology. *Plant Cell, Tissue Organ Cult.* 38, 103–113.

Volk, O.H., and Sinn, M. (1968). Linosid ein neues Flavon aus *Linum maritimum* L. [Linoside: a new flavone from *Linum maritinum* L]. *Z. Naturforsch.* **23b**, 1017.

Wagner, H., Budweg, W., and Iyengar, M.A. (1972). Linoside A and B, two new flavone-C-glycosides from *Linum maritimum* L. *Z. Naturforsch.* **27B**, 809–812.

Wannerberger, K., Nylander, T., and Nyman, M. (1991). Rheological and chemical properties of mucilage in different varieties from linseed (*Linum usitatissimum*). *Acta Agric. Scand.* **41**, 311–319.

Weiss, S.G., Tin-Wa, M., Perdue, R.E., and Farnsworth, N.R. (1975). Potential anticancer agents II. Antitumor and cytotoxic lignans from *Linum album* (Linaceae). *J. Pharm. Sci.* **64**, 95–98.

Westcott, N.D., Hall, T.W., and Muir, A.D. (2000). Evidence for the occurrence of the ferulic acid derivatives in flaxseed meal. *Proceedings of the 58th Flax Institute of the United States. Fargo, ND*, pp. 49–52.

Westcott, N.D., and Muir, A.D. (1996). Variation in the concentration of the flaxseed lignan concentration with variety, location and year. In *Proc. of the Flax Institute of the United States*, Vol. 56, Flax Institute of the United States, Fargo, ND, pp. 77–80.

Westcott, N.D., Muir, A.D., and Diederichsen, A. (2002). Lignan distribution in the genus *Linum. 85th Candian Society of Chemistry Conference. Vancouver, BC.* pp. Abstract 21.

Wichers, H.J., Versluis-De Haan, G.G., Marsman, J.W., and Harkes, M.P. (1991). Podophyllotoxin related lignans in plants and cell cultures of *Linum flavum. Phytochemistry* **30**, 3601–3604.

Xia, Z.Q., Costa, M.A., Proctor, J., Davin, L.B., and Lewis, N.G. (2000). Dirigent-mediated podophyllotoxin biosynthesis in *Linum flavum* and *Podophyllum peltatum. Phytochemistry* **55**, 537–549.

4   Cultivation of flax

Anatoly Marchenkov, Tatiana Rozhmina,
Igor Uschapovsky and Alister D. Muir

Introduction

Significant areas of land are devoted to oilseed and fiber flax cultivation in more than 16 countries distributed throughout the temperate zones of the world, with production increasing or declining as dictated by local and international market conditions and subsidy policies. In this chapter we will review the cultivation practices in the major flax-growing areas of the world, with particular emphasis on Russia as an example of a major fiber flax production area and Canada as a major oilseed flax production area. Mention of a specific product or cultivar is provided for purposes of illustration and does not constitute endorsement. The reader is urged to consult their local agricultural extension services for specific information on cultivation and weed control practices and for current recommendations on cultivars, fertilizers and agricultural chemicals. Use of agricultural chemicals must comply with the label recommendations for rates and methods of application.

Production areas

Flax is grown as a commercial or subsistence crop in over 30 countries representing all five continents of the world. Although growing flax for fiber is an important industry in some countries, most flax is grown primarily for oil. The main flaxseed-producing countries are Argentina, Canada, China, India, Poland, Romania, Russia, Uruguay, and the USA (Gill, 1987b). The area devoted to flax production changes with market demand, particularly in exporting countries such as Canada.

Russia

Flax is the most important industrial fiber crop in Russia with more than 100,000 ha currently devoted to its cultivation. This is down significantly from the 1980s, when over 400,000 ha were under flax cultivation and over 1 m ha in the former USSR (Table 4.1). The 1991–95 average fiber production was 73,000 tonnes compared to an average of 141,000 tonnes for 1981–85 (Table 4.2) and this has continued to fall reaching a low of 23,000 tonnes in 1997 (FAO). The seed is also recovered, with production also falling from approximately 80,000 tonnes of seed in the late 1980s to less than 20,000 tonnes in 2000 (FAO). The main fiber flax-growing areas of Russia are in the western part, from Pskov in the west to Perm on the western slopes of the Ural Mountains

Table 4.1 Fibre and oilseed flax production in the former USSR countries ('000 ha)

	1988		1989–1991		1992–1995		1996–2000	
	Fibre	Oil	Fibre	Oil	Fibre	Oil	Fibre	Oil
USSR	931	1,039	758.6	831.7	457	470.5	238.6	230.4
Russia					181.2	225.5	121.4	96.8
Ukraine					115.6	115.6	32.6	32.7
Belarus					100.5	100.5	75.5	76.1
Lithuania					12.2	12.2	7.1	7.3
Kazakhstan						8.3		2.0
Kyrgyzstan								11.2

Note: The difference between oil and fibre represents areas devoted to oilseed flax. It appears that in the late 1990s seed may not have been recovered from all of the fibre flax crop.
Source: FAO.

Table 4.2 Fibre flax production in the former USSR countries

	Fiber production (1,000 tons)		
	1981–1985 Average	1986–1990 Average	1991–1995 Average
Russia	141	122	73
Ukraine	102	109	75
Belarus	98	88	49
Lithuania	14.2	13.7	N.A.

Source: Anonymous, 1996a.

Table 4.3 The acidity of soils in the non-chernozemic zone of Russia

		Acidity (pH)				
		<4.5	4.6–5.0	5.1–5.5	5.6–6.0	>6.0
Area evaluated for acidity ('000 ha)	38,876	5,449	8,498	10,790	7,739	6,401
%			14	22	28	20	16

in the east (Figure 4.1). Flax production exclusively for oil which occurs on about 10 percent of the area devoted to flax production, including Kazakhstan and Kyrgyzstan, has also declined significantly over the same period. Fiber flax is grown principally on acidic non-chernozemic soils in this region with approximately 60 percent of the area devoted to flax cultivation classified as acidic (pH $\leq$ 5.5) (Table 4.3).

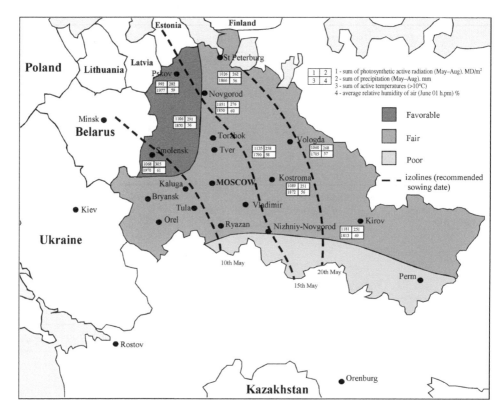

Figure 4.1 Principal fiber flax-growing areas of Russia (adapted from Greenhoff, 1988).

Ukraine and Belarus

Fiber flax is also a significant industrial crop in Ukraine and Belarus, with fiber production in Ukraine similar to that of Russia and slightly less in Belarus (Table 4.2). The decline in oilseed production in the Ukraine has been even more dramatic, declining from 49,000 tonnes in 1992 to 6,000 tonnes in 2000 (FAO).

North America

The main flax-growing areas of North America are the Brown, Dark Brown, Black, and Dark Gray Chernozemic soil zones of the Canadian Prairies and the southern extensions of these soil zones in the northern United States (Figure 4.2). The flax that is grown on the prairies is almost exclusively of the oilseed type (Anonymous, 1996b). In the Brown and Dark Brown zones, the principal limitation is moisture, whereas in the other soil zones moisture is less of a limiting factor. Flax production is concentrated in Saskatchewan and Manitoba involving between 67 and 70 percent of the area (Table 4.4). In 1997 approximately 61,000 ha were devoted to flax production in the USA, of which 91 percent were in North Dakota and the balance in South Dakota, Minnesota and Montana. By 2000, this had risen to over 270,000 ha (FAO; US Agricultural Census, 1997).

Figure 4.2 Principal oilseed flax-growing soil zones of the North American northern plains.

Table 4.4 Flaxseed production in Canada for the years 1999 and 2000

Province	Seeded area 1,000 ha		Production 1,000 tonnes	
	1999	2000	1999	2000
Manitoba	210	176	272	206
Saskatchewan	567	405	711	470
Alberta	32	14	39	18
Total	809	595	1,022	694

Source: Statistics Canada (2000).

India

In India, flax is grown on a variety of different soils with varying moisture (rainfall) and a moderately cool climate being the main factors influencing the growing areas. Flax is grown on black-cotton soils in the Pradesh and Maharashtra states, on lighter gangetic alluvium soils in the Uttar Pradesh and Bihar states and on paddy lands in the Punjab (Figure 4.3) (Gill, 1987a; Anonymous, 1993). The area devoted to flax

Figure 4.3 Principal flax-growing states on the Indian subcontinent.

production has declined from a high of 2.118 m ha in 1976 to a low of under 800,000 ha in 2000 (FAO), although improved yields have to some extent offset the decline in acreage.

China

Although China is one of the largest producers of flax, its production is almost entirely for domestic use and relatively little is known about this industry outside of China (Pan, 1990). The cultivation of flax can be traced back at least 2,000 years in China and possibly as long as 5,000 years. Production is largely concentrated in northern and northwestern China with smaller areas of flax production in the southern provinces of Yunnan and Guangxi. The main production areas are the Gansu, Nei Monggol, Xinjiang and Shanxi provinces (Figure 4.4). Flax production areas in China are characterized by high elevation (1,000–2,500 m), low annual precipitation (50–500 mm) and low annual temperatures (2.5–10°C) (Pan, 1990). Approximately 570,000 ha of land were under flax cultivation in the late 1990s (FAO).

Figure 4.4 Principal flax-growing provinces in China (adapted from Pan, 1990).

Argentina

Flax production in Argentina has declined considerably since the peak production year of 1992–3. The principal flax-growing areas are located in the provinces of Entre Ríos, Buenos Aires and the southern areas of Santa Fe.

Europe

Both fiber and/or oilseed flax is grown in most western and central European countries, although the area devoted to flax cultivation in each country is relatively small and the production in any given year may vary significantly depending on market conditions and the effects of the European Union (EU) agricultural policies. The areas devoted to fiber flax production in France and Belgium, for example, are relatively stable, while the area devoted to oilseed flax can vary significantly from year to year (Pouzet and Sultana, 1991). In the United Kingdom the area devoted to fiber production is relatively stable at around 19,000 ha, while oilseed flax production has varied from a low of 37,000 ha in 1996 to a high of 194,000 ha in 1999 (FAO), largely as a result of changes in the EU policies. Production of oilseed flax in Germany has also increased significantly during the 1990s, from nothing in 1989 to 199,000 ha in 1999–2000 (FAO), while fiber flax production in Poland, Romania, Slovakia and the Czech Republic has all but disappeared. Spain, on the other hand, has become a significant producer of fiber flax, with 85,000 ha under cultivation in 1999–2000 (FAO).

Africa

There is a long history of flax cultivation in North Africa. However, in many areas it is grown only as a subsistence crop and little information is available on production.

Egypt is one of the few countries in Africa with significant commercial production where cultivation has varied from a low of around 3,000 ha in the 1930s to a high of over 28,400 ha in 1980 (El-Hariri, 1998). Between 70,000 and 90,000 ha annually is planted to flax in Ethiopia/Eritrea (FAO).

Cultivars

In areas where there is a long history of flax cultivation (eg. India, China, North Africa), the cultivars grown tend to reflect the evolution of flax in the local area over a long period of time, while in the New World, the cultivars tend to reflect composite germplasm imported from several different locations and refined by concerted plant breeding into well-defined cultivars.

Russia

The majority of the cultivars grown (29 of 31 of the recommended cultivars on the State List) are the results of national breeding work. The average yield in Russia of flax fiber is 600–700 kg of fiber and 300–400 kg of seed per hectare. New cultivars (Alexim, A-93, Lenok etc.) have a high yield potential for fiber (up to 2,800 kg/ha) and seed (up to 1,400 kg/ha) production.

North America

The high α-linolenic acid traditional flax cultivars registered in Canada all have brown seed coats, although at least one yellow seeded variety is grown in both Canada and the USA primarily for human consumption. Flax varieties are classified as medium or late maturing varieties with a range of cultivars developed to suit different regions. Varieties of traditional oilseed flax currently grown in Canada are either the product of the breeding program at Agriculture and Agri-Food Canada (AAFC), or that of the Crop Development Center (CDC) of the University of Saskatchewan. Cultivars adapted for the northern states of the USA are primarily the product of breeding programs at North Dakota State University (NDSU) (Table 4.5).

In 1993, a low α-linolenic acid cultivar "Linola 947" was registered in Canada and subsequently the Canadian Grain Commission Standards were amended to recognize the name "Solin" for varieties with less than 5 percent α-linolenic acid. These standards also require that these cultivars have a yellow seed coat. The only solin varieties currently registered in Canada are the products of United Grain Growers' (UGG) breeding programs although this is expected to change when solin-type varieties under development at CDC are released.

India

Both dual purpose (oil/fiber) and oilseed flax are grown in India (Reddy and Pati, 1998). The varieties are predominantly of Indian origin and much of the production is from home-grown seed, although this is slowly changing with the introduction of new improved cultivars developed as the result of systematic plant breeding activities. Flax cultivars in India either have a deep root system (peninsular type) or shallow root system (alluvial type) (Anonymous, 1993).

Table 4.5 North American oilseed flax cultivars

Cultivar	Maturity	Developer and year of registration
Linott	Medium–Early	AAFC (1966)
McGregor	Late	AAFC (1981)
NorLin	Medium	AAFC (1982)
NorMan	Medium	AAFC (1984)
Vimy	Medium	CDC (1986)
Neche	Medium	NDSU (1988)
Flanders	Medium–Late	CDC (1989)
Omega[1]	Medium	NDSU (1989)
Somme	Medium	CDC (1989)
AC Linora	Medium–Late	AAFC (1991)
AC McDuff	Late	AAFC (1993)
Linola 947[2]	Late	UGG (1993)
AC Emerson	Medium	AAFC (1994)
CDC Normandy	Medium	CDC (1995)
Linola 959[2]	Medium–Late	UGG (1995)
Andro	Medium	CDC (1998)
Cathay	Medium	NDSU (1998)
Pembina	Medium	NDSU (1998)

Notes:
[1] Yellow seed coat.
[2] Yellow seed coat–solin.
Sources: Anonymous, 1996b; Berglund *et al.*, 1999.

China

Five general ecotypes of flax are grown in China, although the distinctive character-istics of Chinese flax cultivars is changing due to the need to incorporate genes for disease resistance, particularly as a result of changes in agricultural practices brought about by the Cultural Revolution and more recently by the privatization that is now occurring in China. The five traditional ecotypes are Loess Plateau, Grassland, Yellow River Valley, Xinjiang and Quighai Highlands. Beginning in the 1950s germplasm was imported into China for incorporation into indigenous cultivars, first to improve yield and subsequently to improve disease resistance (Pan, 1990).

Europe

Like North America, Europe has a long history of systematic flax breeding, both for specialized fiber flax and for oilseed flax adapted to specific regions and for varying maturity periods (Pavelek, 1998).

Africa

Cultivars currently grown in Egypt are mainly long fiber varieties with good seed yields derived by crossing local varieties with imported germplasm from a variety of sources (El-Hariri, 1998).

Crop rotation

Crop rotations are many and varied, reflecting local growing conditions, local custom and the suitability of the soil and climate for other crops grown in the rotation.

Russia

The continuous monoculture of flax has a negative effect on yield and quality of flax fiber. Diseases, insect pests, and weeds increase rapidly. The yields of seeds and fiber, as well as the quality of the fiber, decrease sharply. The optimal rotation period for flax is five to seven years, but in extraordinary situations, a three to four year rotation is acceptable.

The traditional practice in Russia was to grow flax after perennial grasses. But under more intensive agricultural systems with increased soil fertility, flax is now being grown after other crops including potato, grains, and annual grasses (Karpunin *et al.*, 1988).

During the cultivation of different crops in the rotation cycle, different fertilizers are used with consequent effects on soil fertility which, in turn, have an effect on the yield and quality of the flax fiber. It is desirable to determine the elemental and organic content of the soil prior to planting flax to identify any deficiencies that should be corrected before planting.

For example, clover and grains mixes (with 50 percent clover in the mix), used for hay production, enrich the soil with nitrogen (12–15 kg), but at the same time deplete the soil of potassium (18–20 kg) and phosphorus (4–7 kg) for every ton of hay. This situation can lead to potassium and phosphorus deficiencies in a flax crop. Cereals reduce the levels of the main nutritional elements (N, P, K) and others from the soil in approximately equal amounts and a balanced NPK fertilizer should be applied to the fields after cereals and before flax is planted (Petrova, 1989).

The level of weed infestations in fields prior to the planting of flax is also an important consideration. Perennial grass fields should be kept relatively free of weeds for two years prior to the planting of flax. Cereal grain yields of more than 300 kg/ha tend to leave the fields relatively free of weeds compared to other predecessors, with winter crop fields more weedy than spring fields because of the longer effective growing season.

Growing flax after potatoes is less favorable than following cereals. The application of high doses of manure (40–50 tons per ha) desirable for potato cultivation leads to an elevated nitrogen content in the soil which increases the risk of lodging and lowers the flax fiber quality.

An analysis of soil fertility and its organic content and structure, and an assessment of the amount of weeds should both be done to identify fields suitable for flax production. On a rich, well-fertilized soil, a high yield of flax fiber can be obtained in a rotation of perennial grasses followed by winter or spring cereals prior to flax cultivation. On low fertility soils, flax should immediately follow the perennial grasses in the rotation. On light soils, potatoes can be recommended as a flax predecessor.

North America

Crop rotations for oilseed flax are different from those for fiber flax in Russia in that seeding of flax into fields where perennial grasses were grown previously is rare. Usually flax will follow a cereal (wheat or barley), corn, or an annual legume. However, the fungal disease *Rhizoctonia* may be a problem when flax follows a legume in the rotation. Flax does not do well when following potatoes or sugar beet in areas where these are grown, primarily because of difficulties with seedbed preparation and *Rhizoctonia* may also be a problem. In parts of Saskatchewan, growing flax after canola can also cause significant yield reductions that are thought to be due to allelopathic effects of the canola straw from the previous crop and possibly also from spring volunteer canola seedlings. Continuous cropping of flax is rare in North America (Anonymous, 1996b).

India

As flax is a relatively poor competitor with weeds it should be grown in rotation with other cultivated crops. Typical flax rotations in India include maize, sorghum, pearl millet, groundnut and cowpea. The best rotations include a legume in one of the two preceding seasons (Gill, 1987a). Flax is also grown in intercropping systems with chickpea, safflower or lentil (Reddy and Pati, 1998).

China

In China, flax often follows millet, barley, wheat or potatoes (Pan, 1990). In addition to traditional monoculture type flax production in the central areas of Gansu, flax is sometimes planted together with "Wenggai" (*Eruca sativa*), a member of the Brassica family. In this form of culture, both crops are harvested together (Pan, 1990).

Seedbed preparation

Russia

Seedbed preparation affects the yield and quality of flax products. The preparation consists of basic (summer–fall) and spring tillage (Matukhina 1994).

Basic tillage

The main form of basic tillage is plowing with minimizing soil drifting. The optimal time for plowing for the North-West region of Russia is the end of August or the beginning of September. The basic tillage objectives are weed control and incorporation of plant residue which is usually undertaken two to three weeks before plowing. Straw chopping and fall plowing can result in a 40 percent reduction in weed populations.

Another form of basic tillage which could be recommended in the case of the early harvesting of cereals is termed semi-fallow. This tillage can be conducted during the warm weather during the 60 to 80-day period before the first frost. After harvesting and straw chopping the field should be plowed and harrowed at the same time. Before frosts, two or three cultivations (10–12 cm deep) should be performed depending on precipitation and the weediness of the field, with the last cultivation two weeks before the first killing frost (Matukhin, 1990). In experiments at the VNIIL, the number of weeds was decreased by 30 percent and the yield of flax was increased by 10 percent following this form of semi-fallow tillage.

When perennial grass fields are being broken for flax cultivation, field disking is recommended. This operation should be performed in two perpendicular directions to a depth of 8–12 cm. According VNIIL's data, the disking operation conducted ten days before plowing improved the quality and increased the yield of flax fiber by more than 100 kg per ha (Marchenkov *et al.*, 1998).

Spring tillage

The first early spring tillage prevents the loss of water from the plow level, and has a positive effect on soil microflora activity. On sandy and light loamy soils with low perennial weed levels, the early spring tillage can be achieved with a harrow. On heavy soils with high perennial weeds, tillage with a cultivator implement to a depth of 10–12 cm (followed by harrowing) have been recommended. This should be followed four to seven days later by the seedbed tillage.

As a small-seed crop, flax needs a well-prepared seedbed. The soil should be friable, with medium density, and have a level surface. It can usually be achieved by a combination of cultivation, harrowing, leveling and compaction. Presowing cultivation (8–10 cm in depth) should be done with simultaneous harrowing in two cross-over directions. The disc choppers and heavy disk harrow can be used for spring tillage on weed-free fields.

A level field surface leads to an approximately 20 percent increase in the emergence rate and yield increases of up to 100 kg/ha. The final step of field compaction is usually achieved with teeth or plain water-filled rollers. This facilitates a uniform planting depth, increased water supply to the seedbed level from the lower levels of soil and good seed germination. Rolling results in a 10 to 15 percent increase in yield compared to poorly compacted soils. Rolling is recommended on light sandy and light loamy soils, in dry conditions, and is less important on wet and loamy soils with normal soil moisture. To limit deterioration of soil quality during multi-operational tillage it is desirable to combine harrowing, leveling and rolling in one operation.

If fallow tillage has not been undertaken, the field should be plowed, harrowed and rolled in the spring (8–10 cm in depth). The two latter steps can be combined in one operation. At the end of seedbed preparation the soil should be characterized as follows – uniform friable soil (8–10 cm in depth), greater than 90 percent of soil particles should be up to 30 mm in size, with less than 2 percent of the particles greater than 50 mm in size, and the hollows and crests should be less than 2 cm.

North America

In contrast to the situation with fiber flax production in Russia, oilseed flax on the prairies usually follows a cereal or legume crop and therefore the tillage practices are less intensive.

Fall tillage

Fall tillage on the prairies to prepare land for flaxseeding has become less common in recent years and is primarily used as a tool for weed management. This can typically be accomplished by harrowing and care should be exercised to minimize soil erosion and maximize snow trapping (Anonymous, 1996b).

Spring tillage

As with preparation of the seedbed for fiber flax, obtaining a firm well-packed seedbed is desirable and therefore shallow tillage is desirable.

Minimum and zero tillage

In recent years there has been a significant shift toward minimum and zero tillage production of oilseed flax. Specialized seeding equipment developed for zero tillage can be used effectively for minimum and zero tillage seeding of flaxseed. Because moisture is often the main yield-limiting factor on the prairies, minimum and zero tillage production systems not only reduce wind and water erosion but they also reduce the loss of soil moisture.

China and India

Seed preparation requirements are similar for other areas of the world except that fields are generally much smaller and equipment is smaller and less mechanized than in North America.

Soil fertility

Flax is very responsive to the soil nutrient status. Obvious effect of fertilizers on flax plants can appear soon after seedling emergence. The emergence rate, height, size and color of stems depends on the quality of fertilizers and the way they are used.

Russia

The highest yield of flax can be achieved on non-chernozemic soils with 15 mg phosphorus (P_2O_5) and 15 mg potassium (K_2O) per 100 g of soil. In general, the soil fertility in the range of 15–20 mg of P_2O_5 and K_2O per 100 g of soil is considered the optimum for fiber flax.

Uniform placement of fertilizers is usually achieved during fallow (basic) tillage. Fertilizer requirements and rates for flax for different soil types and fertility are listed in Table 4.6.

Rates of application are based on the optimum ratio for fiber for the elements N, P, K in ratio of 1:4:5 respectively. The placement method depends on the type of fertilizer (solid or liquid).

During the first spring cultivation the fertilizers should be placed according to the following methods – broadcasting, seed-placed (2–3 cm below the seed level) or between

Table 4.6 Average rate of fertilizer for fiber flax for different Russian soil types

Soil type	Nitrogen	Phosphorus	Potassium
High fertility (> 100 mg P_2O_5/100 g soil)	10–15	80–100	100–125
Medium fertility (10–15 mg/100)	15	70–85	90–110
Poor fertility (< 10 mg/100 g)	15–20	70	90

seed rows (15 cm) at a depth of 5 cm. In the seed-placed or between rows methods the rate of fertilizer, especially the nitrogen content, should be reduced to half the rate for broadcast spreading (Kuzmenko, 1999). Flax is very sensitive to zinc deficiency in the soil. In soils with low levels of this micro-nutrient, injured (yellowish, thin, small) young plants will be observed. Flax usually outgrows this "physiological disease," but with reduced yield. The effect of zinc deficiency can be reduced by spraying a solution of 500 g $Zn(SO_4)_2$ and 150 g H_3BO_3 per hectare at the seedling stage (Sorokina, 1997).

North America

The soil type and the effect of the previous crop on the soil fertility should always be considered when determining the fertilizer requirements for flax. Oilseed flax often responds well to nitrogen, particularly when soil nitrogen levels are low. Typical application rates for oilseed flaxseeded on stubble in western Canada would be in the range 35–80 kg N/ha, with the higher rates applied when moisture is not limiting. Although flax responds to phosphorus applications of 15–30 kg P_2O_5/ha, the response is less pronounced or consistent compared to other annual crops grown in western Canada. Potassium is frequently deficient on sandier soils of the Dark Gray/Gray soil zone and in organic soils, while sulphur deficiency can be more widespread.

Flax is very sensitive to seed-placed fertilizer. Therefore the preferred practice is to place the fertilizer to the side or below the seed. Pre-plant deep-banding or broadcast applications are less effective with up to four-fold higher rates required to achieve the same result.

As with fiber flax, oilseed flax is sensitive to low levels of zinc and iron. However, experience on the Canadian prairies has suggested that plant and soil testing should be undertaken for these micro-nutrients before their application as the margin between deficiency and toxicity is relatively narrow and high levels can depress yield.

China and India

Traditionally flax has often been planted on marginal land and rarely received adequate fertilizer. Trials in both countries have shown similar beneficial effects on flax yield to those observed in North America, Europe and Russia. In general, flax grown on nitrogen-deficient soils in India will respond well to nitrogen application in proportion to the soil moisture levels (Gill, 1987b).

Growing season and seeding dates

The growing seasons and seeding dates vary from region to region and reflect the annual nature of the crop. Flax is normally planted as a spring crop except in a few locations where it is grown as part of a two-crop annual rotation. In all regions, seeding depth is important, with the optimum depth 2.5–4 cm. Deeper seeding can delay seedling emergence, causing significant reductions in yield.

Russia

The vegetation period of fiber flax is around 85–95 days depending on environmental conditions and cultivar. During this period cumulative heat units (above 10°C) are

1,500 (Greengoff, 1988; Petrova, 1994). The recommended seeding dates are shown in Figure 4.1. For maximum biomass and seed production, seeding at the earliest possible date is desirable. Early seeding leads to full utilization of soil moisture by young plants, with resultant increased tolerance to drought and resistance to lodging, decreased susceptibility to insect damage, and earlier maturity. The optimal time for seeding is when the temperature of soil at a depth of 10 cm is 7–8°C (at 50–60 percent soil moisture). These conditions occur in the first half of May in the central zone of the flax-growing area of Russia (Figure 4.1).

North America

As with most crops on the prairies, seeding dates are influenced by the probability of frost. In Canada, early spring seeding is desirable for the best results. Newly emerging seedlings can withstand temperatures down to −3°C (27°F) (Anonymous, 1996b). Seeding in mid-May usually will allow the crop to reach the flowering stage when there is still sufficient moisture present. If for weed management reasons or because of drought or excessive moisture, seeding is delayed beyond the first week of June, then only early maturing varieties should be seeded. Particular attention should be paid to the cultivars selected since some will perform much better than others under these conditions. Seeding dates and growing seasons are slightly earlier and longer in the northern USA and dates should be adjusted accordingly. Seed yields are strongly influenced by the duration of the flowering period which can be extended by irrigation.

India

The growing season and time of seeding varies considerably over the Indian subcontinent (Gill, 1987a). For example flax is generally sown in the south in early October, in the northern states such as Bihar in November, while in Kashmir, flax is not sown until February or March. As in other flax-growing regions, timing is always important as seeding outside the optimum window has significant negative effects on crop yield (Gill, 1987a).

Most flax is seeded by drilling, but some is still seeded by broadcast, particularly in paddy fields. Seeding rates, as in other areas, are dependent upon the method of planting, fertility and moisture conditions, with typical seeding rates between 25 and 50 kg/ha.

Argentina

In Argentina, flax is primarily grown as a winter crop with optimal seeding dates between May 15 and July 1 with the later dates more appropriate for the northern areas in Buenos Aires province and southern areas of the province of Santa Fe (Acosta and Frutos, 1978).

Seed preparation and seeding rates

In most flax-growing areas of the world it is common practice to treat the seed before planting with either a fungicide or micro-nutrient coating or a combination of treatments. Readers are cautioned to consult local agricultural advisors for specific recommendations for their conditions.

Table 4.7 Recommended micro-nutrient content of seed coating mixes for fiber flaxseed in Russia (per 100 kg of seed)

Boric acid	125–250 g
Ammonium molybdenite	200 g
Zinc sulfate	240–250 g

Russia

Fungicide treatment and coating seed with micro-nutrients are the main ways of increasing yield potential. The chemical agents used in Russia depend on the seed-borne disease and on the soid characteristics. Based on data generated by VNIIL, the application of B, Zn and Mo as a seed treatment increases yield and quality of flax fiber (Sorokina, 1997). The recommended amount of these micro-nutrients for seed coating is shown above (Table 4.7).

A high yield of flax products (seed and fiber) is achieved with a plant density at maturity of 1,600–1,800 plants/m^2. To achieve the optimal plant density a seeding rate of 18–24 million viable seeds per ha is recommended. On heavy soils a higher seeding rate will result in increased resistance to lodging. If the forecast is for a cold wet spring, then the seedling rate should be increased by 10 percent to compensate for the cool temperatures and the crusting effect of the rain.

North America

It is recommended that flaxseed be treated with a fungicide before planting. Suitable fungicidal preparations include the following active ingredients: carbathiin, carbathiin-thiram, maneb and mancozeb. These active ingredients can be formulated into dust, suspension and wettable powders (Anonymous, 1996b).

India and China

As in other major flax-growing regions of the world it is recommended that flax seed be treated with a fungicide before planting to protect against seed-borne pathogens.

Crop management

Russia

The flax flea beetle (*Longitarsus parvulus*) is the most damaging insect that attacks flax in Russia with most damage occurring at the seedling stage. The recommended insecticide for control of this insect is Tigham. A broad spectrum of highly effective herbicides (Shogun, Centurion, Targo, Nabu, Zeleck, Harmony, Bazogran, Lenok) are available for weed control. The right choice of cultivar, good seedbed preparation and the optimal rate of fertilizers are the basis for realization of the seed and fiber yield potential of flax.

North America

The principal insect pests of oilseed flax on the prairies are described in Chapter 6. In western Canada, provincial agricultural departments issue current recommendations for the chemical control of insects.

As the planting density is much lower in oilseed flax than in fiber flax, weed competition, especially in the seedling stage of development, has much greater impact on yield. This makes weed control an important element in the yield equation (Friesen, 1988; Friesen *et al.*, 1990; Lutman, 1991). Weeds can be controlled by either cultural practices such as cultivation or by chemical means or more usually by a combination of both. In Canada and northern USA a range of herbicides are registered for use on flax crops. Registered products fall into one of three categories: pre-emergent, post-emergent and pre-harvest. The pre-emergent herbicides are applied before planting or before flax emerges. Post-emergent herbicides are usually applied in the spring at the seedling stage where they have the greatest beneficial impact on yield. Their use may be restricted until after the seedling stage due to the permissible residue levels in the mature grain. Pre-harvest herbicides can be used to reduce the level of weed contamination in the harvested crop and to control weeds that may impact on the next crop. Desiccants can be used to accelerate the drying of the crop which can facilitate easier harvesting. Pre-harvest herbicides and desiccants may contain the same active ingredient.

Harvesting of oilseed flax

In North America, two approaches to the harvesting of flaxseed are possible. Since most of the cultivars are of the short straw type, if the crop is planted early then it is possible to straight combine the standing crop when 75 percent or more of the bolls have turned brown. The resulting seed may have a higher than desirable moisture content and require further drying to achieve a moisture content (< 10 percent) safe for long term storage (Anonymous, 1990).

Under any conditions that are less than ideal, swathing before threshing with a combine is the standard approach. Swathing allows late maturing crops to dry down to acceptable moisture levels. Swathing may also result in cleaner seed, particularly if the field is weedy. Since frost can cause significant damage to the maturing crop, desiccation and swathing can be used to accelerate the maturation process.

Utilization or correct disposal of the straw from oilseed flax is an important aspect of crop management. The short straw fiber from the oilseed cultivars grown in North America is not suitable for manufacture of spun flax fiber used in traditional linen fabrics. However, it is finding increasing application in non-woven materials and in composite products. The short straw fiber has also been used for many years in the production of specialty paper products. Some flax straw is also used to supplement ruminant livestock feed. If a market is not available for the straw, two options are available. The traditional approach was to burn the straw; however, there are now increasing restrictions on this practice on the prairies. The alternative is to chop the straw as it exits the combine and spread the chopped straw evenly on the field to facilitate rapid decomposition. If the amount of straw is too great for effective decomposition to occur, then the straw must be baled, removed from the field and other options for its disposal explored.

References

Acosta, P.P., and Frutos, E. (1978). *Las epocas de siembra mas convenientes para linos oleaginosos en Pergamino {Most advantageous sowing dates for oil flax in Pergamino cultivars, Argentina.}*. 36 p.

Anonymous (1990). *Official Grain Grading Guide*. Canadian Grain Commission.

Anonymous (1993). Cpt 7. Linseed, *Selected Oils and Oilseeds in India*, The Institute of Economic and Market Research, New Delhi, pp. 7:1–18.

Anonymous (1996a). Agricultural Statistics of the Former USSR Republics and the Baltic States: Grains. http://usda.mannlib.cornell.edu/data-sets/international/93009/

Anonymous (1996b). *Growing Flax: Production, Management and Diagnostic Guide*. Flax Council of Canada. 56 p.

Berglund, D.R., Peel, M.D., and Zollinger, R.K. (1999). *Flax Production in North Dakota*. North Dakota State University Extension Service, Fargo. 7 p.

El-Hariri, D.M. (1998). Varietal development of flax and other blast fibrous plants in Egypt. *Blast Fibrous Plants Today and Tomorrow. St. Petersburg, Russia*, Natural Fibers, pp. 77–83.

FAO. FAOSTAT. http://apps.fao.org/

Friesen, G.H. (1988). Annual grass control in flax (*Linum usitatissimum*) with quizalofop. *Weed Technol.* 2, 144–146.

Friesen, L., Morrison, I.N., Marshall, G., and Rother, W. (1990). Effects of volunteer wheat and barley on the growth and yield of flax. *Can. J. Plant Sci.* 70, 1115–1122.

Gill, K.S. (1987a). Agronomy. In *Linseed*, Indian Council of Agricultural Research, New Delhi, India, pp. 241–268.

Gill, K.S. (1987b). Insect pests of linseed. In *Linseed*, Indian Council of Agricultural Research, New Delhi, India, pp. 342–355.

Greengoff, N. (ed.) (1988). *Reference Book on Agricultural Meteorology*, Gidrometizdat. Leningrad, p. 527, *Справочник агронома по сельскохозяйственной метеорологии, под ред. Н.Грингофа, Л. Гидрометиздат, 1988, с. 527.*

Karpunin, F., Matukhin, A., Tikhomirova, V., Petrova, L., Komarov, A., Bartseva, A., Matukhina, G., and Sorokina, O. (1988). *Flax Growing*. VNIIL. Torzhok. 48p (in Russ.). *Возделывание льна-долгунца (практическое руководство), ВНИИЛ, Торжок, 1988, 48с.*

Kuzmenko, N. (1999). Efficiency of local method of fertilizer application for flax. PhD Thesis, University of Moscow, Moscow, Russia, 20 p. *Кузьменко Н.Н. Эффективность локального способа внесения сложных удобрений под лен долгунец, автореф. к.с.-х.н., Москва, 1999, 20 с.*

Lutman, P.J.W. (1991). Weed control in linseed: A review. In Froud-Williams, R.J., Gladders, P., Heath, M.C., Jenkyn, J.F., Knott, C.M., Lane, A., and Pink, D. (eds), *Production and Protection of Linseed*, Association of Applied Biologists, Wellsbourne, UK, pp. 137–144.

Marchenkov, A., Ponazhev, V., Matiukhin, A., Pavlov, E., Loshakova, N., Pavlova, L., Sorokina, O., Bojarchenkova, M., Uschapovsky, I., Mukhin, V., and Pozdniakov, B. (1998). *Flax Growing Technology in Tver Region, Practical Guidance*. VNIIL, Tver. 78 (In Russ.) p. 78 *Технология возделывания льна долгунца в Тверской области, под ред. А.Н.Марченкова, Торжок, 1998, 78с.*

Matukhin, A. (1990). *Soil Preparation for Flax Crop (Guidance)*. VNIIL, Torzhok. 10p (in Russ.) *Обработка почвы по лен-долгунец, под ред. Матюхина А.П., ВНИИЛ, Торжок, 1990, 7с.*

Matukhina, G. (1994). The basic soil preparation for flax with different devices., *Proc. VNIIL*, VNIIL, Torzhok, pp. 212–218 (in Russ.). *Матюхина Г.Н. Основная обработка почвы различными пчвообрабатывающими орудиями., сб.ВНИИЛ.в.28-29, Торжок, 1994. 212–218.*

Pan, Q. (1990). Flax production, utilization and research in China. *Proc. of the Flax Institute of the United States. Fargo, ND*, Flax Institute of the United States. pp. 59–63.

Pavelek, M. (1998). Analysis of current state of international flax database, *Natural Fibers, Special Edition*, Institute of Natural Fibers, Poznan, Poland, pp. 36–44.

Petrova, L. (1989). *Fertilizers for Flax Crop Rotation Cycle*. VNIIL, Torzhok. 15 p. *Петрова Л.И. Удобрение культур льняного севооборота(рекомендации), ВНИИЛ, Торжок, 1989, 15 с.*

Petrova, L. (1994). Hydro-thermal conditions and flax yield, *Proc. VNIIL*, VNILL, Torzhok, pp. 204–212 (Russ.) *Петрова Л.И. Гидротермические условия и урожайность льна-долгунца, сб.ВНИИЛ.в.28-29, Торжок, 1994. 204–212.*

Pouzet, A., and Sultana, C. (1991). Prospects for linseed and flax in France. In Froud-Williams, R.J., Gladders, P., Heath, M.C., Jenkyn, J.F., Knott, C.M., Lane, A., and Pink, D. (eds), *Production and Protection of Linseed*, Association of Applied Biologists, Wellesbourne, UK, pp. 7–13.

Reddy, P.S., and Pati, D. (1998). Indian oilseeds: Present status and future needs. *Indian J. Agric. Sci.* **68**, 453–459.

Sorokina, O. (1997). Microelement fertilizers applying in flax crop rotation cycle, *Agrochemistry (Russ)* 4, 40–48 (in Russ.). *Сорокина О.Ю. Применение микроудобрений в льняном севообороте, Агрохимия, 1997, 4, 40–48.*

Statistics Canada (2000). *Field Crops Reporting Series No. 8.*

US Agricultural Census (1997). http://www.hort.purdue.edu/newcrop/default.html

5 Principal diseases of flax

Khalid Y. Rashid

Introduction

Flax diseases are mainly caused by fungal pathogens, and by a few viruses and a phytoplasma. No serious diseases in flax are caused by bacteria or nematodes. Fungal pathogens infect all types of flax (*Linum usitatissimum* L.) including linseed (industrial oilseed flax), solin flax (low linolenic acid/vegetable oil flax), and the fiber flax. Differences in the reaction to specific pathogens or races of the same pathogen may exist among varieties of each type of flax. The occurrence and severity, and the importance of flax diseases, vary from one region to another in the flax-growing areas of the world.

Historically, rust and fusarium wilt have been the limiting factors in flax production in most of the flax-growing areas worldwide. Early in the twentieth century, fusarium wilt was recognized as the major disease problem in flax, especially in North America where susceptible varieties were grown. Rust has been another important and widespread disease from Europe to America and China. New virulent races developed over time and attacked commonly resistant varieties. Breeding for resistance to these diseases became the main objective of all major flax-breeding programs in America, Europe and Asia (Millikan, 1951; Singh *et al.*, 1956; Kommedahl *et al.*, 1970). Other diseases affecting flax were more regional and considered to be of minor importance to this crop. The efforts for breeding for resistance against minor flax disease have been limited in most flax-breeding programs.

This chapter describes the major diseases affecting flax worldwide with emphasis on the North American flax crops in western Canada and the north-central United States. The distribution and economic importance of each disease, diagnostic symptoms, causal organisms, physiological races and genetics of resistance, and control measures are discussed. For further detailed information, the reader is referred to descriptions and accounts of flax diseases from around the world that have been published over the years (Muskett and Colhoun, 1947; Millikan, 1951; Hoes, 1975; Gill, 1987b; Turner, 1987; Fitt *et al.*, 1991d; Froud-Williams *et al.*, 1991; Paul *et al.*, 1991; Islam, 1992; Mercer *et al.*, 1992; Fitt and Ferguson, 1993; Mercer *et al.*, 1994; Anonymous, 1996).

Fungal diseases

Rust

Distribution and economic importance

Rust is a common disease reported from all flax-growing regions of the world. Potentially, rusts are the most explosive diseases affecting crops, and this is true of the flax rust. Early infections and rapid disease development favored by weather conditions may completely defoliate flax plants and cause major losses in yield and quality of both seed and fiber (Flor, 1944; Hora *et al.*, 1962; Hoes and Dorrell, 1979; Acosta, 1986; Shukla, 1992). In North America, the build-up of rust during the 1930s and continuing until the early 1950s resulted in severe epidemics and heavy losses in yield (Flor, 1946; Vanterpool, 1949). The frequent development of new races (Flor, 1954; 1958) that attacked hitherto resistant varieties has been the major challenge for breeding resistant cultivars. The latest outbreaks of flax rust in North America were in the 1960s and 1970s when a new group of races were identified, including races 300, 370, and 371 (Hoes and Tyson, 1963; Zimmer and Hoes, 1974; Hoes and Zimmer, 1976). Thirteen flax cultivars licensed in Canada had to be replaced in the 1970s because of susceptibility to the new races of rust (Hoes, 1975; Anonymous, 1996). Presently, this disease is under control in North America where all commercial flax cultivars are immune to local races of rust (Rashid and Kenaschuk, 1992; 1994). Rust remains a constant threat to flax production worldwide because of the ability of the fungus to produce new races that can attack resistant cultivars. The reasonable levels of control of this disease worldwide have been attributed to the availability of the sources of genetic resistance to most flax-breeding programs all over the world and the wise deployment of these genes in commercial flax cultivars.

Symptoms

The main characteristic symptoms of rust are the bright orange and powdery pustules which are developed on leaves (Figure 5.1), on stems and bolls, and all aerial plant parts (Figure 5.2). These pustules are usually circular on leaves and bolls but elongated on the stems. Chlorotic or necrotic zones may appear around the pustules depending on specific race–genotype interactions. Severe epidemics cause the leaves to dry and wither resulting in heavy defoliation of the plants. Rust pustules produce numerous urediospores which are detached and readily dispersed by wind over long distances. As the plants mature, the orange pustules (uredia) turn black and form the telia pustules which produce the overwintering hardy teliospores (Figure 5.3). The black pustules are present on all infected plant parts but are more conspicuous on the stems.

Causal agent

Melampsora lini (Ehrenb.) Desmaz. Flax rust is an autoecious rust and the fungus can complete the four stages of its life cycle, namely pycnia, aecia, uredia, and telia, on the flax plant, unlike many other rusts that require alternate hosts. The pycnia and aecia are usually formed on cotyledons and lower leaves early in the season and are

Figure 5.1 Rust pustules, uredia on a flax leaf.
Figure 5.2 Rust pustules on flax boll, leaf and stem. Some are turning into black telia.
Figure 5.3 Black telia of *M. lini* on flax stems and leaves.

rarely noticed by field inspectors. Pycnia and aecia are the most important stages for completing the sexual life cycle of the fungus and for the development of new races. The third stage of the development of this fungus is the uredia. The uredial stage is the most destructive stage to the crop since uredia produce cycles of urediospores that can create new infections with each cycle throughout the growing season. At maturity, the fourth stage produces the hardy overwintering teliospores which can survive the adverse weather conditions.

Physiological races and host resistance

Physiological specialization is a common phenomenon among rust pathogens. This phenomenon was observed and studied in flax rust early in the twentieth century (Hart, 1926; Flor, 1931). Interactions between flax genotypes and different rust races and the phenotypic expression of these interactions in resistant or susceptible infection types led to the identification of distinct races (Flor, 1954; 1955), and the development of the gene-for-gene theory of pathogenesis (Flor, 1956; 1971). The gene-for-gene theory and the flax/rust system have been the models for studying most of the plant pathogen systems.

In North America, 239 races of *M. lini* were identified (Flor, 1954) using a set of 29 single-gene flax genotypes (flax rust differential set). The number of races identified was up to 400 by the 1970s, and an additional flax differential genotype was added to the differential set (Hoes and Tyson, 1966; Hoes, 1975; Hoes and Zimmer, 1976). Additional races have been identified in India, Australia and Europe (Misra, 1963; Misra and Lele, 1963; Saharan and Singh, 1978; Statler *et al.*, 1981). Further studies on races and genes conditioning resistance to regional races of flax rust have been described (Kerr, 1960; Misra, 1966; Saharan, 1991).

Studying the virulence in the rust *M. lini* has been paralleled by studying the genetics in the host *L. usitatissimum*. Flor (1947; 1954) identified 29 resistance genes which constitute the differential set of cultivars used to differentiate the 239 races at that time. Reviews of the genes and genetics of rust resistance in flax (Islam and Mayo, 1990; Islam and Shepherd, 1991) provided a list of 34 resistance genes grouped into seven groups or loci, namely: 2 at the K locus, 14 at the L locus, 7 at the M locus, 3 at the N locus, 6 at the P locus, and 1 each at D and Q loci. Several genes for resistance have since been identified (Zimmer and Statler, 1976; Hoes and Kenaschuk, 1986; Jones, 1988; Rashid and Kenaschuk, 1996a). Rust resistance genes in flax are dominant traits and are characterized by race-specific interaction. However, various studies have demonstrated the effectiveness of field resistance or slow-rusting and partial resistance in flax cultivars (Hoes and Kenaschuk, 1980; Parish and Statler, 1988; Rashid, 1991; Hoes and Kenaschuk, 1992; Rashid, 1997).

The rust *M. lini* has been reported to infect 16 species of *Linum* (Henry, 1928; Arthur, 1962; Conners, 1967). The importance of the genetic stock in the wild relatives of flax is yet to be explored for future use in flax breeding (Wicks and Hammond, 1978). Studies on the virulence of *M. lini* on the perennial *Linum marginale* in Australia indicate that race-specific/host pathogen interactions occur in the wild population (Burdon and Jarosz, 1992; Jarosz and Burdon, 1992; Burdon, 1994; Burdon and Thompson, 1995; Burdon *et al.*, 1999; Carlsson-Granér *et al.*, 1999).

The recent advancement in the area of molecular biology has provided the tools to study and understand the molecular basis of the gene-for-gene functions and interactions (Ellis *et al.*, 1997). The flax–flax rust system has been the model for studies which have led to isolation, nucleotide sequencing, and the cloning of rust resistance genes at the L and M loci of the flax genome (Anderson *et al.*, 1992; Ellis *et al.*, 1992; Dickinson *et al.*, 1993; Lawrence *et al.*, 1993; Lawrence *et al.*, 1994; Ellis *et al.*, 1995; Lawrence *et al.*, 1995; Anderson *et al.*, 1997; Ayliffe *et al.*, 1999). This new area of research led to the development of molecular markers to identify race-specific rust resistance genes as a tool for developing a wide base of multigenic resistance in flax cultivars (Burdon and Roberts, 1995; Hausner *et al.*, 1999a; Hausner *et al.*, 1999b).

Control

The most adequate and economical control recommendation is the use of resistant cultivars. These are available worldwide and some are characterized by multigenic resistance (Flor and Comstock, 1971; Flor and Comstock, 1972; Kenaschuk, 1974; Zimmer, 1976; Rashid and Kenaschuk, 1992; Rashid and Kenaschuk, 1994). This approach has been successful in North America where the use of highly resistant cultivars have kept this disease under control for the last two decades (Rashid and Platford, 1990; Rashid *et al.*, 1998b; Rashid *et al.*, 2000). The challenge remains to monitor the development of new virulent races and their reaction on common cultivars. The use of susceptible cultivars will not only create epidemics and serious yield loss, but also provides the fungus with opportunities to produce new races that can attack resistant cultivars.

In areas where rust is a problem, destroying the plant debris and volunteer plants, following a three-year crop rotation, and planting flax in a field distant from that of the previous year are measures that can reduce the inoculum pressure, especially early in the season, to avoid severe epidemics. In areas where rust is not well established, avoid using seed contaminated with plant material carrying the black teliospores of this fungus in order to prevent introducing the disease into the area. The use of fungicides has been successful in reducing the severity of rust epidemics and maintaining yield (Gill, 1987a), and can prove a useful measure provided the economic returns are warranted.

Fusarium wilt

Distribution and economic importance

Flax wilt is one of the most widespread diseases affecting flax wherever it has been grown on the same land for a lengthy period of time. Reports of yield losses due to wilt are rare but severe epidemics can result in 80–100 percent yield reduction (Kommedahl et al., 1970; Sharma and Mathur, 1971; Kroes et al., 1999). In North America, flax wilt became a destructive disease by 1890 (Lugger, 1890). This disease has been one of the limiting factors affecting the production of flax in "flax-sick fields" in the United States and Canada (Kommedahl et al., 1970) but has been less destructive to the flax crops in Canada. The development of resistant cultivars worldwide has reduced the negative impact of this disease. In North America, all commercial flax cultivars are characterized by resistance or moderate resistance to fusarium wilt (Kenaschuk and Rashid, 1993; Kenaschuk et al., 1996). Resistance has also been reported in European cultivars of oil and fiber flax (Kroes et al., 1998c; Kroes et al., 1999), and from China (Liu et al., 1993).

Symptoms

Flax wilt can occur at any time during the growing season, and affected plants may be scattered through the field but they occur more commonly in patches. Early wilt can kill the seedlings before or after emergence and up to the pre-flowering stage. Seedlings cease to grow, wilt at the top, turn brown, dry up and die (Figure 5.4). On older seedlings, yellowing and browning usually start on upper leaves and the top parts of the plants droop and die. Roots of dead plants turn ashy gray. The tops of wilted plants often turn downward, and form a "shepherd's crook" (Figure 5.5). Late wilt also affects the plants from the top downwards with leaves turning yellow to brown, leaving the top drooping (Figure 5.6). The stems become necrotic and brittle. Partial wilt can also occur when seedlings are affected but the roots and the buds remain alive. Under cool weather conditions, these buds can produce healthy shoots which remain healthy as long as the cool weather prevails (Figure 5.7). Occasionally, unilateral wilt may develop on plants when all branches on one side of the stem connected to the same vascular elements become infected. Microscopical examination of the stems at the base of infected plants reveals the presence of creamy white pustules (sporodochia) which constitute the reproductive stage of this fungus with massive numbers of sickle-shaped conidia.

Figure 5.4 Symptoms of early wilt on flax seedlings.
Figure 5.5 Typical fusarium wilt "shepherd's crook" on flax.
Figure 5.6 Symptoms of late wilt on flax plants.
Figure 5.7 Partial wilt, main stem infected, with a healthy branch.

Causal agent

Fusarium oxysporum Schlechtend.: Fr. f.sp. *Lini* (Bolley) Snyder and Hansen, is a highly specialized pathogen that causes wilt on flax only. The fungus is primarily soil-borne, and primary infections occur through the roots (Nair and Kommedahl, 1957; Turlier *et al.*, 1994; Kroes *et al.*, 1998a; Kroes *et al.*, 1998b). After the initial invasion of the roots, the fungus continues to grow inside the water-conducting tissue. This interferes with water uptake, and the plants wilt rapidly under warm weather conditions. The fungus produces three distinct kinds of spores; macroconidia with 1–5 septa and 25–40 μ long, microconidia 5–12 μ long, and chlamydospores 5–10 μ in diameter (Kommedahl *et al.*, 1970; Brayford, 1996). The fungus persists in the soil, mycelia and spores survive for five to ten years in the debris of flax and other organic tissue. Dissemination of this fungus can take place by wind, run-off soil, infected seed and plant debris.

Physiological races and host resistance

The first report of pathogenic variability in *F. oxysporum lini* was in 1926 (Broadfoot and Stakman, 1926). This fungus comprises an indefinite number of biotypes with differences in cultural characteristics and pathogenic types (Armstrong and Armstrong, 1968; Kommedahl *et al.*, 1970). Different pathogenic races have been reported from different regions of the world: Argentina, United States, Canada, Australia, India, and Japan (Borlaug, 1945; Millikan, 1945; Millikan, 1948; Houston and Knowles, 1949; Tochinai and Takee, 1950; Sharma and Mathur, 1971; Kroes *et al.*, 1999; Kroes *et al.*, 2002). High pathogenic variability has also been demonstrated in the local population of this fungus in western Canada (Rashid and Kroes, 1999; Mpofu and Rashid, 2000a; 2000b).

Resistant cultivars have been successfully identified, developed and released (Kommedahl *et al.*, 1970; Ondrej, 1977; Kenaschuk and Rashid, 1993; Kenaschuk *et al.*, 1996). Despite several reports on the inheritance of resistance to fusarium wilt and the identification of a few resistant genes (Tisdale, 1917; Burnham, 1932; Knowles and Houston, 1955; Knowles *et al.*, 1956; Jeswani and Joshi, 1964; Kamthan *et al.*, 1981; Goray *et al.*, 1987), the genetics of resistance to this fungus remains as complex as its race composition.

Control

The most important control measure is the use of resistant cultivars which are available in most flax-growing regions of the world, including North America, Europe, and Asia. There are no reports of immunity to this pathogen and all recommended cultivars are resistant or moderately resistant. Crop rotation of at least three years between flax crops, destroying the crop residues, and the use of disease-free seed are recommended in order to maintain low levels of inoculum in the soil. Seed treatments with fungicides may prevent the introduction of this disease into new areas, and reduce the incidence of early wilt in seedlings (Rashid and Kenaschuk, 1996b).

Pasmo

Distribution and economic importance

Pasmo disease, also known as spasm or septoriosis, is a foliar pathogen that infects leaves, stems and bolls. It is widespread in North and South America (Garassini, 1935; Sackston and Gordon, 1945; Rashid *et al.*, 1998b), in Europe (Colhoun and Muskett, 1943; Muskett and Colhoun, 1947; Martelli, 1961; Holmes, 1976), in Africa

(Nattrass, 1943), and in Australiasia (Newhook, 1942; Millikan, 1951). Although early infections start at the seedling stage, the severity of pasmo is not generally recognized until after boll setting and the ripening stage. Contrary to the belief that flax is most susceptible at the ripening stage, epidemics can occur early in the season when favorable environmental conditions prevail. Pasmo can cause defoliation, premature ripening and boll drop which result in reduced yield and poor quality of seed and fiber (Millikan, 1951; Sackston and Carson, 1951; Frederiksen and Culbertson, 1962; Rashid *et al.*, 1998b). Commercial cultivars lack resistance to pasmo, although minor differences among cultivars in susceptibility to this disease have been reported (Millikan, 1951).

Symptoms

The early symptoms of this disease are brown lesions on cotyledons and leaves of seedlings where numerous dark colored pycnidia are formed (Figure 5.8). The disease moves upward throughout the growing season to cover the stem, leaves, branches, and bolls at maturity. Disease lesions on leaves coalesce causing rapid drying of leaves and severe defoliation. On the stems, pasmo is characterized by brown lesions that expand and encircle the stems to form bands of dark brown color which often alternate with the green color of the stems giving the stems a mottled appearance (Figure 5.9). The infections weaken the infected stems and pedicels causing stem breakage and heavy boll-drop by rain and wind. Infected tissues are occupied by tiny black pycnidia which are the fruiting bodies of the fungus.

Causal agent

Pasmo is caused by the fungus *Septoria linicola* (Speg.) Garassini (teleomorph: *Mycosphaerella linorum* Naumov). This fungus is seed-, soil-, and stubble-borne. Numerous pycnidia are present on flax debris which, when exposed to humid conditions, release spores causing the primary infections on the cotyledons, leaves and stems. The lesions from the early infections form pycnidia which also release millions of spores in a globular mass or in a tendril form. These spores are dispersed by splashing rain and wind upward on the same plant and to adjacent plants. Pasmo is favored by high moisture and warm temperatures, dense crop canopy and heavy lodging.

Control

Genetic resistance can provide the best control but in spite of several reports of resistance to pasmo in flax (Convey, 1962; Loshakova and Korneeva, 1979; Loshakova, 1984), no major differences have been observed among commercial cultivars of the different flax types (K.Y. Rashid, unpublished). The control recommendations are to follow a three-year crop rotation between flax crops to minimize the overwintering inoculum on plant debris, seeding early to avoid the early infections and the warm weather conditions favorable for disease development, use clean seed to avoid early seedling infections and early start of epidemics, use recommended seeding rate and control weeds to avoid thick crop canopy and microclimates favoring the disease development. Fungicides can be used to control this disease provided economic returns are warranted (Rashid and Kenaschuk, 1998b).

Figure 5.8   Pasmo lesions on flax leaves.
Figure 5.9   Brown bands of pasmo on flax stems.
Figure 5.10 Powdery mildew on flax leaves.
Figure 5.11 Severe powdery mildew infections and defoliation on susceptible flax; resistant flax
               remains green and healthy.

Alternaria *blight*

Distribution and economic importance

Flax diseases attributed to infection by *Alternaria* spp. have been reported from most
of the flax-growing regions of the world (Dey, 1933; Muskett and Colhoun, 1947; de
Tempe, 1963; Fitt *et al.*, 1991d). It is commonly reported from India (Gill, 1987b),
and from wherever the crop is grown in Europe (Fitt and Vloutoglou, 1992). Disease
affects seedlings causing seedling blight, and also affects leaves and flowers. Yield losses
of up to 90 percent were reported from India (Chauhan and Srivastava, 1975), and up
to 35 percent in the United Kingdom (Mercer *et al.*, 1989). In North America, this
disease has been observed occasionally but no major epidemics have been reported. This
disease reduces the quality and viability of seed harvested from diseased plants.

Symptoms

The earliest symptoms of *A. linicola* are the seedling blight or brown rot from seed-borne inoculum. Seedlings are stunted with dark red lesions developing on the hypocotyls and cotyledons causing the total collapse of seedlings in one to two weeks. The infected and dead seedlings become a rich source of inoculum which spreads and infects healthy plants. Dark brown lesions appear on infected leaves, and usually coalesce to cover the entire leaf which turns chlorotic and dies.

A. *lini* reported from India (Dey, 1933) infects all aerial parts of the flax plant but is more severe on the flowers. Dark brown spots are common on the leaves, the calyx, and the pedicels. Infected buds may produce shriveled seed or fail to produce any seed. Under favorable conditions, these lesions develop and the fungus moves into the stem and the whole plant may be killed. Infected seeds range from dark black to dark brown, and some normal appearing seeds may have enough inoculum to cause seedling blight.

Causal agents

Alternaria linicola, *A. alternata*, and *A. infectoria* Simmons are reported to be the three main species on flax (Fitt *et al.*, 1991b; Fitt and Vloutoglou, 1992). Other species include A. *linicola* Groves and Skolko, and *A. lini* reported from India, which is favored by 26–33°C and humid conditions (Gill, 1987b). A. *linicola* survives as thick-walled chlamydospores in flax debris on and in the soil (Fitt *et al.*, 1991b; Fitt *et al.*, 1991c; Vloutoglou *et al.*, 1995). Conidia produced early in the season infect flax seedlings. The fungus can also move from infected seed to seedling, causing seedling blight (Vloutoglou *et al.*, 1995).

Control

The use of resistant cultivars when available is the most economical approach to disease control. Resistant genotypes were reported (Hossain, 1992) with specific dominant genes (Singh and Chauhan, 1988) and recessive genes (Kalia *et al.*, 1965; Evans *et al.*, 1995; Evans *et al.*, 1997). Klose *et al.*, (1993) reported that resistance to *A. linicola* in flax is a polygenic trait. Crop rotation and destroying crop residue will reduce local inoculum but might have little effect on disease level since inoculum moves in from adjacent fields. Seed treatment with fungicides was reported to be effective in the production of healthy seed in the UK (Turner, 1987). In India, well drained fields are recommended to avoid this disease due to the severe epidemics in poorly drained fields (Gill, 1987b).

Powdery mildew

Distribution and economic importance

Powdery mildew is common on flax in Europe, Australia, Asia, and America. In North America, powdery mildew has been reported only from Minnesota (Allison, 1934). In western Canada, this disease was first reported in 1997 and has been noted to spread widely in a short period of time (Rashid, 1998; Rashid *et al.*, 1998a). Yield losses of up to 38 percent have been reported in susceptible cultivars in India (Pandey and Misra, 1992).

Symptoms

The common symptom of powdery mildew is the superficial whitish powdery colonies of mycelial mass on the host surface. In flax the mycelia of powdery mildew appear on leaves, stems, and sepals (Figure 5.10). A close examination of the colonies reveals the abundance of mycelia and microscopic conidia which are borne singly or in chains on short conidiophores. Small black spherical fruiting bodies (cleistothecia) can be seen embedded in the mycelia on the surface of the host when the perfect state of the fungus occurs. Severe infection of powdery mildew causes drying of leaves and heavy defoliation (Figure 5.11).

Causal agents

Erysiphe polygoni DC from Japan (Homma, 1928), *E. lini*, *E. cichracearum* DC from Siberia (Badayeva, 1930) and Minnesota (Allison, 1934), *Sphacelotheca lini* from Russia (Tvelkov, 1970), *Leveillula taurica* from India (Saharan and Saharan, 1994b), *Oidium lini* Skoric in the UK, India (Saharan and Saharan, 1994a) and Canada. These fungi survive the harsh conditions in dark colored cleistothecia (perfect state) which remain dormant until favorable conditions prevail for the production of spores and the primary infections. In areas where the perfect state has not been found, the fungus probably survives as hyphae in susceptible tissues of perennial secondary hosts.

Physiological races

Although the indications from the reactions of flax cultivars in different flax-growing areas under natural powdery mildew epidemics in Europe and North America point to the presence of different races of powdery mildew (Beale, 1991; Saharan and Saharan, 1992) (K.Y. Rashid, unpublished), there is no scientific confirmation for the existence of races in powdery mildew.

Control

The use of resistant cultivars is the most economic method of controlling powdery mildew. Dominant genes for resistance have been reported from India (Kaushal and Shrivas, 1974; Badwal, 1975). Recent studies (K.Y. Rashid, unpublished) have identified several resistance flax cultivars from Canada and Europe, and three dominant genes for resistance. Early seeding and early maturing cultivars help the crop escape the severe infections toward the end of the season. Powdery mildew can be easily controlled by fungicides (Sharma and Khosla, 1979; Turner, 1987), provided economic returns are warranted.

Verticillium *blight*

Distribution and economic importance

This disease has been reported from Europe (United Kingdom, Germany, The Netherlands) (de Tempe, 1963; Turner, 1987; Fitt *et al.*, 1991a). The disease has also been reported from California (de Tempe, 1963). In Canada there are no reports of this disease on flax, although *Verticillium dahliae* has been observed on some flax accessions tested under controlled conditions, probably originating from seed-borne inoculum (K.Y. Rashid, unpublished). *V. dahliae* and *V. albo-atrum* are pathogenic on other crops, including potatoes and sunflowers. This disease can cause severe epidemics under prolonged wet weather conditions as the crop approaches maturity (Fitt *et al.* 1991a). The severity of this disease is compounded when the nematode *Pratylenchus penetrans* is present in the soil (Coosemans, 1977).

Symptoms

Dark brown stems with purplish appearance starting from the base and extending upward are the typical symptoms of this disease. All plant parts are affected and progressive desiccation results in the death of the plants (Figure 5.12). Close examination of the infected stems reveals the presence of dark black microsclerotia (Figure 5.13) (Fitt *et al.*, 1991a).

Causal agents

Verticillium dahliae Kleb. (Microsclerotia) and *Verticillium albo-atrum*. These two fungi can survive in plant debris, in soil and can be disseminated by plant material, soil particles and by seed contamination. *V. dahliae* is distinguished from *V. albo-atrum* by the formation of microsclerotia which can survive in the soil for a long period of time.

Control

Use of resistant cultivars when available. Follow a long term crop rotation, and avoid any susceptible crops such as sunflowers and potatoes in the rotation. Practice strict sanitation by using disease-free seed and destroy the plant debris after harvesting.

Sclerotinia stem rot

Distribution and economic importance

This disease has been reported to affect a few fields in England (de Tempe, 1963; Mitchell *et al.*, 1986). In Canada, sclerotinia infections were reported in irrigated flax in Alberta (Mederick and Piening, 1982) and from heavily lodged flax in Manitoba and Saskatchewan (Rashid, 2000). The severity of the disease depends on the severity of lodging and the level of field infestation with the sclerotia of the fungus from the previous crops. Differences in flax cultivars to sclerotinia stem rot have been reported (Pope and Sweet, 1991).

Symptoms

Early symptoms are water-soaked lesions on stems which develop rapidly to girdle the stem. The fungus causes bleaching and shredding of the stem (Figure 5.14). Mycelia grow on the surface of the stem and cylindrical sclerotia are formed inside the stem (Figure 5.15). When heavily lodged flax remains lodged, small sclerotia can also be formed on the surface of the stem.

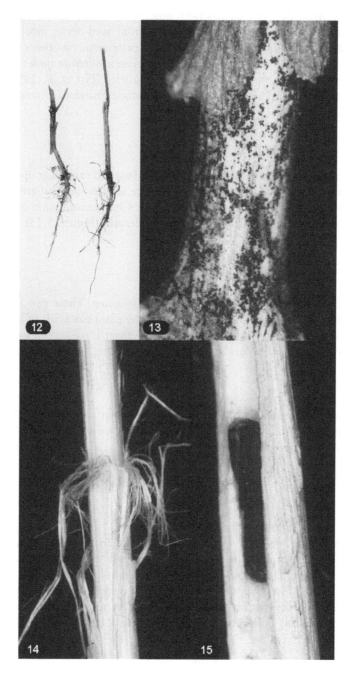

Figure 5.12 Verticillium blight, purplish stems and roots.
Figure 5.13 Black microsclerotia on infected stem.
Figure 5.14 Sclerotinia stem infection, bleached and shredded stem.
Figure 5.15 Sclerotinia stem infection, black cylindrical sclerotia inside the flax stem.

Causal agent

Sclerotinia sclerotiorum (Lib) de Barry is a widespread fungus that infects several hundred plant species worldwide. It overwinters as sclerotia which are compact masses of mycelia that can survive adverse environmental conditions for three to five years in the soil. Sclerotia germinate to produce either mycelia that can infect roots and stems of certain crops such as sunflowers, or to form apothecial fruiting bodies which produce spores that can infect foliar parts of several crops, including canola, beans and sunflowers. Apparently, stems of heavily lodged flax pick up the infection by direct contact with the fungus from the soil.

Control

There is no evidence of any resistance to *S. sclerotiorum* in any host plant species affected by this fungus. In order to minimize the risk of epidemics, it is recommended to use cultivars with lodging resistance, use the proper seeding rates to avoid dense crop canopy and heavy lodging, and avoid fields which are contaminated by sclerotinia from previous years' susceptible crops.

Browning and stem break

Distribution and economic importance

Stem break and browning are phases of a disease caused by a seed- and soil-borne fungus. This disease occurs in Europe, North America, Africa, and Asia. Under heavy infestations in Europe, fiber flax can suffer a significant loss in yield and fiber quality (Muskett and Colhoun, 1947). This disease is of minor importance in western Canada in spite of some damage that has been reported in the parkland regions of Saskatchewan and Alberta in some years (Henry, 1934; Henry and Ellis, 1971). This fungus also occurs on wild *Linum* spp. (Conners, 1967). The fungus can penetrate the bolls as well as the seeds, or may produce spores on the seed surface. However, affected seeds may remain viable.

Symptoms

Dark colored circular lesions are produced on cotyledons and lower stem parts originating from infected seeds during seedling emergence. These lesions develop slowly and may result in seedling blight or stem break. Cankers on the stem base weaken the plant, and the stem may break early at the seedling stage or later on during the season depending on the severity of infection (Figure 5.16). Stem break is a distinguishing sign of this disease. Plants may remain alive after stem-breakage, but produce few seeds which will be lost at harvest.

The browning phase of this disease starts late after the flowering initiated by infections on the upper part of the plant. Lesions which are oval or elongated in shape, dark brown in color, and surrounded by narrow purplish margins are common on the branches, leaves and bolls. These lesions often coalesce giving the leaves and stem a brownish appearance. Patches of heavily infected plants appear brown, giving the disease its name.

Figure 5.16 Browning and stem break in flax.
Figure 5.17 Anthracnose burning effects on flax leaves, stems, and bolls.
Figure 5.18 Seedling blight, shrivelled and dead seedlings.
Figure 5.19 Aster yellows in flax, flowers converted into small yellowish green leaves.

Causal agent

Polyspora lini Laff., Syn. *Aureobasidium lini* (Laff.) Hermanides-Nijof (teleomorph: *Guignardia fulvida* F.R. Sanderson) (Hoes, 1991) can survive in seed coats or in plant debris. The primary infections may start during seedling emergence, or from spores produced on diseased stubble that are spread by wind and rain to infect emerging seedlings (Muskett and Colhoun, 1947). The fungus can also survive and reproduce on secondary hosts and weeds (MacNish, 1963). The fungus produces masses of unicellular spores freely on the surface of infected tissue. The development of the disease is favored by warm and wet conditions in a thick crop canopy.

Physiological races

Differences in pathogenicity among isolates of *P. lini* based on resistance and/or levels of infection on a set of host differentials were reported (Henry, 1934; Colhoun, 1960).

Control

Use of disease-free seed produced by healthy plants is an important control measure. Fungicidal seed treatment controls surface-borne inoculum, but is unlikely to be effective when the fungus has penetrated coat-deep into the seed (Muskett and Colhoun, 1947; Turner, 1987). Crop rotation of two years between flax crops would reduce spread of infection from diseased stubble. Currently recommended varieties may differ in susceptibility to stem break and browning.

Scorch

Distribution and economic importance

This disease is present mainly in Europe with unconfirmed reports from other parts of the world (Muskett and Colhoun, 1947; Wiersema, 1955). It occurs in patches of various sizes in the first part of the growing season. The most severely affected plants in the center of the patch will soon die, while the less affected plants at the periphery of the patch may partially recover and produce some seed.

Symptoms

Infected plants are stunted with short inter-nodes and shriveled brown leaves. Yellowing starts at the lower leaves and moves upward. Infected leaves rapidly turn brown with dry tips. Top leaves grow close together and generally rapidly dry and wither. Infected roots are glossy and brittle.

Causal agent

Pythium megalacanthum De Bary is a soil-borne fungus which infects the roots of young plants. The presence of large echinulate oospores of the fungus in the root tissue is the distinguishing characteristic of this disease. This is primarily a saprophytic fungus which can survive in the soil for many years (Muskett and Colhoun, 1947; Wiersema, 1955).

Control

Late seeding may help in reducing the incidence since the wet and cold spring conditions favor the disease development. Long term crop rotation may help reduce the inoculum in the soil. Seed treatment may also protect the seedlings from early infections especially to cracked or damaged seed (Turner, 1987).

Anthracnose

Distribution and economic importance

Anthracnose in flax has been reported from most flax-growing areas in the world. The disease is primarily seed-borne but can be soil- or stubble-borne since the fungus can overwinter in crop debris and soil. The disease is destructive to the crop, especially when the seed source is contaminated by the fungus. It can spread rapidly under favorable conditions causing severe local epidemics and heavy losses in yield and quality of both fiber and seed.

Symptoms

Seedling blight is the early symptom in seedlings infected from seed-borne inoculum. Spores are abundantly produced on blighted seedlings and spread to adjacent plants. Circular water-soaked lesions appear on the cotyledons and lower leaves early in the growing season. These lesions turn to olive green and reddish brown before the leaves dry out and die. Stem infections appear as cankers of various sizes which may girdle the stem and cause shriveling and stem-breaking at the point of infection. Under favorable conditions of splashing rain and warm weather, the disease develops rapidly through the crop canopy giving the plants a burning appearance and destroying the crop (Figure 5.17). The flower color may change from white to blue in diseased plants (Pospisil, 1971). Microscopic examination of lesions shows the presence of dark colored hair-like protrusions (setae) imbedded in the spore-producing structures (acervuli) which are typical characteristics of this fungus.

Causal agent

Colletotrichum lini (Westerdijk) Tochinai: Syn. *C. linicola* Manns and Bolley. This fungus is commonly seed-borne as mycelia in the outer seed coat. Primary infections normally occur on seedlings from infected seed. Infected seedlings and aerial plant parts produce abundant spores in acervuli with characteristic setae. These spores are spread by rain and wind and cause subsequent infection cycles.

Physiological races

Several races have been identified based on virulence on flax genotypes (Zarzycka, 1973). Zarzycka (1976) identified 11 races of *C. lini*, two of which were predominant. Recently, 25 races of *C. lini* were identified using a set of flax genotypes (Kudriavtseva, 1998).

Control

Resistant varieties when available are the best control, along with the use of clean seed to prevent the introduction of the disease into new areas. Seed treatment with recommended fungicides can control the seed-borne phase of this disease. Adequate crop rotation between flax crops may help in reducing the inoculum in soil and stubble.

Seedling blight

Distribution and economic importance

Seedling blight is commonly present in most flax-growing areas. Severe epidemics reduce the stand and result in heavy yield losses. In spite of seed treatment, seedling blight and root rot can develop, and cause reductions in yield. Yellow-seeded varieties are more affected by blight and root rot than brown-seeded varieties due perhaps to the thin seed coat (Vest and Comstock, 1968; Groth *et al.*, 1970).

Symptoms

Blighted seedlings show red to brown lesions on the roots below the soil surface. These symptoms develop down the roots and upward to the stem, and the seedlings turn yellow, shrivel, wilt and die (Figure 5.18). Infected seedlings may occur singly or in patches. These symptoms may appear similar to those caused by early symptoms of *Fusarium* wilt.

Causal agent

Rhizoctonia solani Kuhn (teleomorph: *Thanatephorus cucumeris* (A.B. Frank) Donk) is the principal agent and can be particularly destructive in soils that are loose, warm and moist. It survives as a composite of strains (Anastomosis Groups (AG)) that differ in host range and pathogenicity. Flax is affected by the same AGs which attack sugar beets and leguminous crops.

Control

Use resistant varieties when available. The variety Linott is reported to be resistant to *Rhizoctonia* blight (Anderson, 1973). The reactions of other varieties are not known. Seedling blight and root rot can be controlled by a combination of farm practices, such as firm seedbed preparation, use of certified seed free of cracks, treating the seed with recommended fungicide (Rashid and Kenaschuk, 1998a), practicing a crop rotation of at least three years between flax crops, and avoiding legumes and sugar beets in the rotation.

Dieback

Distribution and economic importance

This disease has been reported from Saskatchewan, Canada (Vanterpool, 1947), with minor significant importance at that time. Since then, little has been reported on this disease from Canada or other flax-growing areas.

Symptoms

Lesions appear on stems and branches of early maturing flax plants. The upper part of the plants die prematurely. Numerous pycnidia are formed on infected tissue.

Causal agent

Selenophoma linicola Vanterpool overwinters on flax stubble. There is no evidence of a seed-borne phase of this disease (Vanterpool, 1947).

Control

Follow an adequate crop rotation and proper sanitation to reduce the level of fungal inoculum in the soil.

Damping-off

Distribution and economic importance

Damping-off is another common disease affecting flax worldwide (Millikan, 1951). The severity of this disease is more common under cold soil temperatures and water-logging which delay germination and emergence and prolong the seedling exposure to infections.

Symptoms

Seedlings are killed before emergence or shortly after. Reddish brown spots may develop on the cotyledons, stems and roots of affected seedlings. Less severely affected seedlings produce secondary roots and partially recover. Microscopically, the presence of the spherical oospores in the infected root tissue is characteristic of the *Pythium* infection to the seedlings.

Causal agent

Pythium spp. are soil-borne pathogens which can survive on soil organic matter for a long time.

Control

Seed treatment may help protect the seedlings during the early growing stage when they are most susceptible to this disease. Avoiding early seeding when soils are wet and cold will help the crop escape the disease.

Gray mould

Distribution and economic importance

Gray mould is a common disease affecting a wide host range of plants. It reduces the yield and quality of seed harvested from diseased plants. This disease is capable of producing local epidemics and major yield losses in flax fields with heavy lodging or under stress from abiotic factors.

Symptoms

The first sign of the disease is the damping-off of seedlings emerging from infected seed with brownish spots on the base of the stem at the soil surface. Symptoms on mature plants are light brown patches on the stem. These infected stem parts become soft and decayed, resulting in the death of the upper part of the plant.

Causal agent

Botrytis cinerea Fr. can survive on decaying organic matter for a long time. It is seed-borne in flax, and the seed-borne phase acts as a source of primary inoculum. This fungus produces small, hard black sclerotia which can survive adverse weather conditions.

Control

Seed treatment with fungicides protects the seedlings from early infections. Controlling this disease at the mature stage of the plant growth may not be practical since the fungus is widespread in nature and has an abundance of inoculum.

Basal stem blight/foot rot

Distribution and economic importance

This disease is common in most flax-growing areas in Europe (Muskett and Colhoun, 1947). The seedling phase of this disease is of less importance in comparison to damage caused by this disease at the flowering stage. Occasionally severe epidemics occur, resulting in a total loss of the crop. There are no described cultivars with resistance to this disease but there are some indications that linseed flax cultivars are more resistant than fiber flax cultivars (Muskett and Colhoun, 1947).

Symptoms

Elongated brown lesions appear on the stem just above ground level. Seedlings turn yellow and die in a short period of time. Brownish and discolored areas are observed on the base of the stems and top part of the roots. Plants turn yellow, wilt and die. Microscopic examination of the dead tissue reveals the presence of numerous small pycnidia which release masses of tiny unicellular spores in a yellowish tendril. Occasionally, the upper parts of the stems are infected, become discolored and die, while the bases of the stems remain healthy and normal (Muskett and Colhoun, 1947).

Causal agent

Phoma exigua Desmaz. var. *linicola* (Naumov and Vass.) Maas. (Syn. *Ascochyta linicola* Naumov and Vass) (Maas, 1965; Boerema and Verhoeven, 1976). This pathogen is soil- and stubble-borne, and the seed-borne phase can have serious consequences. Primary infections start from mycelia in the seed coat and from spores present on the seed or in the soil. The fungus produces large numbers of pycnidia on infected and dead tissue. These pycnidia release millions of spores which are capable of creating disease epidemics under favorable weather conditions.

Control

There are some indications of resistance in some flax cultivars to this disease (Ondrej and Rosenberg, 1975). Seed treatments failed to protect the crop from this disease, while foliar applications of fungicides may be effective in reducing the disease (Turner, 1987; Decognet *et al.*, 1994). Sanitation is an important approach to protect the crop from this disease; use clean and disease-free seed, follow a crop rotation of at least three years between flax crops, and destroy crop residue after harvest.

Viral and phytoplasma diseases

Aster yellows

Distribution and economic importance

Aster yellows is a minor disease in flax but has a wide host range, which includes canola, sunflower, sugar beet, carrot and many other plant species, including several weeds. The six-spotted leafhopper (*Macrosteles fascifrons* Uhl.) is the insect vector that transmits the phytoplasma (Frederiksen, 1964). In North America, the disease occurs annually at low levels of infection. However, an epidemic in 1957 caused widespread severe yield losses in flax and other crops (Conners, 1967). The phytoplasma overwinters in perennial broad-leafed weeds and crops, but most infections are carried by leaf-hoppers that migrate from the southern United States. The severity of the disease depends on the stage at which plants become infected, the prevailing temperature, and the number of insect vectors that carry the organism (Frederiksen, 1964).

Symptoms

Aster yellows' symptoms include yellowing of the top part of the plant, conspicuous malformation of the flowers and stunted growth. The infection is systemic. All flower parts are converted into small, yellowish green leaves (Figure 5.19). Diseased flowers are sterile and produce no seed.

Causal agent

This disease is caused by a phytoplasma which is transmitted by the six-spotted leaf-hopper (*Macrosteles fascifrons* Uhl.).

Control

Early seeding may help to avoid the migrating leafhoppers in mid- and late-season, and to reduce the incidence and severity of aster yellows. Control of leafhoppers by pesticide application depends on the leafhopper population carrying the inoculum.

Crinkle

Distribution and economic importance

This virus is of minor importance in flax but it infects oats, barley and wheat (Westdal, 1968). Traces of this disease have been observed in western Canada (Hoes, 1975; Rashid *et al.*, 2000).

Symptoms

The distinguishing symptoms of this virus infection are ennations on the axial leaf surface, the puckering of leaves, stunted growth and reduced tillering. Flowers appear normal in contrast to symptoms of aster yellows but seed production is usually reduced.

Causal agent

Oat blue dwarf virus (OBDV) transmitted by the same leafhopper that transmits aster yellows.

Control

Conduct early seeding to avoid the migrating leafhoppers in mid- and late-season. Control of leafhoppers by pesticide application depends on the leafhopper population and the presence of the virus.

Curly top

Distribution and economic importance

This disease is of minor importance in flax but the virus infects sugar beet. It was first reported from California in 1944 (Muskett and Colhoun, 1947).

Symptoms

The distinguishing symptoms are irregular leaves grouped together at the growing point with general chlorosis of the whole plant.

Causal agent

The beet curly top virus (BCTV) is transmitted by the leafhopper (*Eutettix tenellus* Baker).

Control

Late seeded crops escape the disease due to early migration of the insect vector. Control of the insect vector by pesticide application depends on insect population and the presence of the virus.

Non-parasitic disorders

Chlorosis

Distribution and economic importance

These disorders are associated with calcareous soils which are rich in lime but deficient in trace elements such as zinc, manganese and iron (Flor, 1943). Such disorders are most common with severe epidemics in fields with high soil moisture contents and under cool conditions at the seedling stage. Chlorosis may also occur in flax on water-logged soils. Flax cultivars can recover if the soil moisture conditions and temperature return to normal, otherwise plants will start to die from the top down resulting in yield losses depending on the severity of dieback.

Symptoms

Leaf chlorosis that starts from the top downward is the most common symptom (Figure 5.20). If this condition persists for a few weeks, terminal buds start to die and lateral-branch dieback of terminal buds will occur and development of lateral branches are formed at the lower stem. Longer exposure to this condition will result in the death of the whole plant and probably the whole field (Figure 5.21).

Control

Adequate fertility levels in the soil, including the trace elements, are essential for a healthy crop. Late seeding will help the crop escape the high soil moisture conditions and cool temperatures at the seedling stage, especially in areas where this problem persists year after year. Some flax cultivars are tolerant to chlorosis (Kenaschuk *et al.*, 1996) and are recommended for areas where chlorosis is a problem.

Heat and frost canker

Distribution and economic importance

Flax is affected by these disorders wherever the crop is exposed to high temperature conditions or freezing conditions at the seedling stage. Damage is usually most severe in thin stands on light soils. Low spots are more conducive to frost canker. The damage becomes severe if these conditions prevail for a longer period or if the extreme temperatures recur for several days. In such situations, the damage is commonly inconspicuous, stands may be drastically reduced, and fields may have to be re-seeded to ensure acceptable plant populations and reasonable yield.

Symptoms

Heat canker is formed when the surface layer of soil around the stem is over-heated. This will kill the outer tissue of the stem. The affected region of the stem becomes constricted at or near the soil line, and seedlings fall over and die (Figure 5.22). Frost cankers are similar to heat cankers in appearance (Figure 5.23) but are caused by freezing temperatures

Figure 5.20 Flax chlorosis: yellowing starts from the top of the plant and moves downward.
Figure 5.21 Severe chlorosis and poor stand can destroy the whole field.
Figure 5.22 Heat canker, stem girdling and breakage.
Figure 5.23 Frost canker, stem breakage at soil surface.

at the soil surface. The area below the constriction is usually thin and dry, while that above it is swollen and appears rough and cracked. Less severely damaged seedlings produce new shoots and may grow to maturity but will not produce normal yield.

Control

The damage from heat and frost cankers can be reduced by using recommended seeding rates and following proper seeding practices to ensure good and vigorous stands. Early seeding at a higher seeding rate in a north–south direction is recommended in areas where these cankers are expected to occur.

Injury due to herbicides

Distribution and economic importance

Crop injury may occur at any time when herbicides are not applied according to recommended procedures of application, including spray drifts from ground or aerial applications in adjacent fields. Herbicide injury may also result from residual herbicides in the field from previous years. The level of tolerance is usually low and the damage is high, resulting in partial or complete death of affected crops and loss in yield. The damage is compounded under extreme environmental conditions of high humidity, low soil moisture, temperature extremes, or other conditions of stress.

Symptoms

Symptoms of damage vary depending on the herbicide causing the injury. Some of the common symptoms are bending and twisting of stems, yellowing of leaves, delayed maturity, temporary leaf yellowing, thin stand and delayed emergence in the case of soil residues, and suppression of flax growth for a short period.

Control

Following the proper guidelines for the use and application of herbicide will reduce incidence of direct injury, reduce spray drift, and minimize the residual impact in the soil. The use of recommended rates at the correct stage of the crop under normal environmental conditions will also reduce the probability of causing crop injury.

References

Acosta, P.P. (1986). Estimation of losses caused by *Melampsora lini* (Pers.) Lev. using sister lines of linseed. *Boletin Genetico, Instituto de Fitotecnia, Castelar, Argentina* 14, 35–40 (Rev. Plant Pathol. 1988. 67:84).

Allison, C.C. (1934). Powdery mildew of flax in Minnesota. *Phytopathology* 24, 305–307.

Anderson, N.A. (1973). The *Rhizoctonia* complex in relation to seedling blight of flax. *Proc of the Flax Institute of the United States. Fargo, ND*, Flax Institute of the United States, 43, 4–5.

Anderson, P.A., Lawrence, G.J., Morrish, B.C., Ayliffe, M.A., Finnegan, E.J., and Ellis, J.G. (1997). Inactivation of the flax rust resistance gene M associated with loss of a repeated unit within the leucine-rich repeat coding region. *Plant Cell* 9, 641–651.

Anderson, P.A., Lawrence, G.J., and Pryor, A. (1992). The inheritance of restriction fragment length polymorphisms in the flax rust *Melampsora lini*. *Theor. Appl. Genet.* 84, 845–850.

Anonymous (1996). *Growing Flax: Production, Management and Diagnostic Guide*. Flax Council of Canada. 56 p.

Armstrong, G.M., and Armstrong, J.K. (1968). Formae speciales and races of *Fusarium oxysporum* causing a tracheomycosis in the syndrome of disease. *Phytopathology* 58, 1242–1246.

Arthur, J.C. (1962). *Manual of the Rusts in United States and Canada*. Hafner Publ. Co. New York, NY. 438 p.

Ayliffe, M.A., Frost, D.V., Finnegan, E.J., Lawrence, G.J., Anderson, P.A., and Ellis, J.G. (1999). Analysis of alternative transcripts of the flax L6 rust resistance gene. *Plant J.* 17, 287–292.

Badayeva, P.K. (1930). 'Flax diseases in Siberia'. In *Morbi Plantarum*, Vol. 19, Leningrad, pp. 192–199 (Abs. *Rev. Appl. Mycol.* 10:459)

Badwal, S.S. (1975). Inheritance of resistance to powdery mildew in linseed. *Indian J. Gen. Plant Breeding* 35, 432–433.

Beale, R.E. (1991). Studies of resistance in linseed cultivars to *Oidium lini* and *Botrytis cinerea*. *Aspects Applied Biol.* 28, 85–90.

Boerema, G.H., and Verhoeven, A.A. (1976). Series 2a: Fungi on field crops: beet and potato; caraway, flax and oilseed poppy. *Neth. J. Plant Pathol.* 82, 193–214.

Borlaug, N.E. (1945). *Variation and Variability of* Fusarium lini. Minn. Agr. Exp. Sta. Tech. Bull. 168. 40 p.

Brayford, D. (1996). *Fusarium oxysporum* f. sp. *lini. IMI Descriptions of Fungi and Bacteria No. 1267. Mycopathologia* 133, 49–51.

Broadfoot, W.C., and Stakman, E.C. (1926). Physiologic specialization of *Fusarium lini*, Bolley. *Phytopathology* 16, 84.

Burdon, J.J. (1994). The distribution and origin of genes for race-specific resistance to *Melampsora lini* in *Linum marginale. Evolution* 48, 1564–1575.

Burdon, J.J., and Jarosz, A.M. (1992). Temporal variation in the racial structure of flax rust (*Melampsora lini*) populations growing on natural stands of wild flax (*Linum marginale*): local versus metapopulation dynamics. *Plant Path.* 41, 165–179.

Burdon, J.J., and Roberts, J.K. (1995). The population genetic structure of the rust fungus *Melampsora lini* as revealed by pathogenicity, isozyme and RFLP markers. *Plant Path.* 44, 270–278.

Burdon, J.J., and Thompson, J.N. (1995). Changed patterns of resistance in a population of *Linum marginale* attacked by the rust pathogen *Melampsora lini. J. Ecol.* 83, 199–206.

Burdon, J.J., Thrall, P.H., and Brown, H.D. (1999). Resistance and virulence structure in two *Linum marginale–Melampsora lini* host–pathogen metapopulations with different mating systems. *Evolution* 53, 704–706.

Burnham, C.R. (1932). The inheritance of *Fusarium* wilt resistance in flax. *J. Am. Soc. Agron.* 24, 734–748.

Carlsson-Granér, U., Burdon, J.J., and Thrall, P.H. (1999). Host resistance and pathogen virulence across a plant hybrid zone. *Oecologia* 121, 339–347.

Chauhan, L.S., and Srivastava, K.N. (1975). Estimation of loss of yield caused by blight disease of linseed. *Indian J. Farm. Sci.* 3, 107–109.

Colhoun, J. (1960). Physiologic specialization in *Polyspora lini* Laff. *Trans. Br. Mycol. Soc.* 43, 150–154.

Colhoun, J., and Muskett, A.E. (1943). "Pasmo" disease of flax. *Nature, Lond.* 151, 223–224.

Conners, I.L. (1967). *An Annotated Index of Plant Diseases in Canada.* Publ. 1251, Can. Dept. Agric., Res. Br., Ottawa. 381 p.

Convey, R.P. (1962). Field resistance of flax to pasmo. *Phytopathology* 52, 1–34.

Coosemans, J. (1977). Interaction and population dynamics of *Pratylenchus penetrans* (Cobb) and *Verticillium* spp on flax. *Parasitica* 33, 53–58.

de Tempe, J. (1963). Health testing of flaxseed. *ISTA* 28(1), 107–131.

Decognet, V., Cerceau, V., and Jouan, B. (1994). Control of *Phoma exigua* var. *linicola* on flax by seed and foliar spray treatments with fungicides. *Crop. Prot.* 13, 105–108.

Dey, P.K. (1933). An *Alternaria* blight of the linseed plant. *Indian J. Agric. Sci.* 5, 881–896.

Dickinson, M.J., Zhang, R., and Pryor, A. (1993). Nucleotide sequence relationships of double-stranded RNAs in flax rust, *Melampsora lini. Curr. Genet.* 24, 428–432.

Ellis, J., Lawrence, G., Ayliffe, M., Anderson, P., Collins, N., Finnegan, E.J., Frost, D., Luck, J., and Pryor, T. (1997). Advances in the molecular genetic analysis of the flax–flax rust interaction. *Ann. Rev. Phytopathol.* 35, 271–291.

Ellis, J., Lawrence, G., Finnegan, E.J., and Anderson, P.A. (1995). Contrasting complexity of two rust resistance loci in flax. *Proc. Natl. Acad. Sci. USA* 92, 4185–4188.

Ellis, J.G., Finnegan, E.J., and Lawrence, G.J. (1992). Developing a transposon tagging system to isolate rust-resistance genes from flax. *Theor. Appl. Genet.* 85, 46–54.

Evans, N., McRoberts, N., Hitchcock, D., and Marshall, G. (1995). Screening for resistance to *Alternaria linicola* (Groves and Skolko) in *Linum usitatissimum* L. using a detached cotyledon assay. *Ann. Appl. Biol.* 127, 263–271.

Evans, N., McRoberts, N., Hitchcock, D., and Marshall, G. (1997). Identification of the determinants of host resistance and pathogenicity in interactions between *Alternaria linicola*

Groves and Skolko and *Linum usitatissimum* L. accessions using multivariate analyses. *Ann. Appl. Biol.* **130**, 537–547.

Fitt, B.D.L., Cook, J.W., and Burhenne, S. (1991a). *Verticillium* on linseed (*Linum usitatissimum*) in the United Kingdom. *Aspects Applied Biol.* **28**, 91–94.

Fitt, B.D.L., Cosku, H., and Schmechel, D. (1991b). Biology of three *Alternaria* species on linseed: A comparison. *Aspects Applied Biol.* **28**, 101–106.

Fitt, B.D.L., and Ferguson, A.W. (1993). Effects of fungal diseases on linseed (*Linum usitatissimum*) growth and yield, 1988–1990. *J. Agric. Sci.* **120**, 225–232.

Fitt, B.D.L., Ferguson, A.W., Dhua, U., and Burhenne, S. (1991c). Epidemiology of *Alternaria* species on linseed. *Aspects Applied Biol.* **28**, 95–100.

Fitt, B.D.L., Jouan, B., Sultana, C., Paul, V.H., and Bauers, F. (1991d). Occurrence and significance of fungal diseases on linseed and fiber flax in England, France and Germany. *Aspects Applied Biol.* **28**, 59–64.

Fitt, B.D.L., and Vloutoglou, I. (1992). *Alternaria* diseases of linseed. In Chelkowski, J., and Visconti, A. (eds), *Alternaria, Biology, Plant Diseases and Metabolites*, Elsevier, Amsterdam, pp. 289–300.

Flor, H.H. (1931). Physiologic specialisation of *Melampsora lini* in *Linum usitatissimum. J. Agric. Res.* **51**, 119–137.

Flor, H.H. (1943). Chlorotic dieback of flax grown on calcareous soils. *J. Am. Soc. Agron.* **35**, 259–270.

Flor, H.H. (1944). Relation of rust damage in seed flax to seed size, oil content, and iodine value of oil. *Phytopathology* **34**, 348–349.

Flor, H.H. (1946). Genetics of pathogenicity in *Melampsora lini. J. Agric. Res.* **73**, 335–337.

Flor, H.H. (1947). Inheritance of reaction to rust in flax. *J. Agric. Res.* **74**, 241–262.

Flor, H.H. (1954). *Identification of Races of Flax Rust by Lines with Single Rust-Conditioning Genes.* U.S. Dep. Agric. Tech. Bull. 1087. 25 p.

Flor, H.H. (1955). Host–parasite interaction in flax rust–its genetics and other implications. *Phytopathology* **45**, 680–685.

Flor, H.H. (1956). The complementary genic systems in flax and flax rust. *Adv. Genet.* **8**, 29–54.

Flor, H.H. (1958). Mutation to wider virulence in the *Melampsora lini. Phytopathology* **48**, 297–301.

Flor, H.H. (1971). Current status of the gene-for-gene concept. *Ann. Rev. Plant Pathol.* **9**, 275–296.

Flor, H.H., and Comstock, V.E. (1971). Flax cultivars with multiple-rust conditioning genes. *Crop Sci.* **11**, 64–66.

Flor, H.H., and Comstock, V.E. (1972). Identification of rust conditioning genes in flax cultivars. *Crop Sci.* **12**, 800–804.

Frederiksen, R.A. (1964). Simultaneous infection and transmission of two viruses in flax by *Macrosteles fasciforms. Phytopathology* **54**, 1028–1030.

Frederiksen, R.A., and Culbertson, J.O. (1962). Effect of pasmo on the yield of certain flax varieties. *Crop Sci.* **2**, 434–437.

Froud-Williams, R.J., Gladders, P., Heath, M.C., Jenkyn, J.F., Knott, C.M., Lane, A., and Pink, D. (1991). *Production and Protection of Linseed.* Association of Applied Biologists. 160 p.

Garassini, L.A. (1935). El "pasmo" del Lino Phlyctaena? Linicola Speg. Ensayo a campo de resistencia varietal y estudio morfol gico y fisiol gico del parasito. *Rev. Fac. Agron. La Plata* **20**, 170–261.

Gill, K.S. (1987a). 'Diseases of linseed'. In Linseed (ed.), Indian Council of Agricultural Research, New Delhi, India, pp. 317–341.

Gill, K.S. (1987b). *Linseed.* Indian Council of Agricultural Research. New Delhi. 386 p.

Goray, S.C., Khosla, H.K., Upadhyaya, Y.M., Naik, S.L., and Mandloi, S.C. (1987). Inheritance of wilt resistance in linseed. *Indian J. Agric. Sci.* **57**, 625–627.

Groth, J.V., Comstock, V.E., and Anderson, N.A. (1970). Effect of seed color on tolerance of flax to seedling blight caused by *Rhizoctonia solani. Phytopathology* **60**, 379–380.

Hart, E. (1926). Factors affecting the development of flax rust. *Phytopathology* 16, 185–205.

Hausner, G., Rashid, K.Y., Kenaschuk, E.O., and Procunier, J.D. (1999a). The development of codominant PCR/RFLP based markers for the flax rust resistance alleles at the L locus. *Genome* 42, 1–8.

Hausner, G., Rashid, K.Y., Kenaschuk, E.O., and Procunier, J.D. (1999b). The identification of a cleaved amplified polymorphic sequence (CAPS) marker for the flax rust resistance gene M3. *Can. J. Plant Pathol.* 21, 187–192.

Henry, A.W. (1928). Reaction of *Linum* species of various chromosome numbers to rust and powdery mildew. *Sci. Agr.* 8, 460–461.

Henry, A.W. (1934). Observations on the variability of *Polyspora lini* Lafferty. *Can. J. Res.* 10, 409–413.

Henry, A.W., and Ellis, C. (1971). Seed infestation of flax in Alberta with the fungus causing browning or stem-break. *Canadian Plant Disease Survey* 51, 76–79.

Hoes, J.A. (1975). Diseases of flax in western Canada. *Oilseed and Pulse Crops in Western Canada: A Symposium. Calgary*, Western Co-operative Fertilizers. pp. 415–423.

Hoes, J.A. (1991). Common names for plant diseases: flax (*Linum usitatissimum* L. and other *Linum* spp.). *Plant Disease* 75, 228.

Hoes, J.A., and Dorrell, D.G. (1979). Detrimental and protective effects of rust in flax plants of varying age. *Phytopathology* 69, 695–698.

Hoes, J.A., and Kenaschuk, E.O. (1980). Postseedling resistance to rust in flax. *Can. J. Plant Pathol.* 2, 125–130.

Hoes, J.A., and Kenaschuk, E.O. (1986). Gene K1 of Raja flax: a new factor for resistance to rust. *Phytopathology* 76, 1043–1045.

Hoes, J.A., and Kenaschuk, E.O. (1992). Host–pathogen specificity in postseedling reaction of *Linum usitatissimum* to *Melampsora lini*. *Can. J. Bot.* 70, 1168–1174.

Hoes, J.A., and Tyson, I.H. (1963). A naturally occurring North American race of *Melampsora lini* attacking flax variety Ottawa 770B. *Plant Dis. Rep.* 47, 836.

Hoes, J.A., and Tyson, I.H. (1966). Races of flax rust in the Canadian prairies in 1963 and 1964. *Plant Dis. Rep.* 50, 62–63.

Hoes, J.A., and Zimmer, D.E. (1976). New North American races of flax rust, probably products of natural hybridization. *Plant Dis. Rep.* 60, 1010–1013.

Holmes, S.J.I. (1976). Pasmo disease of linseed in Scotland. *Plant Path.* 25, 61.

Homma, Y. (1928). On the powdery mildew of flax. *Bot. Mag. (Tokyo)* 42, 331–334.

Hora, T.S., Chenulu, V.V., and Munjal, R.L. (1962). Studies on assessment of losses due to *Melampsora lini* on linseed. *Indian Oilseeds J.* 6, 196–198.

Hossain, M.D. (1992). Screening fifteen linseed genotypes for leaf blight resistance. *Bangladesh J. Bot.* 21, 297–298.

Houston, B.R., and Knowles, P.F. (1949). Fifty years survival of flax *Fusarium* wilt in the absence of *Fusarium* culture. *Plant Dis. Rep.* 33, 38–39.

Islam, M.R. (1992). Control of flax diseases through genetic resistance. *J. Plant Dis. Prot.* 99, 550–557.

Islam, M.R., and Mayo, G.M.E. (1990). A compendium on host genes in flax conferring resistance to flax rust. *Plant Breeding* 104, 89–100.

Islam, M.R., and Shepherd, K.W. (1991). Present status of the genetics of rust resistance in flax. *Euphytica* 55, 255–267.

Jarosz, A.M., and Burdon, J.J. (1992). Host–pathogen interactions in natural populations of *Linum marginale* and *Melampsora lini*. *Oecologia* 89, 53–61.

Jeswani, L.M., and Joshi, A.B. (1964). Inheritance of resistance to wilt in linseed. *Indian J. Gen. Plant Breeding* 24, 92–94.

Jones, D.A. (1988). Genes for resistance to flax rust in the flax cultivars Towner and Victory A and the genetics of pathogenicity in flax rust to the L8 gene for resistance. *Phytopathology* 78, 338–341.

Kalia, H.R., Chand, J.N., and Ghai, B.S. (1965). Inheritance of resistance to *Alternaria* blight of linseed. *J. Res. Ludhiana* 2, 104–105.

Kamthan, K.P., Misra, D.P., and Shukla, A.K. (1981). Independent genetic resistance to wilt and rust in linseed. *Indian J. Agric. Sci.* 51, 556–558.

Kaushal, P.K., and Shrivas, S.R. (1974). Inheritance of resistance to powdery mildew (*Oidium lini*) in linseed, *Linum usitatissimum. Curr. Sci.* 43, 353–354.

Kenaschuk, E.O. (1974). Parentage of North American flax varieties. *Proc. of the Flax Institute of the United States. Fargo, ND*, Flax Institute of the United States. 44, 21.

Kenaschuk, E.O., and Rashid, K.Y. (1993). AC Linora flax. *Can. J. Plant Sci.* 73, 839–841.

Kenaschuk, E.O., Rashid, K.Y., and Gubbels, G.H. (1996). AC Emerson flax. *Can. J. Plant Sci.* 76, 483–485.

Kerr, H.B. (1960). The inheritance of resistance of *Linum usitatissimum* L. to the Australian *Melampsora lini* (pers.) Lev. race complex. *Proc. Linn. Soc. N.S.W.* 85, 273–321.

Klose, A., Bauers, F., and Paul, V.H. (1993). Pathogenicity of two isolates of *Alternaria linicola* on 16 cultivars of *Linum usitatissimum. Bulletin OILB/SROP* 16, 100–108.

Knowles, P.F., and Houston, B.R. (1955). Inheritance of resistance to *Fusarium* wilt of flax in Dakota selection 48–94. *Agron. J.* 47, 131–135.

Knowles, P.F., Houston, B.R., and McOnie, J.B. (1956). Inheritance of resistance to *Fusarium* wilt of flax in Punjab 53. *Agron. J.* 48, 135–137.

Kommedahl, T., Christensin, J.J., and Freederikson, R.A. (1970). *A Half Century of Research in Minnesota on Flax Wilt Caused by Fusarium oxysporum*. Minnesota Agr. Exp. Sta. Tech. Bull. 34 p.

Kroes, G.M.L.W., Baayen, R.P., and Lange, W. (1998a). Histology of root rot of flax seedlings (*Linum usitatissimum*) infected by *Fusarium oxysporum* f.sp. *lini. Eur. J. Plant Pathol.* 104, 725–736.

Kroes, G.M.L.W., Baayen, R.P., and Lange, W. (1998b). Infection and colonization of flax seedling roots (*Linum usitatissimum*) by *Fusarium oxysporum* f.sp. *Lini. Proc. of the Flax Institute of the United States. Fargo ND*, Flax Institute of the United States. 57, 129–134.

Kroes, G.M.L.W., Loffler, H.J.M., Parlevliet, J.E., Keizer, L.C.P., and Lange, W. (1999). Interactions of *Fusarium oxysporum* f.sp. lini, the flax wilt pathogen, with flax and linseed. *Plant Path.* 48, 491–498.

Kroes, G.M.L.W., Rashid, K.Y., Hammond, J., and Lange, W. (1998c). Assessment of resistance to *Fusarium oxysporum* f.sp. *lini* in flax, race specific interactions and environmental factors. *Proc. of the Flax Institute of the United States. Fargo, ND*, Flax Institute of the United States. 57, 118–124.

Kroes, I., K., Rashid, K.Y., and Lange, W. (2002). Variation in *Fusarium* wilt (*Fusarium oxysporum lini*) in Europe and North America. *Eur. J. Plant Pathol.* (in press).

Kudriavtseva, L.P. (1998). Intraspecific differentiation in flax anthracnosis pathogen by virulence. *Mikologiya I Fitopatologiya* 32, 62–64.

Lawrence, G.J., Ellis, J.G., and Finnegan, E.J. (1994). Cloning a rust-resistance gene in flax. *Advances in Molecular Genetics of Plant–Microbe Interactions.* 3, 303–306.

Lawrence, G.J., Finnegan, E.J., Ayliffe, M.A., and Ellis, J.G. (1995). The L6 gene for flax rust resistance is related to the Arabidopsis bacterial resistance gene RPS2 and the tobacco viral resistance gene N. *Plant Cell* 7, 1195–1206.

Lawrence, G.J., Finnegan, E.J., and Ellis, J.G. (1993). Instability of the L6 gene for rust resistance in flax is correlated with the presence of a linked Ac element. *Plant J.* 4, 659–669.

Liu, X.Y., Chen, S.L., Sun, Q.A., He, D.T., and Wu, Y.N. (1993). Evaluation of *Fusarium* wilt resistance of flax varieties. *Scientia Agricultura Sinica* 26, 44–49.

Loshakova, N.I. (1984). Methods of appraising pasmo resistance in flax. *Len Konop.* 20, 28–29 (Rev. Pl. Path. 65, 6061).

Loshakova, N.I., and Korneeva, E.M. (1979). Resistance of fiber flax cultures to pasmo disease. *Rev. Plant Path.* 59, 4199.

Lugger, O. (1890). *A Treatise on Flax Culture.* Minn. Agr. Exp. Sta. Bull. 13. 38 p.

Maas, P.W.T. (1965). The identity of the footrot fungus of flax. *Neth. J. Plant Pathol.* 71, 113–121.

MacNish, G.C. (1963). Diseases recorded on native plants, weeds, fields and fiber plants in Western Australia. *J. Agric. W. Aust. Ser.* 4, 401–408.

Martelli, G.P. (1961). Septoriosi (pasmo) of flax in Italy. *Phytopathol. Medit.* 1, 66–70.

Mederick, F.M., and Piening, L.J. (1982). *Sclerotinia sclerotiorum* on oil and fiber flax in Alberta. *Canadian Plant Disease Survey* 62, 1.

Mercer, P.C., Hardwick, N.V., Fitt, B.D.L., and Sweet, J.B. (1994). Diseases of linseed in the United Kingdom. *Plant Variety and Seeds* 7, 135–150.

Mercer, P.C., McGimpsey, H.C., and Ruddock, A. (1989). Effect of seed treatment and sprays on the field performance of linseed. Tests of Agrochemicals and Cultivars, *Annals of Applied Biology 114 (Supplement)* 10, 50–51.

Mercer, P.C., Ruddock, A., Fitt, B.D.L., and Harold, J.F.S. (1992). Linseed diseases in the UK and their control. *Brighton Crop Protection Conference–Pests and Diseases* 3, 921–930.

Millikan, C.R. (1945). Wilt disease of flax. *J. Dept. Agr. Victoria* 43, 305–313, 354–361.

Millikan, C.R. (1948). Studies of strains of *Fusarium lini*. *Proc. Roy. Soc. Victoria* LXI, 1–24.

Millikan, C.R. (1951). *Diseases of Flax and Linseed*. Dept. of Agric., Victoria, Australia. 140 p.

Misra, D.P. (1963). A new physiologic races of linseed rust in India. *Indian Phytopathol.* 16, 102–103.

Misra, D.P. (1966). Genes conditioning resistance of *Linum* species to Indian races of linseed rust. *Indian J. Gen. Plant Breeding* 26, 63–72.

Misra, D.P., and Lele, V.C. (1963). Physiologic races of linseed rust in India during 1960, 1961, and 1962. Prevalence and distribution. *Indian Oilseeds J.* 7, 336–337.

Mitchell, S.J., Jellis, G.J., and Cox, T.W. (1986). *Sclerotinia sclerotiorum* on linseed: New or unusual records. *Plant Path.* 35, 403–405.

Mpofu, S.I., and Rashid, K.Y. (2000a). Assessment of genetic variation among *Fusarium oxysporum* f. sp. *lini* isolates from western Canada based on vegetative compatibility grouping. *Can. J. Plant Pathol.* 22, 176.

Mpofu, S.I., and Rashid, K.Y. (2000b). Virulence of *Fusarium oxysporum* f. sp. *lini* on flax. *Can. J. Plant Pathol.* 22, 190.

Muskett, A.E., and Colhoun, J. (1947). *The Diseases of the Flax Plant* (Linum usitatissimum Linn.). W. and G. Baird, Ltd. Belfast. 112 p.

Nair, P.N., and Kommedahl, T. (1957). The establishment and growth of *Fusarium lini* in flax tissues. *Phytopathology* 47, 25.

Nattrass, R.M. (1943). The pasmo disease of flax in Kenya, *Sphaerella linorum* Wollenweber. *East African Agric. J.* 8, 223–226.

Newhook, F.J. (1942). Pasmo (*Sphaerella linorum*) on flax in New Zealand. *N. Z. J. Sci. A.* 24, 102–106.

Ondrej, M. (1977). New ideas about fusariosis of flax. *Rev. Plant Path.* 57, 5521.

Ondrej, M., and Rosenberg, L. (1975). Phoma wilt of flax. *Rev. Plant Path.* 55, 1–6.

Pandey, R.N., and Misra, D.P. (1992). Assessment of yield loss due to powdery mildew of linseed. *Indian Bot. Reptr.* 11, 62–64.

Parish, D.L., and Statler, G.D. (1988). Slow rusting in flax. *Proc. of the Flax Institute of the United States. Fargo, ND,* Flax Institute of the United States. 52, 63–66.

Paul, V.H., Sultana, C., Jouan, B., and Fitt, B.D.L. (1991). Strategies for control of diseases on linseed and fiber flax in Germany, France and England. *Aspects Applied Biol.* 28, 65–70.

Pope, S.J., and Sweet, J.B. (1991). Sclerotinia stem rot resistance in linseed cultivars. *Aspects Applied Biol.* 28, 79–84.

Pospisil, B. (1971). The effect of artificial infection by anthracnose on the variability of flax. *Rev. Plant Path.* 51, 1547.

Rashid, K.Y. (1991). Evaluation of components of partial resistance to rust in flax. *Can. J. Plant Pathol.* 13, 212–217.

Rashid, K.Y. (1997). Slow-rusting in flax cultivars. *Can. J. Plant Pathol.* 19, 19–24.

Rashid, K.Y. (1998). Powdery mildew on flax: A new disease in western Canada. *Can. J. Plant Pathol.* 20, 216.

Rashid, K.Y. (2000). Sclerotinia on flax in western Canada – warning for a potential disease problem. *Can. J. Plant Pathol.* 22, 175–176.

Rashid, K.Y., and Kenaschuk, E.O. (1992). Genetics of resistance to rust in the flax cultivars Vimy and Andro. *Can. J. Plant Pathol.* 14, 207–210.

Rashid, K.Y., and Kenaschuk, E.O. (1994). Genetics of resistance to flax rust in six Canadian flax cultivars. *Can. J. Plant Pathol.* 16, 266–272.

Rashid, K.Y., and Kenaschuk, E.O. (1996a). New genes for resistance to flax rust. *Proc. of the Flax Institute of the United States. Fargo, ND*, Flax Institute of the United States. 56, 182–186.

Rashid, K.Y., and Kenaschuk, E.O. (1996b). Seed treatment for fusarium wilt control in flax. *Proc. of the Flax Institute of the United States. Fargo, ND*, Flax Institute of the United States. pp. 158–161.

Rashid, K.Y., and Kenaschuk, E.O. (1998a). Control of rhizoctonia seedling blight in flax by seed treatment. *Can. J. Plant Pathol.* 20, 128.

Rashid, K.Y., and Kenaschuk, E.O. (1998b). Flax yield loss by Pasmo and efficacy of fungicides for disease control. *Can. J. Plant Pathol.* 20, 337.

Rashid, K.Y., Kenaschuk, E.O., and Menzies, J.G. (1998a). Powdery mildew on flax, first encounter in western Canada. *Proc. of the Flax Institute of the United States. Fargo, ND*, Flax Institute of the United States. 57, 125–128.

Rashid, K.Y., Kenaschuk, E.O., and Platford, R.G. (1998b). Diseases of flax in Manitoba in 1997, and first report of powdery mildew on flax in western Canada, Can. Plant Dis. Surv. 78, 99–100. http://res2.agr.gc.ca/london/pmrc/english/catalog.html

Rashid, K.Y., Kenaschuk, E.O., and Platford, R.G. (2000). Diseases of flax in Manitoba in 1999, Can. Plant Dis. Surv. 80, 92–93. http://res2.agr.gc.ca/london/pmrc/english/catalog.html

Rashid, K.Y., and Kroes, G.M.I.W. (1999). Pathogenic variability in *Fusarium oxysporum* f. sp. *Lini* on flax. *Phytopathology* 89, S65.

Rashid, K.Y., and Platford, R.G. (1990). Survey of flax diseases in Manitoba in 1989. *Canadian Plant Disease Survey* 70, 76.

Sackston, W.E., and Carson, R.B. (1951). Effect of pasmo disease of flax on the yield and quality of linseed oil. *Can. J. Bot.* 29, 339–351.

Sackston, W.E., and Gordon, W.L. (1945). Twenty-fourth Ann. Rep. of 1944 Survey. *Canadian Plant Disease Survey*, 29.

Saharan, G.S. (1991). Linseed and flax rust. *Indian J. Mycol. Pl. Pathol.* 21, 119–137.

Saharan, G.S., and Saharan, M.S. (1992). *Studies on Physiological Specialization and Inheritance of Resistance in Linseed to Powdery Mildew Disease*. 2nd Ann. Rept. Dept. Pl. Pathol. CCS HAU, Hisar, India. 47 p.

Saharan, G.S., and Saharan, M.S. (1994a). Conidial size, germination and appressorial formation of *Oidium lini* Skoric, cause of powdery mildew of linseed. *Indian J. Mycol. Pl. Pathol.* 24, 176–178.

Saharan, G.S., and Saharan, M.S. (1994b). Studies on powdery mildew of linseed caused by *Leveillula taurica* (Lev.) Arnaud. *Indian J. Mycol. Pl. Pathol.* 24, 107–110.

Saharan, G.S., and Singh, B.M. (1978). New physiological races of *Melampsora lini* in India. *Indian Phytopathol.* 31, 450–454.

Sharma, H.C., and Khosla, H.K. (1979). Fungicidal control of powdery mildew of linseed (*Linum usitatissimum*) in relation to losses. *JNKVV Res. J.* 10, 161–162.

Sharma, L.C., and Mathur, R.I. (1971). Variability in first single spore isolates of *Fusarium oxysporum* f.sp. Lini. Rajasthan. *Indian Phytopathol.* 24, 698–704.

Shukla, A.K. (1992). Assessment of yield losses by various levels of rust infection linseed. *Plant Dis. Res.* 7, 157–160.

Singh, D., Mital, S.P., and Gangwar, L.C. (1956). Breeding for wilt resistance in linseed (*Linum usitatissimum* L.) in Uttar Pradesh. *Indian J. Gen. Plant Breeding* 16, 29–31.

Singh, N.D., and Chauhan, Y.S. (1988). Genetics of resistance to *Alternaria lini* in linseed (*Linum usitatissimum*). *Indian J. Agric. Sci.* 58, 550–551.

Statler, G.D., Hammond, I.J., and Zimmer, D.E. (1981). Hybridization of *Melampsora lini* to identify rust resistance in flax. *Crop Sci.* 21, 219–221.

Tisdale, W.H. (1917). Flax wilt: A study of the nature and inheritance of wilt resistance. *J. Agric. Res.* 11, 573–606.

Tochinai, Y., and Takee, G. (1950). Studies on the physiologic specialization in *Fusarium lini* Bolley. *J. Fac. Agr. Hokkaido Univ.* 47, 193–266.

Turlier, M.-F., Eparvier, A., and Alabouvette, C. (1994). Early dynamic interactions between *Fusarium oxysporum* f.sp. *lini* and the roots of *Linum usitatissimum* as revealed by transgenic GUS-marked hyphae. *Can. J. Bot.* 72, 1605–1612.

Turner, J. (1987). *Linseed Law: A Handbook for Growers and Advisers.* Alderman Printing and Bookbinding, BASF UK Ltd. Hadleigh. Ipswich. 356 p.

Tvelkov, S.G. (1970). *S. lini* a new species on fiber flax in the Novogorod region. *Mikol i. Fitopatol.* 4, 484.

Vanterpool, T.C. (1947). *Selenophoma linicola* Sp. Nov. on flax in Saskatchewan. *Mycologia* 39, 341–348.

Vanterpool, T.C. (1949). Flax diseases in Saskatchewan in 1948. *28th Ann. Rep. Can. Plant Dis. Surv.* 28, 22–24.

Vest, G., and Comstock, V.E. (1968). Resistance of flax to seedling blight caused by *Rhizoctonia solani*. *Phytopathology* 58, 1161–1163.

Vloutoglou, L., Fitt, B.D.L., and Lucas, J.A. (1995). Survival and seed to seedling transmission of *Alternaria linicola* on linseed. *Ann. Appl. Biol.* 127, 33–47.

Westdal, P.H. (1968). Host range studies of oat blue dwarf virus. *Can. J. Bot.* 46, 1431–1435.

Wicks, Z.W., and Hammond, J.J. (1978). Screening of flax species for new sources of genes resistant to *Melampsora lini* (Ehrenb.) Lev. *Crop Sci.* 18, 7–10.

Wiersema, H.T. (1955). Flax scorch. *Euphytica, Neth. J. Plant Breeding.* 4, 197–205.

Zarzycka, H. (1973). Biological composition of the fungus *Colletotrichum lini* (Westend). *Rev. Plant Path.* 54, 342.

Zarzycka, H. (1976). Physiological races of *Colletotrichum lini* (Westend). Tochinai in Poland. *Rev. Plant Path.* 56, 3581.

Zimmer, D.E. (1976). Genes conditioning rust resistance of the flax varieties Dufferin, Foster and Raja. *Proc. of the Flax Institute of the United States. Fargo, ND*, Flax Institute of the United States. 46, 12–14.

Zimmer, D.E., and Hoes, J.A. (1974). Race 370, a new and dangerous North American race of flax rust. *Plant Dis. Rep.* 58, 311–313.

Zimmer, D.E., and Statler, G.D. (1976). Genetics of rust resistance of six Argentinian flax introductions. *Phytopathology* 66, 661–663.

6   Principal insect pests of flax

Ian L. Wise and Juliana J. Soroka

Introduction

Only a small number of the major cosmopolitan insect pests of crops attack flax, *Linum usitatissimum* L., and these are considered to be minor in their economic impact on the crop. Nevertheless, insect pests can cause serious yield losses wherever flax is grown. Most of the main insect pests of flax are indigenous to specific regions, subcontinents, or continents. Thus, each geographic area of the world has its own unique complex of flax insect pests. No differences in insect diversity are known to occur between flax grown for seed or fiber. Therefore, the relative importance of each insect pest depends on its feeding site on the plant, the stage of the plant when fed on, and on the economic value of fiber versus linseed oil.

Most insects feeding on flax are polyphagous, consuming plants in a variety of plant families. Fewer than a dozen insect species are monophagous, restricted in their feeding to flax and its close relatives. Major insect pests of flax, which include potato aphids, cutworms, flea beetles, flax thrips and linseed midges, occur in both feeding groups.

This chapter discusses the host range, biology, diagnostic methods, economic impact, and control methods for each of the major insect pests of flax listed above. It also provides a brief description of minor insect pests of flax, including climbing cutworms, plant bugs, crane flies, grasshoppers, leaf rollers, stem borers, and leaf miners.

Major insect pests of flax

Potato aphid, Macrosiphum euphorbiae *(Thomas)*

Pest distribution and host range

The potato aphid originated in North America and now has a worldwide distribution (Blackman and Eastop, 1984). It can be found on many commercial vegetable and field crops but until the 1990s was considered to be a serious pest of only tomatoes and potatoes (Hodgson *et al.*, 1974; Lange and Bronson, 1981; Walker *et al.*, 1984). Outbreaks of the potato aphid on flax were first reported in western Canada in the 1980s (Lamb, 1989), and field studies have since shown that this aphid can cause serious yield losses (Wise *et al.*, 1995). The potato aphid is now considered to be the most serious insect pest of flax in western Canada.

Biology and ecology

Potato aphids overwinter as eggs on woody perennials, particularly wild and domest-icated roses (Shands *et al.*, 1962). The first generation of wingless females arises from eggs hatched in the spring. These females give birth to nymphs which develop into both winged and wingless adults. In western Canada, the winged females migrate from the winter hosts to summer hosts such as flax in late June to early July when the crop is flowering and developing seeds (Wise *et al.*, 1995). Wingless females develop on the summer hosts and produce living young by parthenogenesis (Figure 6.1). Most of these females are 3–4 mm long, have long legs, prominent cornicles or siphunculi on the sides of their abdomens, and are entirely green, although a pink phase may occur. A number of generations of wingless females develop on flax until early August, when winged females and males appear in response to diminishing daylight (Lamb *et al.*, 1997). These winged adults migrate back to the winter hosts, resulting in rapid drops in aphid populations on the summer hosts. Adults on the winter host mate and females produce eggs for overwintering.

Inspection and diagnostic methods

The easiest way to detect aphids in flax is to gently sample the upper portion of the plant with an insect sweep net before the crop is flowering in late June or early July. Yellow sticky traps, yellow colored cards covered with an adhesive to which flying insects become stuck, or water pan traps, containers of water that drown insects, can also capture aphids, but are generally less effective at low populations than is sweeping. If aphids are found, fields need to be more closely inspected by randomly collecting plants when the crop reaches the full bloom or early green boll crop stages. To collect plants, the stem should be severed at the base, and the plant placed separately in a plastic bag. Aphids can readily be counted soon after sampling by lightly tapping the severed plants on a white surface to dislodge the insects.

Economic impact

The potato aphid is the most serious insect pest of flax grown in the plains region of western Canada and the northwestern United States. Adults and nymphs damage flax by extracting plant fluids from the stems, leaves and developing bolls. Although the potato aphid can vector plant diseases (Boyce and McKeen, 1967; Halbert *et al.*, 1980), it is not known to transmit any diseases of flax. Aphid feeding causes yield losses mainly by lessening the plant's ability to set healthy seed. The weight of individual seeds is only slightly reduced and oil content and quality is not affected by the aphids. Plants under drought stress may be prematurely desiccated by high aphid densities. Yield losses of 20 percent or more in oilseed flax may occur when aphid densities exceed 50 or more per plant (Wise *et al.*, 1995). Aphid densities generally do not get high enough to kill plants because of the impact of natural enemies on aphid populations.

Control methods

The most effective method to control potato aphids in flax is with a single insecti-cide application at full bloom or at the early green boll stage (Wise *et al.*, 1995).

Table 6.1 Upper and lower cumulative potato aphid counts at full bloom and early green pod crop stages of oilseed flax (from Wise and Lamb, 1995)

Number of plants	Sampling at full bloom		Sampling at early green pod	
	Lower	*Upper*	*Lower*	*Upper*
20			96	224
25	44	106	129	271
30	56	124	163	317
35	69	141	197	363
40	81	159	231	409
45	94	176		
50	107	193		

Treatments applied at these crop stages provide season-long control because there is insufficient time left for aphid populations to recover. To determine if an insecticide treatment is needed, a minimum of 25 plants at full bloom and 20 plants at early green boll should be collected randomly in the field to provide an accurate estimate of aphid densities (Wise and Lamb, 1995). If aphid populations exceed the cumulative aphid total listed in the sequential decision plan table for each crop stage (Table 6.1), control measures need to be applied within 48 hours. Aphid densities below the lower limit should not be sprayed, while those between the upper and lower limits should continue to be sampled until a decision on whether or not to spray can be made (Wise and Lamb, 1995). The upper and lower counts in the table take into account population variances at an economic threshold of three aphids/plant at full bloom or eight aphids/plant at early green boll crop stages (Wise and Lamb, 1995).

The potato aphid is highly susceptible to attack by fungi in the Order Entomophthorales (Shands *et al.*, 1962), especially in years of high rainfall and humidity in late June and July. In years when aphids are attacked by fungal diseases, populations are often decimated before they reach economic thresholds, but high populations can also be drastically reduced by the pathogen. Aphid populations sampled at full bloom that have many diseased insects should be sampled again at the early green boll stage to determine the effect of the disease on aphid densities.

A number of predators such as ladybird beetles, lacewings, hoverfly larvae and parasitic wasps attack the potato aphid, but they are largely ineffective in controlling aphid populations during years of high levels of adult emigration and rapid colonization on flax. Potato aphid populations are known to be suppressed by predators and parasitoids during years of slow colonization of some crops (Walker *et al.*, 1984), but this has yet to be proven in flax.

Cutworms

Pest distribution and host range

The larvae of many cutworm species (Lepidoptera: Noctuidae) attack flax crops in nearly all areas of the world where flax is grown. Localized outbreaks generally are composed of only one species, but several species often occur in a flax-growing area. The pale western cutworm, *Agrotis orthogonia* Morrison, and the red-backed cutworm, *Euxoa ochrogaster* (Guenée), are the most common in North America (Parker *et al.*, 1921; King, 1926; Philip, 1977), and several other species, primarily within the genera *Euxoa*, *Feltia*, *Laphygma*, and *Agrotis*, are found in Europe (Filipyev, 1929), South America (Wille and García Rada, 1942), India (Isaac, 1936; Narayanan, 1962; Malik *et al.*, 1998), and China (Chen, 1985; Chen *et al.*, 1990; Chen, 1992; Anonymous, 1998). All species are highly polyphagous and can feed on most cereal, oilseed, and legume crops as well as on many different weed species. The presence of flax flowers and those of most other dicotyledonous crops can exacerbate outbreaks of cutworms by providing food for ovipositing females (King and Atkinson, 1926).

Biology and ecology

The adults of all pestiferous cutworms are moths that have dark gray or brown forewings, often with mottled patterns of dark and light spots, and lightly colored hind wings. The moths vary in size but most have a wing span of 25–40 mm. Adults generally emerge in late summer to mate and lay eggs. Fields with loose soil and small hills or knolls are selected for oviposition or egg laying. Eggs are laid singly or in small clusters at or just below the soil surface or on the underside of host plant leaves. In some parts of India, female greasy cutworm (*Agrotis ypsilon* Rottemberg. (= black cutworm *A. ipsilon* Hufnagel) moths prefer to lay their eggs on wet or muddy soil just after flood waters have receded (Narayanan, 1962). Most species overwinter as eggs and hatch the following spring; however, species such as the early cutworm *E. tristicula* Morrison overwinter as immature larvae and commence feeding in early spring.

Young larvae begin feeding early in the growing season, usually on weeds before the flax seedlings emerge. Larvae stay in the soil during the day and feed during the evening near the soil surface until they become full grown in about three to five weeks. The larvae of most species are greenish or reddish brown with a brown head capsule, and reach lengths of up to 40 mm. When disturbed, the larvae curl up into a typical C configuration. At maturity, larvae form pupation cells or small chambers at 5 to 7 cm depths in the soil and pupate there.

Inspection and diagnostic methods

The first signs of cutworm outbreaks are the sudden presence of severed or partially severed plants in localized areas on the soil surface. In fields with small hills or knolls, plant damage typically first appears as bare spots or partially severed plants in the highest areas of the field. Plant damage usually progresses within seed rows and is concentrated near the perimeter of the outbreak areas. Areas damaged by cutworms should be inspected early in the morning to determine the severity of the damage before severed or damaged plants desiccate and become dispersed by the wind. Plants that are partially severed usually become more visible later in the day, particularly if conditions are hot and windy.

Larval densities in the soil can be estimated by sifting the top five cm of soil near the margin of the damaged area, using a hand trowel and a #5 to #8 mesh sieve. The larvae can reach 20 to 35 mm in length by the time that damage to flax becomes visible. Sampling can be done at any time during the day, but is easiest later in the day when the soil surface is dry. This sampling method should capture all larvae in the sampled area. A second method (Ayre, 1990), in which only loose soil around plants exhibiting cutworm damage is inspected, is about 75 percent effective in collecting larvae. This method is useful when equipment mentioned earlier is not at hand. While the first method enables a ready conversion of larval densities to a per m² total, the second method requires a measuring of row spacing by length of row inspected to determine larval density.

Economic impact

Cutworms damage flax plants by totally or partially severing the stems of the seedlings at the soil surface. Damaged plants are either completely destroyed or are severely weakened and made susceptible to further damage by wind or disease. Plant injury usually occurs too late in the season to reseed, and results in partial or complete yield loss in the affected areas. The black cutworm *A. ypsilon* Rott. (= *ipsilon* Hufn.) typically causes 10 to 30 percent damage to flax in northern China (Anonymous, 1998). A population of 12 redbacked cutworm (*Euxoa ochrigaster* (Guenée)) larvae per m² can cause a 10 percent reduction in flaxseed yields (Anonymous, 1996). Ayre (1990) in Manitoba, Canada, found a density of 32 redbacked cutworm larvae per m² destroyed all flax plants that had been seeded at a rate of 45 kg/ha in rows 20 cm apart. Since the cutworm species infesting flax have similar feeding habits, they are considered as one pest when determining their potential for economic damage to the crop.

Control methods

The most effective control for cutworm infestations is the application of an insecticide to the base of plants in the damaged areas. Treatments late in the afternoon or early evening will control larvae which come to the soil surface to feed. The insecticide should be applied as soon as damaged plants and larvae are detected, using high water volumes (100 l/ha or more) to improve coverage and penetration of the insecticide into the soil.

Cultural methods such as leaving a protective crust on fallowed fields near the end of the growing season will reduce the attractiveness of these fields as oviposition sites. Cultivation of fallow fields harboring cutworm pupae will bury them or expose them to predators. Also, fallowed fields kept weed free early in the season will prevent the fields from being a source of future infestations. Delaying seeding will effectively prevent feeding injury by some early cutworm species, but is not effective against the most common cutworm species.

Many species of birds can help to reduce cutworm feeding injury by predation on larvae and pupae. Early season cultivation can encourage predation by bringing these stages to the soil surface. In India, damage to flax by the black cutworm *A. ipsilon* was reduced by about 70 percent from predation by the pond heron (Malik, 1998).

Flea beetles

Pest distribution and host range

The large flax flea beetle *Aphthona euphorbiae* (Schrank) and the flax flea beetle *Longitarsus parvulus* (Paykull) are serious pests of flax cultivars grown for fiber and seed in the British Isles (Rhynehart, 1922; Ferguson *et al.*, 1997) and throughout mainland Europe (Grandori, 1946; Fritzsche and Lehmann, 1975; Lewartowski and Piekarczyk, 1978;

Cate, 1984; Sultana, 1984; Beaudoin, 1989; Voicu *et al.*, 1997), Russia (Yaroslavtzev, 1931), and Turkey (Lodos *et al.*, 1984). The two species are usually found together (Jourdheuil, 1960), except in the southern areas of Russia where the flax flea beetle is largely absent (Popov, 1941). In areas where the species are sympatric, their relative abundance can vary considerably from year to year because of marked differences in their ecological preferences. As many as 15 other species of flea beetles in five genera (Yaroslavtzev, 1931), including *Phyllotreta undulata* (Kutschera), *Psylliodes affinis* Paykull, *Ps. chrysocephala* L., and *Aphthona abdominalis* (Duftschmid) (Lakhmanov, 1970) have been found on flax, but their damage is either very localized or not significant.

Cultivated flax is a favored host of *L. parvulus* (Popov and Firsova, 1936) and *A. euphorbiae* (Kurdiumov, 1917), but clovers, cereal crops, beets, and numerous weeds are also fed upon. Beetles of both species feed on the pollen of many flowering plants, especially those in the Cruciferae, Compositae, and Umbelliferae families (Pluzhnichenko, 1963).

Biology and ecology

The adults of the large flax flea beetle *A. euphorbiae* are metallic green in color and 1.5–2.0 mm in length. The flax flea beetle *L. parvulus*, also known as the springing black beetle, is 1.0–1.2 mm long and has long first tarsal segments on its hind legs (Figure 6.2). The adults of both species begin to emerge from winter hibernacula when air temperatures average 9–12°C (Popov and Firsova, 1936) and temperatures in the top 2 cm of soil reach 11°C (Fritzsche, 1958). Emergence is hastened if the soil is dry in spring (Popov and Firsova, 1936). Adults commence feeding as spring temperatures near the soil surface reach 15–20°C (Fritzsche and Hoffmann, 1959; Jourdheuil, 1960). Their numbers and peak flight activity occur in May at the time seedlings of spring-sown cultivars begin to emerge from the soil and air temperatures exceed 20°C (Fritzsche, 1958). Adults feed on the cotyledons, vegetative buds and stems of newly emerged seedlings. Females lay eggs in soil surface cracks on or near the lateral roots of the plants in June to early July. The larvae of both species emerge within about three weeks. They feed on the lateral roots and tunnel into the tap roots of young plants, or feed on the root cortex and lateral roots of older plants (Popov and Firsova, 1936; Jourdheiul, 1963). New generation beetles emerge from August to early September, although summer adults of *L. parvulus* have been found as early as mid-July in Ireland (Lafferty *et al.*, 1922). Adults feed on the leaves, stems and seed capsules of flax until the plants mature and on other food sources before seeking dry sites in wooded or grassy areas, in crevices of walls, or other protected places in late September and October to overwinter.

The large flax flea beetle thrives within a wider temperature and humidity range than the flax flea beetle. Although it prefers warm, moist conditions, the large flax flea beetle can withstand hot, dry weather that could be fatal to pre-adult stages of the flax flea beetle (Popov, 1941). Conversely, the flax flea beetle prefers cool, moist conditions and is often the dominant species in cool, wet summers.

Detection, inspection and diagnostic methods

Insect sweep nets, yellow pan traps and sticky traps are effective means of detecting the migration of overwintered flea beetles into a field. The traps should be placed at

Figure 6.1 Potato aphid, *Macrosiphum euphorbiae* (Thomas), wingless adult and nymphs on a flax leaf. The aphid is the most serious pest of oilseed flax in western Canada. (Photo courtesy of Agriculture and Agri-Food Canada, Cereal Research Center, Winnipeg, Canada).

Figure 6.2 Flax flea beetle, *Aphthona euphorbiae* (Schrank) adult. The beetle is common on flax throughout most of Europe. (Photo courtesy of Dr. T. Rozhmina, The Flax Research Institute (VNIIL), Torzhok, Russia).

Figure 6.3 Linseed blossom midge or bud fly, *Dasyneura lini* Barnes, adult female. The midge is one of the most destructive insects on flax in India. (Photo from Richharia, 1962).

Figure 6.4 Flax bollworm, *Heliothis ononis* (Denis and Schiffermüller) larva on a flax boll. The bollworm, a climbing cutworm, is usually held in check by natural forces, but on occasion has reached damaging levels in western Canada. (Photo courtesy of R. Underwood, Agriculture and Agri-Food Canada, Saskatoon Research Center, Saskatoon, Canada).

the edges of fields immediately after seeding, and checked weekly until the end of May or once the crop is past the seedling stage. Sampling with a sweep net is most effective when the crop emerges from the soil; sweep sampling provides a quicker sample method and catches greater numbers of beetles than pan or sticky traps.

The initial symptom of flea beetle damage is a reduced or lack of seedling emergence, particularly along field borders. Feeding by the adults can kill the seedlings before the plants emerge from the soil. Upon close inspection, the adults can readily be seen on the seedlings, especially during hot sunny days. The leaves and cotyledons on damaged plants may be notched on the edges and have a "shot hole" appearance where the upper and lower surfaces of the plant tissue have been consumed. In addition, stems may be fed upon, causing heavily damaged plants to wilt or die on sunny days, or to become stunted and develop an increased number of basal tillers. Damage to older plants by new generation beetles appears as semicircular notches on leaf margins and sepals or as lesions to the stem. In flax grown for fiber, the lesions may indicate severe injury to the fiber cells and a lowering of fiber production.

Economic impact

Flea beetles are a serious insect pest of flax in many parts of Europe. In 1994 widespread feeding damage by flea beetles to linseed seedlings was reported for the first time in the United Kingdom (Haydock and Pooley, 1997); flea beetles have since become the most serious pest of flax in England (Ferguson *et al.*, 1997). The economic impact of flea beetles is influenced by the differences in yield components of linseed cultivars grown for seed versus flax cultivators grown for fiber. Feeding by the overwintered adults on seedlings can result in complete crop losses (Carpenter, 1920), particularly when mild winters favor a high overwintering survival of beetles and conditions are unsuitable for rapid seedling growth. Plants with severe damage to the cotyledons (> 75 percent of surface area) and to the first true leaves are shorter and produce less and lower quality fiber and less seed than plants defoliated even more severely later in their growth (Sokolov and Bezrukova, 1939). Later attack can slow down vegetation growth (Sultana, 1984). Although Fritzsche (1958) reported plant losses of 30 percent due to larval feeding, larvae are not considered to cause significant damage to the crop. Adults that emerge in the summer are not known to cause significant damage to oilseed flax but their feeding on upper and middle stems of fiber flax, the most important parts in the production of fiber, can cause fiber cells to become corky and decrease the value of the crop (Durnovo, 1935).

Control methods

The use of seed dressing insecticides is the most common means of preventing flea beetle injury to the emerging seedlings (Gheorghe and Doucet, 1987; Gheorge *et al.*, 1990; Horak, 1991; Trotus *et al.*, 1994), particularly to seedlings that may be attacked before they are fully emerged. Foliar treatments will also protect young seedlings (Oakley *et al.*, 1996; Haydock and Pooley, 1997) if applied when flea beetles invade fields after seedlings have emerged. A sampling method to estimate flea beetle densities and an economic threshold to determine the need to apply a foliar insecticide have not been developed.

To date, no genetic resistance to flax flea beetles has been found. Trichomes or hairs on the leaf surfaces of some wild flax species may offer potential deterrence to feeding by flea beetles (Rozhmina, 1999, personal communication).

Generally, cultural control methods have been less effective than the use of insecticides in areas of high flea beetle populations. However, in areas of lower infestations, cultural practices such as good seedbed preparation, the use of faster establishing cultivars, and very early (late April) seeding (Rhynehart, 1922) can allow the seedlings to outgrow flea beetle damage. The effectiveness of these measures depends on weather, chiefly rainfall, that encourages rapid growth of the seedlings (Durnovo, 1927), and on the relative abundance of the two main beetle species. Early sown crops that develop under cool conditions are less susceptible to damage in areas where the large flax flea beetle dominates, while later sowings are more effective where the flax flea beetle is more common (Popov, 1941). Historical reports showed that fields with increased seeding rates or narrower row spacings are often less attacked (Kurdiumov, 1917).

Changes in seeding dates may enable both seedlings and ripening flax to escape feeding damage. In regions with relatively mild winters, spring seedlings from autumn-sown plants are usually large enough to withstand flea beetle attack. The effectiveness of fall planting also depends on the timing of seedling emergence relative to the autumn flight of the new generation adults. Early spring sowings in colder regions can help to reduce the damage by new generation adults late in the season to flax grown for fiber.

Early in the growing season adult flea beetles are susceptible to parasitism by braconid wasps, attack from entomopathic nematodes, and infection by fungi, *Entomophthora* spp. (Jourdheiul and Chansigaud, 1961). In warm, humid weather the fungi spread rapidly throughout the flea beetle population and may affect up to 90 percent of the beetles (Fritzsche and Hoffmann, 1959). The effectiveness of these organisms in controlling flea beetle populations is generally low because considerable egg-laying occurs before adult mortality.

Flax thrips

Pest distribution and host range

The adult and larvae of *Thrips linarius* Ladureau [syn. *Thrips lini* Ladureau] and *T. angusticeps* (Uzel) can seriously damage flax grown in western Europe (Bonnemaison and Bournier, 1964; Czencz, 1985; Brudea and Gheorghe, 1989) and Russia (Uvarov and Glazunov, 1916). *T. linarius* feeds primarily on flax (Czencz, 1985) and is commonly known as flax thrips. *T. angusticeps*, known as flax thrips, cabbage thrips, or field thrips, prefers flax but is common on many crops, particularly peas (Doeksen, 1938), beets, onions, radish, wheat and barley (Franssen, 1955; Franssen and Mantel, 1961). Over 20 other species of thrips have been found on flax (Franssen and Mantel, 1961; Walters and Lane, 1991; Abrol and Kotwal, 1996), including *Aeolothrips fasciatus* L., which preys on the larvae of *T. linarius* (Ermolaev, 1940). However, no more than three or four species breed on flax (Morison, 1943), and most species, except for the two *Thrips* spp. listed above and the ubiquitous onion thrips *T. tabaci* Lindeman (Bonnemaison and Bournier, 1958), are relatively rare and are not known to cause noticeable crop damage.

Biology and ecology

Thrips are very small insects, 0.5 to 2.0 mm long, with narrow bodies and prominent legs and antennae. All adults of *T. linarius* and the summer generation of *T. angusticeps* have four long narrow wings that are fringed with long hairs, and are strong fliers, whereas the adults of the overwintering generation of *T. angusticeps* have short wings and are flightless. The eggs laid by fertilized females produce mostly female offspring. Males develop parthenogenetically from unfertilized eggs (Zawirska, 1963). Except for their smaller size and lack of wings, larvae are similar in appearance to adults. Both larval and adult stages of thrips actively move about on the host when disturbed and are most active on calm, clear days. Distribution of thrips tends to be general across a field, with little congregation at the edge (Walters and Lane, 1991).

In most areas of Europe, *T. linarius* is univoltine, having only one complete generation per year, while *T. angusticeps* produces two generations: the short-winged colorless females of the overwintering generation lay eggs which develop into the long-winged summer adults (Franssen and Huisman, 1958). However, thrips voltinism appears to be plastic: in Finland only one generation of *T. angusticeps*, which is entirely short winged, is produced (Hukkinen and Syrjanen, 1940), while *T. linarius* is thought to have a partial second generation in Romania (Brudea, 1990) and at least two generations in Poland (Zawirska, 1960).

Adults of both thrips species overwinter in the soil at depths of 20 cm or more and emerge in the spring when soils begin to warm (Ermolaev, 1940; Franssen and Huisman, 1958) and daytime temperatures average 8 to 10°C (Gheorghe, 1987), or remain to hibernate another year in the soil (Franssen and Mantel, 1961). Males of *T. linarius* are few in number and usually remain on weeds at the overwintering site, while females move to flax after mating. The short-winged adults of *T. angusticeps* often emerge well before flax is seeded, and seek out other plants on which to feed and lay eggs. The larvae remain on these plants until they become long-winged adults. Then the adults of both sexes either remain or move to flax where they feed on young leaves or growing points at the top of the plant. Females of both species lay eggs in June and July near the growing points on young plants or on the inside of flower buds on more mature plants. Larvae feed for about four weeks and then form a small cell in the soil at depths of 10–25 cm to pass their prepupal and pupal stages (Ermolaev, 1940). The young adults remain in the soil and move to greater depths with the onset of cold weather.

Detection, inspection and diagnostic methods

The movement of flying thrips into flax can readily be detected with water or sticky traps. Water traps should be about 6 cm deep and have a surface area of 250–500 cm^2 (Lewis, 1973). The traps should be filled to within 2 cm of the rim and a drop of detergent added to break the surface tension of the water. The traps are best used at ground level and should be placed around the perimeter of the field. Sticky traps with flat surfaces can be positioned vertically or horizontally at various heights in the field. Cylindrical traps have been found to be even more efficient than flat traps because they reduce air turbulence around the trap (Gregory, 1951). All traps for thrips should be white as opposed to yellow for flea beetles.

Detecting the non-flying overwintering generation of *T. angusticeps* is more difficult. Young seedlings should be inspected for signs of feeding damage, often manifested as silver spots on the stalks or wilting growing tips. Older plants infested by larvae or adults initially develop a yellowish gray color with dark spotted leaves and have terminals that stand up instead of hanging to one side. Thrips can be sampled in older plants by beating the plants against a white surface or by sweeping a damp board through the top of the crop (Ferguson and Fitt, 1991).

Economic impact

Thrips are the most serious insect pest of flax in many flax-growing areas of Europe. Flax seedlings or young plants sown in or near fields infested by thrips the previous year can be severely damaged or killed by the first generation of *T. angusticeps*. Seedlings whose growth is retarded by cool, dry weather are particularly susceptible (Franssen and Huisman, 1958; Czencz, 1985; Beaudoin, 1989). Feeding by larvae and adults of *T. linarius* and second generation *T. angusticeps* on the growing points and young leaves of older plants either kills the growing points or causes abnormal cell division which distorts growth. This can result in profuse branching on the main shoot or the production of flowering tillers which could delay crop maturation and make harvesting difficult. Feeding by thrips can cause flax foliage to assume a red, spotted appearance. Heavily infested plants are often short, which can reduce the length of their fiber but usually does not affect its quality.

Thrips' transfer of pollen from flower to flower on flax plants is the main and often sole cause of self-sterility in fiber flax, leading to reduced seed yields (Rataj, 1974). Enzymes excreted during feeding and the extraction of sap from buds and flowers can cause flowers and leaves at the top of the plant to wither. The damaged flowers often drop prematurely or produce drastically reduced numbers of seeds. Capsules on infested plants frequently burst open before ripening, causing a reduction in their weight and in number of seeds.

Control methods

Thrips are very susceptible to drowning by heavy rains when they are burrowing into the soil to overwinter, when they emerge in the spring, or if they are dislodged from plants by rains in the summer. Cool and rainy summers suppress thrips reproduction and can result in over ten-fold differences in overwintering populations from year to year (Franssen and Mantel, 1961; Bonnemaison and Bournier, 1964).

Cultural practices such as sowing flax early in the growing season can reduce thrips injury, particularly where *T. linarius* is common. Conversely, cultivating fields early in the spring to control weeds and delaying seeding until after adults have begun to emerge from the soil can increase mortality of young adults through starvation and, subsequently, reduce the buildup of the summer adult populations of *T. angusticeps*. Autumn and winter cultivation that redistributes the thrips to both shallower and deeper soil depths can increase the overwintering mortality of thrips in the soil.

Damage to flax by the overwintering generation of *T. angusticeps* also can be prevented by sowing flax in fields not sown with a susceptible crop the previous year. Flax should not be grown after peas, mustard, or cole crops, but by following a rotation that includes cereal crops, red clover or potatoes (Franssen and Huisman, 1958).

Foliar application of insecticides, especially those with systemic activity, can effectively prevent feeding injury by thrips. Application timing should coincide with the most susceptible stage of the thrips and before significant feeding damage has occurred. In most flax fields insecticides should be applied in early June or at the onset of flowering when larvae begin to hatch or are early instars (Brudea and Gheorghe, 1989). However, timing may be complicated by rapid population increases caused by airborne immigration of adults which may necessitate earlier applications in May. For example, application of an insecticide when the flax is about 2.5 cm high followed by a second treatment about two weeks later may be needed in fields being attacked by overwintering populations of *T. angusticeps* and their offspring (Franssen and Kerssen, 1962).

Linseed blossom midge or bud fly

Pest distribution and host range

The linseed blossom midge or linseed bud fly, *Dasyneura lini* Barnes (Diptera: Cecidomyiidae), is found in all flax-growing regions of the Indian subcontinent, and seriously limits flax production in central and northern India. It also attacks pigeon pea, *Cajanus cajan* Millsp. (Pruthi and Bhatia, 1937), and sesame, *Sesamum orientale* L. (Narayanan, 1962), but has not been found on any noncrop hosts. One other midge, *D. sampaina* Tav., produces galls on the terminal leaf buds of flax in Europe and Algeria, but it is of no consequence as a pest (Barnes, 1949).

Biology and ecology

The adult midges are small (1–1.5 mm long), narrow-bodied orange flies that have long legs and hairs on the back edge of the wings (Figure 6.3). The adults emerge early in the morning during late December or early January and can be seen hovering from plant to plant in calm weather (Pruthi, 1936). During the day females lay their eggs singly or in clusters of three to five within the calyx of young flower buds (Pruthi and Bhatia, 1937). A single female can lay up to 100 eggs (Narayanan, 1962). Eggs hatch in one to two days, depending on temperature, and larvae enter the buds and feed on the internal parts of the flower (Prasad, 1967). Typically two to four larvae develop in infested buds, although as many as ten may be found. The light orange-colored larvae feed for about seven days, causing greatest damage in the second and third instar stages. They drop from the flower as orange-red fourth instar larvae (about 2.3 mm long), and immediately begin to pupate 5–7 cm below the soil surface (Prasad, 1967). Three generations of the fly are often completed from January to mid-March (Pruthi and Bhatia, 1937). Infestation levels can remain high throughout this period and are maximized by mean temperatures of 16–20°C and relative humidity of 60–75 percent (Malik *et al.*, 1998). Rainfall in January and March has little effect on midge populations (Shrivastava *et al.*, 1994). After the flax crop matures, most larvae in the soil or under debris form a silken cover and enter a period of quiescence.

Economic impact

The linseed bud fly is the major limiting factor in the production of flax for seed in India (Jakhmola and Yadev, 1983). Yield losses by the midge can be expected in most areas each year, and in the central and northern plains of India losses in some years make the growing of flax uneconomical. During years of severe bud infestation, yield losses by the bud fly can exceed over 90 percent (Malik *et al.*, 1996c). Damage to the crop is caused by the larvae feeding on buds which prevents the flower from opening or producing seed.

Detection, inspection and diagnostic methods

In India, flax should be inspected for the presence of the midge once plants start to develop buds. The adult females are most active during the brightest time of the day. Their small size, tendency to hover near the tops of plants, and relatively slow flight speed distinguish them from other flies. The adults are attracted to light, making their collection with light traps in the evening simple. The larvae or maggots can be found readily by gently breaking open the immature capsules soon after initial bud development.

Control methods

Because midge larvae need to begin feeding on buds immediately after they hatch, early seeding, no later than the first week of November for most flax cultivars, or the growing of early flowering cultivars can reduce the severity of damage by the fly (Jakhmola *et al.*, 1973; Pal *et al.*, 1978). Early maturing plants not only are less susceptible to attack by first-generation larvae but are also more likely to escape damage by the later generations. However, the effectiveness of either method at reducing bud fly infestation should be weighed against their yield disadvantages. For example, crops sown in late October or early November often provide higher yields despite higher levels of midge damage than crops sown earlier (Singh *et al.*, 1991; Malik *et al.*, 1996b). While no flax cultivars have been found to be completely resistant to the midge (Jakhmola and Yadev, 1983; Kumar *et al.*, 1992), attributes of many varieties allow growers to further reduce larval infestation through varietal selection. Flax varieties with such phenotypic traits as short flowering periods, thin buds and thin sepals are less susceptible to the midge (Sood and Pathak, 1990; Malik *et al.*, 1991; Mishra *et al.*, 1996), as are varieties with higher polyphenol contents (Malik *et al.*, 1996d). Supplementing earlier seeding dates and varietal selection with higher amounts of phosphorus (Singh *et al.*, 1991), and irrigating the crop at both branching and capsule formation can also help to reduce bud fly infestation (Malik *et al.*, 1996c). Increasing the amount of nitrogen fertilizer will increase bud fly infestation, but usually not to the extent of preventing higher yields (Malik *et al.*, 1996c).

After bud formation, a foliar insecticide treatment should be applied as soon as adult flies are noticed. Treatments should not be delayed because this allows newly hatched larvae to enter the buds, making their control with most contact insecticides more difficult. Two or three insecticide applications at biweekly intervals can effectively control later generations of the midge (Jakhmola, 1974; Singh and Pandey, 1980). In most years insecticides should be used only in areas where bud fly infestation is likely to exceed 7 percent, in order for growers to realize an economic return (Malik *et al.*, 1996a). To maximize yields, insecticides should not be used as the only means of control but should be combined with a variety that has low susceptibility to the midge and that is most suited to the area (Sood and Pathak, 1986).

Eight species of parasitoids of larval bud fly have been recorded (Narayanan, 1962), and play a role in fly population reduction. Application of pesticide treatments late in the season should be avoided because of potential damaging effects on parasites such as the chalcid wasp *Systasis dasyneurae* Mani. The larva of this wasp parasitizes midge larvae in the buds, attaching to the midge and extracting body fluids from its host. Each wasp larva can destroy three to four midge larvae (Pruthi, 1937). Levels of parasitism of late instar midge larvae can exceed 50 percent (Ahmad and Mani, 1939).

Minor pests of oilseed flax

Foliar caterpillars (climbing cutworms)

A second group of noctuid moth larvae or caterpillars, known as climbing cutworms, occasionally attacks flax, often causing extensive damage in a short period. Climbing cutworms differ from the cutworms discussed previously in that the larvae feed almost exclusively above ground and attack all parts of the plant. The adults are mainly dark-colored, stout-bodied moths with mottled forewings and wing spans of up to 40 mm. The females are prolific egg producers, laying hundreds of eggs over a period of about two weeks. Damage to flax capsules is generally done by later instar caterpillars after they have fed on flowers, buds or leaves.

The most serious of the climbing cutworms in flax is the flax caterpillar *Rachiplusia nu* (Guenée). It is frequently found in many of the flax-growing areas of Brazil and Argentina (Griot, 1944; da Silva, 1987), and can produce as many as five generations per year. Larvae are mostly green with longitudinal dorsal markings, and can reach a length of 4 cm. They walk in a typical inchworm manner by arching their back upwards to bring the posterior part of their abdomen forward. Larval populations are highest in October, when seeds in the capsules are developing, and can cause extensive defoliation. Insecticides should be sprayed at this time if densities exceed ten larvae greater than 1.5 cm per 20 net sweeps (da Silva, 1987). The flax caterpillar also feeds on lucerne (*Medicago* spp.), which can serve as an alternative host until flax becomes suitable for oviposition in the spring.

The larvae of the semi-looper *Plusia orichalcea* F., which is widely distributed in India, feed on a variety of fruit and vegetable crops, of which linseed is one (Narayanan, 1962).

In North America the army cutworm, *Euxoa auxiliaris* (Grote), and the bertha armyworm, *Mamestra configurata* Walker (King, 1928), in Europe and Russia the flax worm or silver Y moth, *Autographa gamma* L. (Boldirev, 1923; Ruszkowski, 1928), and in northern China the cabbage armyworm *Barathra brassicae* L. (Li, 1980; Chen *et al.*, 1990; Anonymous, 1998) are noctuids that feed sporadically on flax after moth populations have increased in previous years. The army cutworm and bertha armyworm are univoltine, having one generation a year, while the flax worm and cabbage armyworm have at least two generations, of which the second is of greatest concern in flax. These species overwinter either as a partially developed larva or as a pupa. Larvae of the army cutworm and the flax worm become active in early spring and feed on various weeds. The late instar larvae of the army cutworm and of the flax worm are highly migratory and can move into flax after they have destroyed nearby food plants. This generally concentrates damage on young plants along the edges of fields. Army cutworm populations greater than ten larvae per m^2 should be treated with an insecticide (Anonymous, 1996). Larvae of the bertha armyworm hatch from eggs laid on the stem and leaves in July. Yield losses occur when the late instar larvae feed on the flowers and immature capsules.

The larvae of a pyralid moth, the beet webworm *Loxostege sticticalis* L., occasionally damage flax in localized areas of western Canada (Strickland and Criddle, 1920), Russia (Esterberg, 1932; Berezhkov, 1936), and northern China (Lei, 1981). Early instar larvae are dark green, becoming black near maturity, with two white lines along the length of their backs. Two and sometimes a partial third generation of the beet webworm develop each year, but only the second generation is of concern to flax producers. Larvae of the second generation feed on leaves, flowers, and stems of flax in July to August, often consuming many different dicotyledonous weeds before beginning to feed on flax. The larvae are highly migratory and may travel over one kilometer in search of new food sources after they have destroyed the local food supply (Esterberg, 1932). The webworm overwinters in the soil as a late instar larva or pupa. Outbreaks and mass movement of webworm larvae generally occur in years when weed growth is reduced by hot, dry summer weather.

Flax is not a preferred host for any of these caterpillar species, and it usually is attacked only after other plant hosts in the flax or nearby fields have been consumed. Common broadleaf weeds such as the goosefoot *Chenopodium album* L. either act as an attractant for ovipositing females such as the bertha armyworm (King, 1928), are a favored food source for early instars, such as the flax worm and the beet webworm (Esterberg, 1932), or enable the larvae to reach a stage capable of attacking flax (Merzheeskaya, 1963). Controlling weeds in flax largely eliminates problems with these insect pests, except during years when larval populations may move into flax from nearby infested fields. In Canada flax has recently become more susceptible to damage because of the rapid expansion of canola or rapeseed – a preferred food source of the bertha armyworm (Bailey, 1976). Cultivation of weeds soon after the soil has warmed in spring is an important means of cutworm control; such cultivation acts to reduce army cutworm larval populations by removing their food supply and starving them before they are large enough to migrate into nearby flax fields.

Outbreaks of these polyphagous pests on flax are usually confined to very damaging but small infestations (Kanervo, 1947; Merzheeskaya, 1963; Mason *et al.*, 1998). Populations of all climbing cutworm species are held in check by parasites and diseases, and both the bertha armyworm and the flax worm are highly susceptible to nuclear polyhedrosis viruses during outbreaks (Vago and Cayrol, 1955; Erlandson, 1990).

Bollworms

Bollworms differ from other climbing cutworms in that young larvae feed on buds and flowers for a short time before they enter a capsule and feed until its contents are exhausted. Larvae then exit the capsule and either feed on foliage or enter another capsule.

Most bollworm species that attack flax are generalist feeders such as *Helicoverpa armigera* Hübner (May, 1949; Chiarelli de Gahan and Touron, 1954; Goyal and Rathore, 1998), *Heliothis virescens* (F.), *Heliothis gelotopoeon* (Dyar) (Velasco de Stacul *et al.*, 1969) and *Helicoverpa punctigera* Wallengren (Kirkpatrick, 1961). Flax is not the preferred host for any of these species, and the flowers and capsules are damaged by their larvae only during years when flowering flax plants attract a large number of ovipositing adults during their migration from other hosts. The moths can be identified in the field by their peculiar habit of flying low and rapidly for a short distance and then alighting after being disturbed. The females produce large numbers of eggs which are laid singly on bud terminals or flowers. The larvae hatch in three to four days and feed for about three weeks (Rossiter, 1969). Larval densities of 3/m row of crop can reduce yields by 30 percent (Passlow *et al.*, 1960). To reduce feeding injury greatly, seeding should be timed so as to avoid having plants in flower during peak moth flights. Fields should be inspected at flowering for the presence of adult moths. An insecticide treatment applied ten days after peak flowering will minimize feeding damage and prevent reinfestation by late emerging larvae (Passlow *et al.*, 1960). Predation and parasitism should be encouraged. In India, preying on the larvae of *H. armigera* by the black drongo bird reduced flax capsule damage by more than 50 percent (Malik, 1998). Damage may have been further reduced by the placement of bird perches near or in flax fields.

Two bollworm species, *Heliothis dipsacea* L. in Russia and the Ukraine (Shchegolev, 1928; Paramonov, 1953), and the flax bollworm *H. ononis* in western Canada (Twinn, 1944; Twinn, 1945; Putnam, 1975) (Figure 6.4), also have other host plants but prefer flax over other crops. The moths move to flax and lay their eggs in open flowers. Young larvae eat the developing seeds within the flax boll, while older larvae leave the boll in which they emerged and feed on the seeds in surrounding bolls. Both species are very sporadic pests in flax as populations are usually kept low by parasites and diseases, but populations are capable of quickly increasing within a year and causing serious yield losses.

Plant bugs

The larvae and adults of a number of polyphagous plant bugs (Heteroptera: Miridae), principally *Calocoris norvegicus* (Gmelin), *Lygocoris pabulinus* L., and the European tarnished plant bug, *Lygus rugulipennis* Poppius, in Europe (Cate, 1984; Ferguson and Fitt, 1991) and the tarnished plant bug, *Lygus lineolaris* L., in North America (Wise and Lamb, 2000), extract fluids from the stems, leaves and vegetative and flower buds of flax from June to August. All species may migrate to flax after completing at least one generation on an alternate host. The *Lygus* spp. overwinter as adults while *C. norvegicus* and *L. pabulinus* overwinter as eggs on woody plants.

Plant bug adults move into flax from headlands as flower buds develop. Females lay eggs in the stems and leaves of the plants soon thereafter, and larvae emerge in July and feed through to mid-August. New generation adults begin to appear in late July to mid-August, and remain on flax for a short period before leaving to seek other food sources. Usually only one generation of mirids is found on flax each year (Wise and Lamb, 2000). However, two generations have been observed in areas where *L. rugulipennis* is common (Ferguson *et al.*, 1997).

Flax plant injury by mirids is most severe before the development of the capsules when nymphal populations typically are at their peak. Feeding can cause large swellings where the epidermis has been punctured, which render the plant liable to *Botrytis* infection (Pethybridge *et al.*, 1921), and cause buds to become necrotic and abscise and flowers to abort (Walters and Lane, 1991). Severely damaged plants may develop multiple branching, malformed growing points and have arrested development (McKay and Loughnane, 1946; Cate, 1984) – symptoms very similar to thrips damage. However, unlike thrips, which tend to be distributed evenly throughout a field, plant bugs tend to be clumped or aggregated in their distribution (Ferguson *et al.*,

1997), and are often in greater numbers close to field edges and hedgerows. Thus, flax grown near woody areas or fields with known alternate hosts such as alfalfa should be sampled with a sweep net during the bud and flowering crop stages. Yield losses by mirids are confined mostly to the boundaries of fields and, thus, control measures should be concentrated on field margins (Ferguson and Fitt, 1991). The value of using insecticides to control mirids awaits further study (Wise, 1992; Wise and Lamb, 2000). Under good growing conditions, oilseed flax appears to be tolerant of feeding damage by plant bugs, but whether tolerance extends to flax crops grown under less favorable conditions is uncertain (Wise and Lamb, 2000).

Say's stink bug *Chlorochroa sayi* Stål in North America (Munro and Butcher, 1940; Gardiner, 1946), *Piezodorus hybneri* F. in India (Joseph, 1953; Gill, 1987), and the southern green stink bug, *Nezara viridula* L. (Heteroptera: Pentatomidae) in many countries often become very abundant on several crops before attacking flax. Feeding by nymphs and adults during flowering and seed set can reduce seed quality and production. Stink bug populations can be readily measured at flowering with a sweep net, and generally become a concern only in locations where flax is sown near other susceptible crops. Attack by parasitoids such as the tachinid fly *Trichopoda giacomellii* (Bolach.) Guimeres on the pentatomid *N. viridula* can significantly reduce pest populations (la Porta, 1990).

Crane flies or tipulids

Periodic outbreaks of several species of crane flies (*Tipula oleracea* L., *T. paludosa* Meigen) (Diptera: Tipulidae), especially during years of high rainfall, can cause severe damage to flax in northern Europe and western Russia (Silantyev, 1930; Rawlinson and Dover, 1986). Adult females lay eggs from July to September with larvae emerging 10–15 days later (Silantyev, 1930). The larvae or "leatherjackets," so called because of their tough cuticle, are cylindrical maggots up to 4 cm long. They develop in the soil until July of the next year, and may attack and destroy flax seedlings by severing the stems and dragging the young plants below the ground. Larval densities of about 50 per m^2 can reduce seedling populations by over one-half (Rawlinson and Dover, 1986). Damage to flax is most severe in fields that had perennial grasses or cereals the previous year or when larvae move *en masse* from clover or meadows into flax (Yakushev, 1930). The sowing of resistant crops before flax, the construction of trap ditches (25 cm), and the preservation of avian habitat to increase predation by birds can help to reduce damage to flax. Treatment thresholds of one larva per 20 cm row length for cereals generally are applicable for flax. Spraying the soil surface with an insecticide at dusk soon after seedling emergence will control larvae in the field (Beaudoin, 1989).

Grasshoppers

A number of grasshopper species in the prairie regions of North America, Russia and Argentina attack flax late in the growing season. These species typically are highly migratory and feed on flax only after other food sources become scarce. The two-striped grasshopper *Melanoplus bivittatus* (Say) in the northern great plains of North America (Shotwell, 1941), *Trigonophymus arrogans* Stål (Schiuma, 1938) in Argentina, and *Chorthippus* spp. in eastern Russia (Kopaneva *et al.*, 1983) are the most common species that periodically damage flax by feeding on flowers and buds or by cutting off the capsules. Flax is best protected by controlling young grasshoppers in surrounding crops or rangeland before they begin to migrate. If grasshoppers do start to clip bolls from the stems late in the season, the crop should be harvested as quickly as possible (McLeod, 1997).

Leaf rollers

The leaf-rolling tortricids or leaf tiers *Cnephasia pasiuana* (Hübner), *C. pumicana* (Zeller), *C. asseclana* (Denis and Schiffermüller), *C. incertana* (Treitschke), *C. interjectana* (Haworth), and *Cochylis epilineata* Dup. in Europe (Banita and Peteanu, 1982; Cate, 1984; Beaudoin, 1989; Ferguson *et al.*, 1997), *C. longana* (Haworth) in North America (Edwards *et al.*, 1934; Rosenstiel *et al.*, 1944), and *Eulia loxonephes* Meyr. in Argentina (Chiarelli de Gahan, 1945) occasionally cause damage to the growing tips of shoots of young plants (Fritzsche, 1959) and to terminal buds, flowers, and capsules later in the season. Damage to the terminal growing points can cause lower shoots to develop, and reduce the length of the fiber. Most of these species have at least two generations per year, both of which may attack flax or a number of other host plants. The larvae of all species are highly visible in flax because of their habit of spinning terminal shoots together.

Control measures are rarely needed against these pests, except for the flax tortrix *C. pumicana* in the western Ukraine. Here feeding by this pest, mainly by second generation larvae, is known to reduce seed production by 10–15 percent, and occasionally up to 50 percent on medium or late sown crops (Fomenko, 1965b). Seeding in late winter or very early spring when the ground is thawing (Fomenko, 1965a), followed by harvesting the crop at yellow ripeness, and then quickly drying and threshing the crop will reduce feeding damage. In flax, larvae of this species emerge over a long period, which can make the timing of insecticide treatments both difficult and uneconomical because of the possible need to make more than one application. For other species, particularly *C. longana*, crop rotations that include a cereal crop instead of legumes before flax will greatly reduce feeding injury (Fritzsche, 1959). The braconid wasp *Dacnusa areolaris* (Nees) is reported to parasitize *C. interjectana* (Anonymous, 1982), and other parasitoids may act as biocontrol agents of these moths.

Stem borers

Larvae of the flax weevil *Ceuthorrrynchus sareptanus* Schultze in Siberia and the long-horned beetle *Phytoecia coerulea* (Scop.) in Hungary are known to attack flax (Semenov and Nikiforov, 1937; Dolinka, 1958). The distribution of the weevil in Siberia closely follows that of the wild perennial flax *Linum perenne*, its primary host (Lukyanovich, 1937). Both insect species damage flax by boring into the stem, destroying the fiber and causing plants to either break off or become deformed. Plants severely attacked by the flax weevil can be stunted to less than 20 percent of their normal height (Semenov and Pankova, 1938).

The flax weevil is the more serious pest, with infestations in fields sown repeatedly to flax exceeding 90 percent (Semenov and Nikiforov, 1937). The weevil overwinters in the soil as an adult, emerges in May to June, and feeds on the leaves of early developing flax or on perennial grasses such as quack or couch grass (*Agropyron repens* (L.) Beauv.). Eggs are laid in the stems of flax over a period of two months from the seedling to early flowering crop stages. The larvae mine the stems for about three weeks, usually upward, during June and July before dropping to the soil to pupate (Semenov and Nikiforov, 1937; Semenov and Pankova, 1938). Newly developed adults appear in August, feed on any available flax plants, and then enter the soil to overwinter.

Autumn tillage of flax fields, crop rotation to avoid planting flax in the same field for more than one year, and early sowing can reduce damage by the weevil (Semenov and Nikiforov, 1937). Insecticide treatments are most effective if applied to control the young adult weevils on flax plants in August. This will not reduce damage to flax in the year of application, but should do so in subsequent years.

Leaf miners

Larvae of the leaf-mining flies *Phytomyza horticola* (Gourea) and *P. atricornis* (Meigen) (Diptera: Agromyzidae) are polyphagous pests of linseed in India, mining up to 25 percent of leaves and adversely affecting linseed production (Kumar *et al.*, 1992). *P. atricornis* is also recorded as a pest of flax in Russia (Filipyev, 1929). The larvae of leaf miners feed between the upper and lower surfaces of host leaves, creating serpentine mines or tunnels; under heavy feeding, the tunnels may coalesce to form blotches. In Egypt, numbers of *P. atricornis* on flax reach a maximum in February and March; rates of leaf miner infestations are lower on flax cultivars planted in early compared to late November (El-Sheikh, 1990). Varietal resistance has been investigated as a means of reducing leaf miner damage to flax (El-Sheikh, 1990; Kumar *et al.*, 1992), and flax strains with moderate resistance to the pest have been found (Kumar *et al.*, 1992).

References

Abrol, D.P., and Kotwal, D.R. (1996). Insect pollinators of linseed (*Linum usitatissimum* Linn.) and their effect on yield components. *J. Anim. Morphol. Physiol.* 43, 157–161.

Ahmad, T., and Mani, M.S. (1939). Two new chalcidoid parasites of the linseed midge, *Dasyneura lini* Barnes. I. Biology and morphology of *Systasis dasyneurae* Mani. II. Description of the parasites. *Indian J. Agric. Sci.* 9, 531–539. Abs. in *Rev. Appl. Entomol. 1940 Series A*, 28, 204–205.

Anonymous (1982). *Annual Report on Research and Technical Work of the Department of Agriculture for Northern Ireland 1982.* Dept. Agriculture, Northern Ireland. 360 p.

Anonymous (1996). *Growing Flax: Production, Management and Diagnostic Guide.* Flax Council of Canada. 56 p.

Anonymous (1998). Control of flax pests and diseases. Chapter 5. In Wang, J.Z., and Zhou, M.C. (eds), *Flax Cultivation and Processing*, China Agricultural Press, Beijing, pp. 64–90.

Ayre, G.L. (1990). The response of flax to different population densities of the redbacked cutworm, *Euxoa ochrogaster* (Gn.) (Lepidoptera: Noctuidae). *Can. Entomol.* 122, 21–28.

Bailey, C.G. (1976). A quantitative study of consumption and utilization of various diets in the bertha armyworm *Mamestra configurata* (Lepidoptera: Noctuidae). *Can. Entomol.* 108, 1319–1326.

Banita, E., and Peteanu, S. (1982). Aspects of attack by thrips (*Thrips linarius* Uzel) and moth (*Cochylis epilinana* Dup.) in flax crops. (In Romanian). *Probleme de Protectia Plantelor* 10, 289–299. Abs. *CAB Abstracts 1984–1986.*

Barnes, H.F. (1949). *Gall Midges of Economic Importance.* Lockwood and Sons, Ltd, London, 142–145.

Beaudoin, X. (1989). Disease and pest control. In Marshall, G. (ed.), *Flax: Breeding and Utilisation*, ECSC, Brussels, pp. 81–88.

Berezhkov, R.P. (1936). *Loxostege sticticalis* L. in the forest zone of western Siberia. (In German). *Trav. Inst. Sci. Biol. Tomsk* 2, 98–131. Abs. *Rev. Appl. Entomol. 1938 Series A*, 26, p. 361.

Blackman, R.L., and Eastop, V.F. (1984). *Aphids on the World's Crops*. Wiley Interscience. Chichester, UK. 466 p.

Boldirev, V.F. (1923). *Instructions for the Control of Phytometra gamma L., and its Larva, the Flax Worm*. (In Russian). 29 pp. Abs. *Rev. Appl. Entomol. 1924 Series A*, **12**, p. 23.

Bonnemaison, L., and Bournier, A. (1958). Note preliminaire sur les thrips nuisibles au lin. *Comptes Rendus des Seances (Academie d'Agriculture de France)* **44**, 828–831.

Bonnemaison, L., and Bournier, A. (1964). The flax thrips: *Thrips augusticeps* Uzel and *Thrips linarius* Uzel. *Ann. Épiphyt.* **15**, 97–169. Abs. *Rev. Appl. Entomol. 1965 Series A*, **53**, 327–328.

Boyce, H.R., and McKeen, C.D. (1967). Some observations on vectors and transmission of tobacco etch virus. *Proc. Entomol. Soc. Ontario* **97**, 68–71.

Brudea, V. (1990). Research on the bioecology of the flax thrips (*Thrips linarius* Uzel) in northern Moldova. (In Romanian). *Analele Institutului de Cercetari pentru Cereale si Plante Technice Fundulea* **58**, 309–316.

Brudea, V., and Gheorghe, M. (1989). Results of experiments on control of the flax thrips (*Thrips linarius* Uzel) in Suceava district. (In Romanian). *Cercetari Agronomice in Moldova* **22**, 67–70. Abs. *CAB Abstracts CD-ROM 1993–1994*.

Carpenter, G.H. (1920). Injurious insects and other animals observed in Ireland during the years 1916, 1917, and 1918. *Econ. Proc. R. Dublin Soc.* **ii**, no **15**, 259–272. Abs. *Rev. Appl. Entomol. 1921 Series A*, **9**, 152–153.

Cate, P. (1984). The pests of flax. (In German). *Der Pflanzenarzt* **37**, 27–28. Abs. *Rev. Appl. Entomol. 1984 Series A*, **72**, p. 453.

Chen, H.F. (1985). List of insect pests of fiber flax in China. (In Chinese). *China's Fiber Crop* **1985**, 42–47.

Chen, H.F. (1992). List of insect pests of fiber flax in China. (In Chinese). *China's Fiber Crop* **1992**, 35–38.

Chen, H.F., Lu, P.W., and Zhang, J.W. (1990). Insect pests of bast fiber crops. (In Chinese). *Encyclopedia of Chinese Agriculture*, China Agricultural Press, Beijing, p. 247.

Chiarelli de Gahan, A. (1945). A microlepidopterous larva attacking flax, *Eulia loxonephes* Meyr. (Tortricidae). (In Spanish). *Publ. Inst. Sanid. Veg. (A), 1*, 11 pp. Abs. *Rev. Appl. Entomol. 1946 Series A*, **34**, p. 298.

Chiarelli de Gahan, A., and Touron, E.A. (1954). Biology and taxonomy of *Heliothis armigera* (Hbn.) (In Spanish). *Rev. Invest. Agric.* **8**, 111–148. Abs. *Rev. of Appl. Entomol. 1956 Series A*, **44**, p. 385.

Czencz, K. (1985). Thrips pests of cultivated flax. (In Hungarian). *Növényvédelem* **21**, 293–298. Abs. *Rev. Appl. Entomol. 1987 Series A*, **75**, 766–767.

da Silva, M.T.B. (1987). Bioecology, damage and control of *Rachiplusia nu* (Guenée, 1852) in flax. (In Spanish). *Revista do Centro de Ciências Rurais* **17**, 351–367.

Doeksen, J. (1938). "Bad heads" of flax caused by *Thrips lini*. (In Dutch). *Ladureau. Tijdschrift Pl. Ziekten* **44**, 1–44. Abs. *Rev. Appl. Entomol. 1939 Series A*, **27**, p. 233.

Dolinka, B. (1958). *Phytoceia coerulea* Scop. (Coleoptera, Cerambycidae), a new pest of flax in Hungary. (In Hungarian). *Növénytermelés* **7**, 79–84. Abs. *Rev. Appl. Entomol. 1959 Series A*, **47**, 358–359.

Durnovo, Z.P. (1927). Flax flea beetles in connection with the time of sowing of flax. (In Russian). *Trud. Opuitno-Issled. Uchastka Stantz. Zashch. Rast. Vred. Moskovsk. Zemel. Otd.* **pt 1**, 21–43. Abs. *Rev. Appl. Entomol. 1928 Series A*, **16**, p. 412.

Durnovo, Z.P. (1935). Character of damage caused to ripening flax by *Aphthona euphorbiae* Schr. (In Russian). *Plant Protection* **fasc. 2**, 104–106. Abs. *Rev. Appl. Entomol. 1936 Series A*, **24**, p. 5.

Edwards, W.D., Gray, K., and Mote, D.C. (1934). Observations on the life habits of *Cnephasia longana* Haworth, *Monthly Bulletin, California Department of Agriculture*, 23, pp. 328–333. Abs. *Rev. Appl. Entomol. 1935 Series A*, **23**, 134–135.

El-Sheikh, M.A.K. (1990). Effect of varieties and sowing dates of flax on the rate of infestation with the leaf miner, *Phytomyza atricornis* Mg. *Bulletin of the Faculty of Agriculture, University of Cairo* 41, 665–674.

Erlandson, M.A. (1990). Biological and biochemical comparison of *Mamestra configurata* and *Mamestra brassicae* nuclear polyhedrosis virus isolates pathogenic for the bertha armyworm, *Mamestra configurata* (Lepidoptera: Noctuidae). *J. Invert. Pathol.* 56, 47–56.

Ermolaev, M.F. (1940). The biology of *Thrips linarius* Uzel and control measures against it. (In Russian). *Bulletin of Plant Protection* 3, 23–34. Abs. *Rev. Appl. Entomol. 1942 Series A*, 30, p. 234.

Esterberg, L.K. (1932). Sugar beet web worm (*Loxostege sticticalis* L.) in the district of Nizhni-Novgorod in 1929–1930. *Plant Protection* 8, 275–292. Abs. *Rev. Appl. Entomol. 1933 Series A*, 20, p. 262.

Ferguson, A.W., and Fitt, B.D.L. (1991). Insect infestations in linseed: Plant injury and distribution within fields. *Aspects Applied Biol.* 28, 129–132.

Ferguson, A.W., Fitt, B.D.L., and Williams, I.H. (1997). Insect injury to linseed in south-east England. *Crop Protection* 16, 643–652.

Filipyev, I.N. (1929). Annual report of the Division of Applied Entomology. (In Russian). *Ann. Inst. Exp. Agron.* 7, 94–106. Abs. *Rev. Agric. Entomol. 1929 Series A*, 17, p. 476.

Fomenko, L.D. (1965a). The effects of sowing times on damage to flaxseed by the flax tortricid. (In Ukrainian). *Zashch. Roslyn., pt. 2,* 37–41. Abs. *Rev. Appl. Entomol. 1969 Series A*, 57, 544–545.

Fomenko, L.D. (1965b). The flax tortrix. (In Russian). *Zashch. Rast. Vredit. Bolez, pt. 8,* 19–20. Abs. *Rev. Appl. Entomol. 1967 Series A*, 55, p. 425.

Franssen, C.J.H. (1955). The bionomics and control of *Thrips augusticeps.* (In Dutch). *Tijdschrift Pl. Ziekten* 61, 97–102.

Franssen, C.J.H., and Huisman, P. (1958). *The Bionomics and Control of* Thrips angusticeps *Uzel on Flax in Holland.* (In Dutch). 103 pp. Abs. *Rev. Appl. Entomol. 1846 Series A*, 48, 356 p.

Franssen, C.J.H., and Kerssen, M.C. (1962). Aerial control of thrips on flax in the Netherlands. (In Dutch). *Agricultural Aviation* 4, 50–54. Abs. *Rev. Appl. Entomol. 1964 Series A*, 52, p. 6.

Franssen, C.J.H., and Mantel, W.P. (1961). The damage cause to flax by thrips and its prevention. (In Dutch). *Tijdschrift Pl. Ziekten* 67, 39–51.

Fritzsche, R. (1958). Contributions to the biology, ecology and control of the flax flea beetles. (In German). *Nachrichtenblatt für den Deutschen Pflanzenschutzdienst* 12, 121–133.

Fritzsche, R. (1959). *Cnephasia wahlbomiana* L. as a pest of flax and hemp. (In German). *Wiss. Z. Univ. Halle* 8, 1117–1119. Abs. *Rev. Appl. Entomol. 1961 Series A*, 49, p. 331.

Fritzsche, R., and Hoffmann, G.H. (1959). Infestation of adults of *Aphthona euphorbiae* Schrk. and *Longitarsus parvulus* Payk. by *Entomophthora* sp. (In German). *Beitr. Ent.* 9, 517–523.

Fritzsche, R., and Lehmann, H. (1975). Effect of micro-climate on the feeding activity of flax flea beetles. (In German). *Arch. Phytopath. Pflanzen.* 11, 153–159.

Gardiner, J.G. (1946). Entomology Section, *Report of the Minister of Agriculture for the Dominion of Canada for the Year Ended March 31, 1946*, pp. 63–83. Abs. *Rev. Appl. Entomol. 1949 Series A*, 37, 268–269.

Gheorghe, M. (1987). Aspects concerning the ecology and control of the flax thrips *Thrips linarius* Uzel. (In Romanian). *Analele Institutului de Cercetari pentru Cereale si Plante Technice Fundulea* 54, 355–361.

Gheorge, M., Brudea, V., Bigiu, L., and Popescu, F. (1990). Elements of integrated control of diseases and pests of flax. (In Romanian). *Analele Institutului de Cercetari pentru Protectia Plantelor, Academia de Stiinte Agricole si Silvice* 23, 203–207.

Gheorghe, M., and Doucet, I. (1987). Behaviour of some flax cultivars to seed treatments with carbamate insecticides. (In Romanian). *Probl. Protect. Plant.* 15, 91–3.

Gill, K.S. (1987). Insect pests of linseed. In: *Linseed* 1987, Indian Council of Agricultural Research, New Delhi, India, pp. 342–355.

Goyal, S.P., and Rathore, V.S. (1998). Patterns of insect plant relationship determining susceptibility of different hosts to *Heliothis armigera* Hubner. *Indian J. Entomol.* 50, 193–201.

Grandori, R. (1946). An experiment on the control of flax flea beetles. (In Italian). *Boll. Zool. Agric. Bachic* 13, 3–7. Abs. *Rev. Appl. Entomol. 1948 Series A*, 36, p. 268.

Gregory, P.H. (1951). Deposition of air-borne *Lycopodium* spores on cylinders. *Ann. Appl. Biol.* 38, 357–376.

Griot, M. (1944). A caterpillar that eats out the capsules of flax. (In Spanish). *Rev. Argent. Agron.*, 44–57. Abs. *Rev. Appl. Entomol. 1944 Series A*, 32, p. 340.

Halbert, S.E., Irwin, M.E., and Goodman, R.M. (1980). Alate aphid (Homoptera: Aphididae) species and their relative importance as field vectors of soybean mosaic virus. *Ann. Appl. Biol.* 97, 1–9.

Haydock, P.P.J., and Pooley, R.J. (1997). Evaluation of insecticides for control of flax beetle in linseed. *Tests of Agrochemicals and Cultivars 1997* **No. 18**, 4–5.

Hodgson, W.A., Pond, D.D., and Munro, J. (1974). *Diseases and Pests of Potatoes*. Canada Department of Agriculture Publication 1492. 70 p.

Horak, A. (1991). Strategies for control of flax flea beetles (*Aphthona euphorbiae*, *Longitarsus parvulus*) in linseed in Czechoslovakia. *Aspects Applied Biol.* 28, 133–136.

Hukkinen, Y., and Syrjanen, V. (1940). Contribution to knowledge of the Thysanoptera of Finland. (In German). *Annales Ent. Fenniae* 6, 115–128.

Isaac, P.V. (1936). Report of the Imperial Entomologist. *Sci. Rep. Inst. Agric. Res. Pusa 1933–1934*, 168–174. Abs. *Rev. Appl. Entomol. 1936 Series A*, 24, 667–668.

Jakhmola, S.S. (1974). Chemical control of linseed bud-fly, *Dasyneura lini* Barnes (Diptera: Cecidomyiidae). *Indian J. Agric. Sci.* 43, 1078–1080.

Jakhmola, S.S., Kaushik, U.K., and Kaushal, P.K. (1973). Note on the effect of sowing and nitrogen levels on the infestation of linseed bud-fly, *Dasyneura lini* Barnes (Diptera: Cecidomyiidae). *Indian J. Agric. Sci.* 43, 621–623.

Jakhmola, S.S., and Yadev, H.S. (1983). Susceptibility of linseed cultivars to budfly, *Dasyneura lini* Barnes. *Indian J. Entomol.* 45, 165–170.

Joseph, T. (1953). On the biology, bionomics, seasonal incidence and control of *Piezodorus rubrofasciatus* F., a pest of linseed and lucerne at Delhi. *Indian J. Entomol.* 15, 33–37.

Jourdheuil, P. (1960). Observations sur les altises du lin; remarques sur la biologie et les méthodes de lutte. *C.R. Acad. Agric. France* 46, 477–480, Abs. *Rev. Appl. Entomol. 1961 Series A*, 49, p. 658.

Jourdheuil, P. (1963). Famille des Chrysomelidae, sous-famille des Halticinae. In Balachowsky, A.S. (ed.), *Entomologie Appliquée à l'Agriculture*, pp. 762–854.

Jourdheuil, P., and Chansigaud, J. (1961). The parasites of adult flea-beetles infesting cultivated flax. (In French). *Bull. Soc. Ent. France* 66, 219–225.

Kanervo, V. (1947). On the mass occurrence of *Autographa gamma* in the summer of 1946 in Finland. (In German). *Annales Ent. Fenniae* 13, 89–104. Abs. *Rev. Appl. Entomol. 1950 Series A*, 38, p. 87.

King, K.M. (1926). *The Redbacked Cutworm and its Control in the Prairie Provinces*. Canada Department of Agriculture Pamphlet 69. 13 pp. Abs. *Rev. Appl. Entomol. 1926 Series A*, 14, 430–431.

King, K.M. (1928). *Barathra configurata*, Wlk., an armyworm with important potentialities on the northern prairies. *J. Econ. Entomol.* 21, 279–293.

King, K.M., and Atkinson, N.J. (1926). The relation of the redbacked cutworm to diversified agriculture in western Canada. *Sci. Agri.* 7, 86–91.

Kirkpatrick, T.H. (1961). Comparative morphological studies of *Heliothis* (Lepidoptera: Noctuidae) in Queensland. *Queensland J. Agric. Sci.* 18, 179–194. Abs. *Rev. Appl. Entomol. 1962 Series A*, 50, p. 88.

Kopaneva, L.M., Efimova, L.F., and Ivanova, I.V. (1983). The species composition and abundance of grasshoppers in various crops in the system of rotation in the non-chernozem zone and Siberia. (In Russian). *Noveishie Dostizheniya Sel'Skokhozyaistvennoi entomologii*, 95–98. Abs. *Rev. Appl. Entomol. 1983 Series A*, 71, p. 193.

Kumar, D., Singh, B., Singh, S.V., and Tuhan, N.C. (1992). Screening of linseed strains against major insect pests. *J. Insect Sci.* 5, 190–192.

Kurdiumov, N.V. (1917). *Blue Flax Flea Beetle* Aphthona euphorbiae *Schrank.* (In Russian). Proc. Poltava Agric. Expt. Sta., No. 30. 26. Abs. *Rev. Appl. Entomol. 1923 Series A*, 11, p. 154.

la Porta, N.C. (1990). Evaluation of field parasitism by *Trichopoda giacomellii* (Blach.) Guimeres, 1971 (Diptera: Tachinidae) on *Nezara viridula* (L.) 1758 (Hemiptera: Pentatomidae). (In Spanish). *Revista Chilena de Entomologia* 18, 83–87.

Lafferty, H.A., Rhynehart, J.G., and Pethybridge, G.H. (1922). Investigations on flax diseases (Third report). *J. Dept. Agric. Tech. Instr. Ireland* 22, 103–120. Abs. *Rev. Appl. Entomol. 1922 Series A*, 10, p. 589.

Lakhmanov, V.P. (1970). The injuriousness of the yellow spurge flea beetle. (In Russian). *Zashchita Rastenii* 15, 10. Abs. *Rev. Appl. Entomol. 1973 Series A*, 61, p. 758.

Lamb, R.J. (1989). Aphids in flax. *Canadian Agricultural Insect Pest Review* 67, 20.

Lamb, R.J., Wise, I.L., and MacKay, P.A. (1997). Photoperiodism and seasonal abundance of an aphid, *Macrosiphum euphorbiae* (Thomas), in oilseed flax. *Can. Entomol.* 129, 1049–1058.

Lange, W.H., and Bronson, L. (1981). Insect pests of tomatoes. *Ann. Rev. Entomol.* 26, 345–371.

Lei, Z.M. (1981). Preliminary observation on *Loxostege sticticalis*. (In Chinese). *Shanxi Nongye Kexue No. 10*, 14–15.

Lewartowski, R., and Piekarczyk, K. (1978). Characteristics of development, appearance, intensity and noxiousness of more important diseases and pests of industrial plants in Poland in 1976. (In Polish). *Biuletyn Instityutu Ochrony Roslin* 62, 151–221.

Lewis, T. (1973). *Thrips: Their Biology, Ecology and Economic Importance.* Academic Press Inc., London. 349 p.

Li, Z.D. (1980). *The theory and practice of mast crops.* Shanghai Science and Technology Publishing House, pp. 302–303.

Lodos, N., Onder, F., and Simsek, Z. (1984). Study of overwintering insect fauna and research on flight activity and migration behavior of some other species at the spring emergence during the migration period of the sunn pest (*Eurygaster integriceps* Put.) to the plain of Diyarbakir (Karacadag). (iii). Species of Coleoptera: Chrysomelidae. (In Turkish). *Bitki Koruma Bulteni* 24, 113–118. Abs. *CAB International; Flax and Insects.*

Lukyanovich, F.K. (1937). Geographical distribution of the flax weevil, *Ceuthorrynchus sareptanus* Schultze. (In Russian). *Plant Protection* 14, 25–39. Abs. *Rev. Applied Entomol. 1938 Series A*, 26, p. 351.

Malik, Y.P. (1998). Birds: An eco-friendly approach for insect-pests management in linseed. *Insect Environment* 3, 104.

Malik, Y.P., Singh, B., and Pandey, N.D. (1991). Role of blooming period in linseed bud fly (*Dasyneura lini* Barnes) resistance. *Indian J. Entomol.* 53, 276–279.

Malik, Y.P., Singh, B., Pandey, N.D., and Singh, S.V. (1996a). Assessment of economic threshold level of bud fly, *Dasyneura lini* Barnes, in linseed. *Indian J. Entomol.* 58, 185–189.

Malik, Y.P., Singh, B., Pandey, N.D., and Singh, S.V. (1996b). Management of linseed bud fly through non monetary inputs. *Indian J. Entomol.* 58, 136–139.

Malik, Y.P., Singh, B., Pandey, N.D., and Singh, S.V. (1996c). Role of fertilizer and irrigation in management of the linseed bud fly. *Indian J. Entomol.* 58, 132–135.

Malik, Y.P., Singh, S., Singh, B., Pandey, N.D., and Singh, S.V. (1996d). Determination of physico-chemical basis of resistance in linseed for bud fly. *Indian J. Entomol.* 57, 267–272.

Malik, Y.P., Singh, B., Pandey, N.D., and Singh, S.V. (1998). Infestation dynamics of linseed bud fly *Dasyneura lini* Barnes in relation to weather factors under irrigated conditions in central Uttar Pradesh. *Ann. Plant Prot. Sci.* 6, 80–83.

McKay, R., and Loughnane, J.B. (1946). A survey of flax diseases in Eire in 1945. *J. Dept. Agric., Eire* 43, 24–30. Abs. *Rev. Appl. Entomol. 1948 Series A*, 36, p. 208.

McLeod, M. (1997). *Insect Control for Oilseed Crops 1997: Sunflower, Soybeans, Flax, Canola, Sunflower.* South Dakota State University, Co-operative Extension Service, FS 888-OS. 5 p.

Mason, P.G., Arthur, A.P., Olfert, O.O., and Erlandson, M.A. (1998). The bertha armyworm (*Mamestra configurata*) (Lepidoptera: Noctuidae) in western Canada. *Can. Entomol.* 130, 321–336.

May, A.W.S. (1949). The control of *Heliothis* in linseed. *Queensland Agric. J.* 68, 216–220. Abs. *Rev. Appl. Entomol. 1950 Series A*, 38, 481–482.

Merzheeskaya, O.I. (1963). Noctuids (Lepidoptera) reducing the yield of agricultural crops in the Byelorussian S.S.R. (In Russian). *Zool. Zh.* 42, 359–367. Abs. *Rev. Appl. Entomol. 1965 Series A*, 53, 83–84.

Mishra, P.R., Sontakke, B.K., Mukherjee, S.K., and Dash, P.C. (1996). Field evaluation of some linseed cultivars for resistance to bud-fly, *Dasyneura lini* Barnes. *Indian J. Plant Protect.* 24, 119–121.

Morison, G.D. (1943). Notes on Thysanoptera found on flax. *Ann. Appl. Biol.* 30, 251–259.

Munro, J.A., and Butcher, F.G. (1940). Say's stinkbug. *Bimonthly Bulletin of N. Dakota Agric. Experim. Stn.* 3, 11–13. Abs. *Rev. Appl. Entomol. 1942 Series A*, 30, p. 97.

Narayanan, E.S. (1962). Insect pests of linseed and methods of their control. In Richharia, R.H. (ed.), *Linseed*, Monograph. Indian Central Oilseeds Committee, Hyderabad, India, pp. 107–113.

Oakley, J.N., Corbett, S.J., Parker, W.E., and Young, J.E.B. (1996). Assessment of risk and control of flax flea beetles. *Proc. Brighton Crop Protection Conference*, Pests and Diseases 1996. pp. 191–196.

Pal, S., Srivastava, J.L., and Pandey, N.D. (1978). Effect of different dates of sowing on the incidence of *Dasyneura lini* Barnes (Diptera: Cecidomyiidae). *Indian J. Entomol.* 40, 433–434.

Paramonov, S. (1953). The principal pests of oil-seed plants in the Ukraine. (In German). *Z. Angew. Ent.* 35, 63–81. Abs. *Rev. Appl. Entomol. 1955 Series A*, 43, 97–98.

Parker, J.R., Strand, A.L., and Seamans, H.L. (1921). Pale western cutworm (*Porosagrotis orthogonia* Morrison). *J. Agric. Res.* 22, 289–321. Abs. *Rev. Appl. Entomol. 1922 Series A*, 10, 111–112.

Passlow, T., Hooper, G.H.S., and Rossiter, P.D. (1960). Insecticidal control of *Heliothis* in linseed. *Queensland J. Agric. Sci.* 17, 117–120.

Pethybridge, G.H., Lafferty, H.A., and Rhynehart, F.G. (1921). Investigations on flax diseases. *J. Dept. Agric. Tech. Instr. Ireland* 21, 167–187. Abs. *Rev. Appl. Entomol. 1922 Series A*, 10, 446–447.

Philip, H.G. (1977). *Insect Pests of Alberta.* Alberta Agriculture Agdex No. 612–1. 77 p.

Pluzhnichenko, T.F. (1963). Types of feeding by flax flea beetles (Coleoptera, Chrysomelidae). (In Russian). *Entomologicheskoe Obozrenie* 42, 273–279.

Popov, K.I. (1941). The effect of ecological and agrotechnical conditions on the behaviour and injuriousness to flax of the flea beetles *Aphthona euphorbiae* Schrank and *Longitarsus parvulus* Payk. (In Russian). *Trud. Obshch. Estestvoisp. Kazan, Gosud. Univ.* 55, 157–203. Abs. *Rev. Appl. Entomol 1941 Series A*, 29, p. 181.

Popov, K.I., and Firsova, A.V. (1936). The influence of environmental conditions on the biology and injuriousness of *Aphthona euphorbiae* Schrank and *Longitarsus parvulus* Payk. (In Russian). *Plant Protection* 11, 94–102. Abs. *Rev. Appl. Entomol. 1937 Series A*, 25, p. 575.

Prasad, S.N. (1967). Biology of the linseed blossom midge *Dasineura lini* Barnes. *Cecidologia Indica* 2, 31–41. Abs. *Review Appl. Entomol. 1970 Series A*, 58, p. 864.

Pruthi, H.S. (1936). *Report of the Imperial Entomologist.* Scientific Report Agricultural Research Institute, Pusa, India 1934–35. 141–152. Abs. *Review. Appl. Entomol. 1937 Series A*, 25, 349–350.

Pruthi, H.S. (1937). *Report of the Imperial Entomologist.* Scientific Report Agricultural Research Institute, New Delhi, India 1936–37.

Pruthi, H.S., and Bhatia, H.L. (1937). A new cecidomyid pest of linseed in India. *Indian J. Agric. Sci.* 7, 797–808.

Putnam, L.G. (1975). Insect pests of *Brassica* seed crops and of flax. In Harapiak, J.T. (ed.), *Oilseed and Pulse Crops in Western Canada*, Western Co-operative Fertilizers, Ltd., Calgary, Canada, pp. 455–474.

Rataj, I.K. (1974). The influence of *Thrips linarius* Uzel on self-sterility. (In Russian). *Len i Konoplya* 12, 91–105.

Rawlinson, C.J., and Dover, P.A. (1986). Pests and diseases of some new and potential altern-ative arable crops for the United Kingdom. *Proc. 1986 Brighton Crop Protection Conference: Pests and Diseases. Brighton*, pp. 731–732.

Rhynehart, J.G. (1922). On the life-history and bionomics of the flax flea beetle (*Longitarsus parvulus* Payk.) with descriptions of the hitherto unknown larval and pupal stages. *Scientific Proc. Royal Dublin Society* 16, 497–541. Abs. *Rev. Appl. Entomol. 1922 Series A*, 10, p. 339.

Richharia, R.H., ed., (1962). *Linseed.* Indian Central Oilseeds Committee. Hyderabad, India.

Rosenstiel, R.G., Ferguson, G.R., and Mote, D.C. (1944). Some ecological relationships of *Cnephasia longana. J. Econ. Entomol.* 37, 814–817.

Rossiter, P.D. (1969). *Heliothis* in linseed. *Queensland Agric. J.* 95, 610–611.

Ruszkowski, J.W. (1928). *Phytometra gamma* L., a serious field and vegetable pest. (In Polish). *Poradnik Gospod.* 39, 730–732. Abs. *Rev. Appl. Entomol. 1929 Series A*, 17, p. 129.

Schiuma, R. (1938). Report on grasshoppers. (In Spanish). *Publ. Misc. Minist. Agric. Argentina, No.* 43, 119 pp. Abs. *Rev. Appl. Entomol. 1939 Series A*, 27, 213–214.

Semenov, A.E., and Nikiforov, A.M. (1937). A new pest of flax. (In Russian). *Len i Konoplya* 14, 9–11. Abs. *Rev. Appl. Entomol. 1938 Series A*, 26, 6–7.

Semenov, A.E., and Pankova, I.A. (1938). The effect of the larvae of the flax weevil on the structure of the flax fiber. (In Russian). *Len i Konoplya* 15, 45–48. Abs. *Rev. Appl. Entomol. 1938 Series A*, 26, p. 601.

Shands, W.A., Hall, I.M., and Simpson, G.W. (1962). Entomophthoraceous fungi attacking the potato aphid in northeastern Maine in 1960. *J. Econ. Entomol.* 55, 174–179.

Shchegolev, V. (1928). Pests of oil-producing plants in the northern Caucasus. (In Russian). *Maslob.-Zhirov. Delo* 9, 32–37. Abs. *Rev. Appl. Entomol. 1929 Series A*, 17, p. 263.

Shotwell, R.L. (1941). *Life Histories and Habits of Some Grasshoppers of Economic Importance on the Great Plains.* 47 pp. Abs. *Rev. Appl. Entomol. 1941 Series A*, 29, 603–604.

Shrivastava, N., Katiya, O.P., Shrivastava, S., and Das, S.B. (1994). Population dynamics of lin-seed budfly *Dasyneura lini* Barnes (Diptera: Cecidomyiidae) on linseed (*Linum usitatissimum*). *J. Oilseeds Res.* 11, 160–164.

Silantyev, I. (1930). Contributions to a monograph of *Tipula oleracea* L. (In Russian). *Plant Protection* 7, 29–45. Abs. *Rev. Appl. Entomol. 1932 Series A*, 20, p. 87.

Singh, B., Katiyar, R.R., Malik, Y.P., and Pandey, N.D. (1991). Influence of sowing dates and fertilizer levels on the infestation of linseed budfly (*Dasyneura lini* Barnes). *Indian J. Entomol.* 53, 291–297.

Singh, S.P., and Pandey, N.D. (1980). Chemical control of linseed gall midge, *Dasyneura lini* Barnes, attacking linseed. *Indian J. Entomol.* 42, 786–787.

Sokolov, A.M., and Bezrukova, V.F. (1939). The injuriousness of the flax flea. (In Russian). *Plant Protection* 18, 150–154. Abs. *Rev. Appl. Entomol. 1939 Series A*, 27, p. 684.

Sood, N.K., and Pathak, S.C. (1986). Stability of effectiveness of insecticides and varieties in the control of linseed budfly (*Dasyneura lini* Barnes). *Indian J. Entomol.* 46, 53–59.

Sood, N.K., and Pathak, S.C. (1990). Thickness of sepals – a factor for resistance in linseed against *Dasyneura lini* Barnes. *Indian J. Entomol.* 52, 28–30.

Strickland, E.H., and Criddle, N. (1920). *The Beet Webworm* (Loxostege sticticalis L.). Canada Dept. Agric., Crop Protection Leaflet 12. 2 p.

Sultana, C. (1984). Lin et autres oléagineux. *Cultivar. Juin 1984. Spécial Oléaoprotéagineux. No. 173*, 97–101.

Trotus, E., Mincea, C., and Alexandrescu, S. (1994). Studies on the control of the flax flea beetle (*Aphthona euphorbiae* Schrank) by seed treatment. *Cercetari Agronomice in Moldova 1994* 27, 187–189.

Twinn, C.R. (1944). A summary of the more important insect pests in Canada in 1943, *74th Report Entomol. Soc. Ontario 1943*, pp. 54–59. Abs. *Rev. Appl. Entomol. 1945 Series A*, 33, 143–144.

Twinn, C.R. (1945). A summary of insect conditions of importance or special interest in Canada in 1944, *75th Report Entomol. Soc. Ontario 1944*, pp. 45–49. Abs. *Rev. Appl. Entomol. 1947 Series A*, 35, p. 13.

Uvarov, B.P., and Glazunov, V.A. (1916). *A Review of Pests.* (In Russian). Report on the work of the Entomological Bureau of Stavropol for 1914. Dept. Agric. of the Ministry of Agric., Petrograd. 13–54. Abs. *Rev. Appl. Entomol. 1917 Series A*, 5, 458–459.

Vago, C., and Cayrol, R. (1955). A polyhedral virus of the gamma noctuid *Plusia gamma* L. (Lepidoptera). (In French). *Ann. Épiphyt.*, 6, 421–432. Abs. *Rev. Appl. Entomol. 1956 Series A*, 44, p. 320.

Velasco de Stacul, M., Barral, J.M., and Orfila, R.N. (1969). Taxonomy, specificity and distinguishing biological characters of the complex of species known as "cotton bollworm", "corn earworm", "tobacco budworm", and "flaxworm". (In Spanish). *Revta Invest. Agropec.* 6, 19–68. Abs. *Rev. Appl. Entomol. 1970 Series A*, 58, p. 539.

Voicu, M., Popov, C., Ioan, I., and Luca, E. (1997). Halticines species from oil flax (*Linum usitatissimum* L.) crops in the Moldavian plain. (In Romanian). *Cercetari Agronomice in Moldova* 30, 295–300. Abs. *CAB Abstracts CD-ROM 1998–1999.*

Walker, G.P., Madden, L.V., and Simonet, D.E. (1984). Spatial dispersion and sequential sampling of the potato aphid, *Macrosiphum euphorbiae* (Homoptera: Aphididae), on processing-tomatoes in Ohio. *Can. Entomol.* 116, 1069–1075.

Walters, K.F.A., and Lane, A. (1991). Incidence and severity of insects damaging linseed in England and Wales 1988–1989. *Aspects Applied Biol.* 28, 121–128.

Wille, J.E., and García Rada, G. (1942). *The Insect Pests and Diseases of Flax.* (In Spanish). 19 pp. Abs. *Rev. Appl. Entomol. 1943 Series A*, 31, 515–516.

Wise, I.L. (1992). The status of lygus bugs as a pest in flax. In Sims, R.W. (ed.), *Research Update '92*, Agriculture Canada. Winnipeg Res. Station, pp. 13–14.

Wise, I.L., and Lamb, R.J. (1995). Spatial distribution and sequential sampling methods for the potato aphid, *Macrosiphum euphorbiae* (Thomas) (Homoptera: Aphididae), in oilseed flax. *Can. Entomol.* 127, 967–976.

Wise, I.L., and Lamb, R.J. (2000). Seasonal occurrence of plant bugs (Hemiptera: Miridae) on oilseed flax (Linaceae) and their effects on yield. *Can. Entomol.* 132, 369–371.

Wise, I.L., Lamb, R.J., and Kenaschuk, E.O. (1995). Effects of the potato aphid *Macrosiphum euphorbiae* (Thomas) (Homoptera: Aphididae) on oilseed flax, and stage-specific thresholds for control. *Can. Entomol.* 127, 213–224.

Yakushev, M.R. (1930). Observations on *Tipula* in the Government of Smolensk. (In Russian). *Plant Protection* 7, 219–225. Abs. *Rev. Appl. Entomol. 1932 Series A*, 20, 228–229.

Yaroslavtzev, G.M. (1931). A brief report on the pests of field cultures in 1930 according to the data of the State Service of dynamics and distribution of the injurious insects. (In Russian). *Plant Protection* 8, 375–413. Abs. *Rev. Appl. Entomol. 1932 Series A*, 20, 263–264.

Zawirska, I. (1960). Development of a population of *Thrips lini* Lad. on flax during the vegetative period. (In Polish). *Biuletyn Instytutu Ochrony Roslin* 10, 51–67.

Zawirska, I. (1963). A contribution to the bionomics of *Thrips linarius* Uzel. (In Polish). *Biuletyn Instytutu Ochrony Roslin* 19, 1–10.

7 The contribution of α-linolenic acid in flaxseed to human health

Stephen C. Cunnane

Introduction

α-linolenate (ALA, 18:3ω3) is one of several ω3 polyunsaturated fatty acids which cannot be synthesized *de novo* in mammals. ALA was demonstrated to be dietarily essential in rats 70 years ago but its metabolism and function in humans are still poorly understood. As a result of its long, enigmatic and controversial history, the potential health attributes of ALA in humans are often overlooked. This review aims to provide an update on ALA metabolism, function and relevance to human health with reference to flaxseed as a source of dietary ALA. There are several previously published reviews on ALA intake, metabolism and health attributes, that provide additional background (Tinoco, 1982; Cunnane, 1991; Cunnane, 1995; Gerster, 1998).

As in many animal studies, human studies have shown that ALA is potentially important as a source of ω3 polyunsaturates, especially in vegetarians and those preferring not to eat fish. The efficacy of its conversion to eicosapentaenoate (EPA; 20:5ω3) and docosahexaenoate (DHA; 22:6ω3) is an ongoing controversy. Another controversy relates to whether ALA has health attributes without being converted to EPA or DHA. Research with traditional high-ALA flaxseed and flaxseed oil suggests that ALA may have an important but, as yet, ill-defined protective role against degenerative Western diseases, particularly cancer and coronary heart disease. This role may be independent of its conversion to EPA and DHA.

Extensive animal studies had previously left little doubt, but it was controversial until the 1980s as to whether human infants could develop adequately without dietary ALA. Indeed, a dietary source of ω3 polyunsaturates is clearly necessary for optimal post-natal development. ALA alone may be sufficient in healthy term infants but its conversion to DHA in infants seems inadequate to meet the needs of development in premature infants. This is yet another controversial area because the gross pathology of ALA deficiency is subtle and takes a long time to develop.

Animal models of nutrient metabolism do not necessarily relate well to effects in humans but they are necessary to investigate new or specific metabolites in individual organs. Work using ALA tagged with a tracer isotope has recently shown that ALA catabolism and carbon recycling into *de novo* lipogenesis is much more extensive than previously understood. In fact, β-oxidation and carbon recycling appear to play a quantitatively significant role in ALA metabolism, especially in brain lipid accumulation during early postnatal development. These recent observations have contributed to a growing body of evidence in mammalian neonates showing that ALA is not a dependable precursor to DHA, therefore indicating the conditional essentiality of pre-formed dietary DHA in neonates.

History

ALA was reported by Burr and Burr (1930) to be dietarily essential for the rat. As the technology of fatty acid analysis improved, the ability to measure individual long-chain polyunsaturates derived from ALA also improved and the sequential metabolic link between these fatty acids emerged in the 1960–70s. Since the time of Burr and Burr's studies (1930), it was understood that ALA and the ω3 polyunsaturates were biochemically and metabolically distinct from linoleate (18:2ω6, LA), arachidonate (20:4ω6, AA) and the other ω6 polyunsaturates. Nevertheless, claims that ALA was not essential in mammals persisted but were vigorously rebutted for lack of adequate precision in tissue measurements (Crawford and Sinclair, 1972). Curiously, and despite their independent effects and metabolism, studies of LA and ALA deficiency have remained to this day a combined entity – "essential fatty acid" deficiency – but not for any good scientific reason.

It was recognized in the 1970s that deficiency of ω3 polyunsaturates could be induced using dietary oils rich in LA, i.e. corn, sunflower or safflower oils. This led to the classic animal studies in rats and monkeys that firmly established the link between ALA intake and DHA accumulation in tissues, and the neurological and visual deficits during deficiency of ω3 polyunsaturates (Sinclair and Crawford, 1972; Lamptey and Walker, 1976; Bourre *et al.*, 1984; Neuringer *et al.*, 1984; Bourre *et al.*, 1989).

Flax or flaxseed oil, instead of "linseed" oil, is now the more common way of describing the edible version of this oil rich in ALA. In the 1960s there were occasional reports of beneficial effects of flaxseed oil. However, these early studies became controversial (Owren *et al.*, 1964) and some of the claimed effects on platelet aggregation were later withdrawn. More sustained interest in the nutritional attributes of ALA in humans arose after the cardiovascular benefits of fish and fish oils became evident in the 1970s and 1980s. Several small comparative studies of changes in blood fatty acid profiles after supplementation with flaxseed or fish oil were performed (Dyerberg *et al.*, 1980; Renaud and Nordøy, 1983). Controversy again emerged with conflicting interpretations over whether these studies indicated that humans "effectively" converted ALA to EPA and DHA or whether it was misleading to recommend ALA intake if it was so poorly converted to the more active longer chain ω3 polyunsaturates (Johnston, 1986). This controversy about whether a diet can be healthy with only ALA as a source of ω3 polyunsaturates continues today.

Controversy also dogged the claims for cases of human ALA deficiency (Holman *et al.*, 1982; Bjerve *et al.*, 1987; Bjerve *et al.*, 1988) and it is likely that these human cases were complicated by additional nutrient deficiencies, especially LA deficiency (Koletzko and Cunnane, 1988). Human research on the potential health benefits of flaxseed began in earnest in the late 1980s with the realization that flaxseed offered several additional components besides ALA that were potentially beneficial and possibly synergistic, including the lignans and soluble fiber mucilage (Cunnane and Thompson, 1995).

Metabolism of ALA

ALA is widely regarded as the "parent" ω3 polyunsaturated fatty acid in the mammalian diet, i.e. it is the only ω3 polyunsaturated that cannot be synthesized *de novo* by mammals and from which the longer chain ω3 polyunsaturates are derived. In actual fact, however, mammals including humans can synthesize ALA by chain elongation of its 16-carbon homologue, hexadecatrienoate (16:3ω3) which exists in several edible, leafy green plants in the human diet (Cunnane *et al.*, 1995b; Periera *et al.*, 2001). Indeed, if 14:3ω3 existed in the human diet, it, too, would be an efficient precursor to ALA (Sprecher, 1968). In reality, unless fish is consumed in large amounts, ALA is the dominant ω3 polyunsaturated fatty acid in the human diet and is present in the same edible plants as hexadecatrienoate. Nevertheless, dietary hexadecatrienoate could contribute as much as 3–4 percent of ALA in the body (Cunnane *et al.*, 1995b).

The pathway of desaturation and chain elongation of ALA to DHA is well known, including the relatively recent discovery of a peroxisomal step involving chain shortening of 24:6ω3 to obtain DHA (Figure 7.1). The competitive interaction between the ω6 and ω3 polyunsaturates is an important attribute of their metabolism, (a) at most steps in desaturation/chain elongation, (b) for acylation into membrane lipids, and (c) for cyclooxygenases leading to eicosanoid (prostaglandin) synthesis. In microsomal desaturase preparations, ALA displays greater affinity for the enzyme than does LA but there is usually more LA in the diet and in tissue lipids, thereby offsetting the substrate preference at the enzyme level.

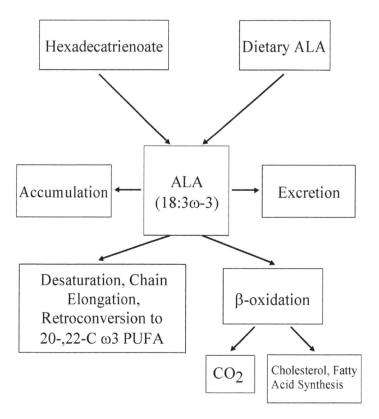

Figure 7.1 Pathways of α-linolenate (ALA) input and metabolism. Hexadecatrienoate (16:3ω3) is present in the human diet and can be chain elongated to ALA. Whole body balance studies show that ALA oxidation is the predominant pathway of its utilization. Tracer studies show that considerable β-oxidized ALA is recycled into cholesterol and fatty acids synthesized *de novo*.

These interactions probably play an important role in the physiological effects of ALA and the influence of other dietary polyunsaturates, especially LA, on ALA requirement but will not be discussed in detail here since they have been reviewed elsewhere (Cunnane, 1991; Cunnane, 1995; Gerster, 1998). This competitive interaction between LA and ALA means that differences in the dietary ratio of LA to ALA may also contribute to the lack of agreement over the sufficiency of ALA for infant nutrition, and the effects of ALA on platelet aggregation and serum lipids in adults. This competition probably contributes to the way in which the comparatively high levels of dietary LA in Westernized societies exacerbate the risk of deficiency of ω3 polyunsaturates (Lands, 2001).

Other important attributes of ALA metabolism are sometimes overlooked, thereby oversimplifying ALA metabolism within the pathway to DHA synthesis. These overlooked attributes include the (a) age- and species-dependent differences in the rate of desaturation, (b) largely unknown whole body capacity to synthesize DHA, (c) substrate inhibition by ALA itself, (d) stimulation of desaturation by precursor deficiency, (e) dependence of the desaturation/chain elongation pathway on co-factor nutrients, and (f) interconvertibility of ω3 polyunsaturates between 14 and 24 carbons long.

Age- and species-dependent differences

Desaturation-chain elongation varies with age and between species. Young, growing rats appear to have the most efficient desaturation-chain elongation compared to all other mammals studied (Ackman and Cunnane, 1992). As a result, extrapolation from the numerous studies done in young rats to other species including humans, or even to older rats, has given rise to a widespread but misleading assumption about the ease with which DHA can be synthesized in humans. As a result of this fairly narrow focus on desaturation/chain elongation, other roles of ALA and other pathways of its metabolism, especially β-oxidation and carbon recycling, have tended to be overlooked, despite many year's knowledge of these prominent pathways (see β-oxidation and carbon recycling, p. 160).

Capacity to synthesize DHA

Although human infants and adults can synthesize DHA from ALA, in fact, DHA changes very little in human plasma or cellular lipids, even after relatively high amounts of ALA supplementation (Bryan *et al.*, 2001) (Figures 7.2 and 7.3) (see detailed review in Cunnane, 1995). Recent tracer studies suggest that the minimal change in blood DHA in humans supplemented with ALA is due to conversion of ALA not exceeding 0.2 percent (Palowsky *et al.*, 2001). Percent conversion does not tell us what capacity we have to synthesize DHA, only that a certain capacity exists and that it could presumably go up or down under various conditions.

The capacity to synthesize DHA and limitations on that capacity are not well understood in any species, especially humans. Indeed, this information is not easy to obtain for any metabolic pathway or product. Several strategies have been taken to get surrogate information. For example, a preparation of liver microsomes can be shown to desaturate ALA at a certain rate. Alternatively, the fatty acid profile of plasma lipids can be shown to have higher EPA or DHA after ALA supplementation. However, the microsomal assay is done under standardized, isolated, *in vitro* conditions that are suitable for characterizing the enzyme's properties, and the change in fatty acid profile

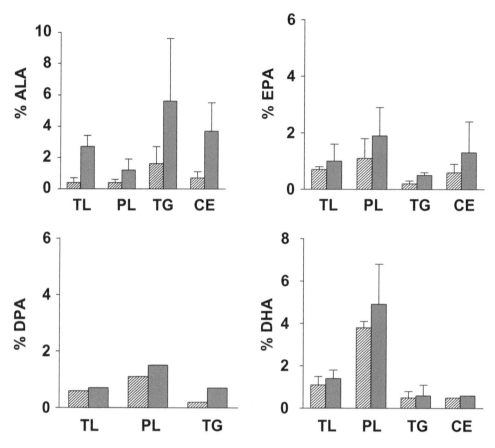

Figure 7.2 α-linolenate (ALA), eicosapentaenoate (EPA), ω3 docosapentaenoate (DPA) and docosahexaenoate (DHA) levels in human plasma lipids before (stippled bars) and after (gray bars) dietary supplementation with ALA (9–21 g/d) for 4–6 weeks. (Data are compiled from Beitz *et al.*, 1981; Kestin *et al.*, 1990; Cunnane *et al.*, 1993a; Kelley *et al.*, 1993; Mantzioris *et al.*, 1994; Cunnane *et al.*, 1995a; Layne *et al.*, 1996.) Plasma lipid classes: TL – total lipids, PL – phospholipids, TG – triglycerides, CE – cholesteryl esters.

is a composite of many pathways and rates of enzyme activity. Neither of these parameters has direct relevance to determining the ability of the liver, let alone the whole body, to make DHA. Nevertheless, these two examples characterize the problem of estimating the capacity to synthesize DHA and hence the necessity or lack thereof to provide dietary DHA. That ability is the sum of desaturation and chain-elongation in many organs, as well as rates of esterification of all the ω3 polyunsaturates, and β-oxidation affecting all the ω3 polyunsaturates.

Knowing the whole body capacity to make DHA or the whole body utilization of ALA (percent conversion to DHA, storage as ALA, or β-oxidation) tells us whether ALA intake is meeting requirements for DHA accumulation. Whole body analysis tells us whether certain diets, metabolic states or ratios of ALA to LA compromise ALA utilization (including DHA synthesis) and, therefore, raise ALA requirement. Whole

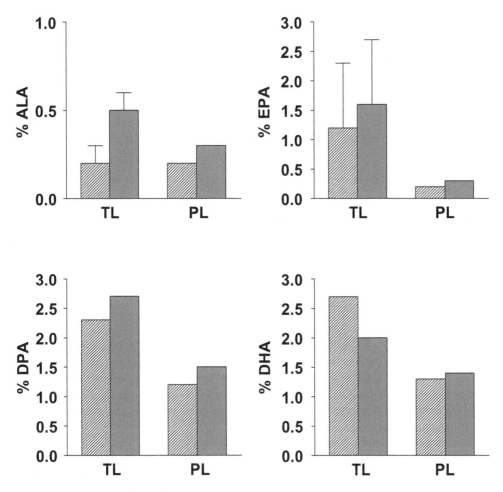

Figure 7.3 α-linolenate (ALA), eicosapentaenoate (EPA), ω3 docosapentaenoate (DPA) and
DHA levels in human platelet lipids before (stippled bars) and after (gray bars) dietary
supplementation with ALA (9–21 g/d) for 2–6 weeks. (Data are compiled from Sanders
and Roshanai, 1983; Adam *et al.*, 1986; Chan *et al.*, 1993; Kelley *et al.*, 1993; Ferrier
et al., 1995; Freese and Mutanen, 1997.) Platelet lipid classes: TL – total lipids,
PL – phospholipids.

body measurements done in animals offer a means of validating the results of tracer
studies done in animals, thereby permitting greater confidence in tracer studies done
in humans (see Whole body fatty acid balance analysis, p. 157).

Substrate inhibition

When present in excess, ALA and LA inhibit their own desaturation-chain elongation.
This is apparent using both *in vitro* microsomal desaturation assays (Ackman and Cunnane,
1992) and in fatty acid profiles of human plasma after supplementation with large
amounts of ALA which does not effectively raise plasma or platelet DHA (Cunnane,
1995) (Figures 7.2 and 7.3), and in recent tracer studies using ^{13}C-ALA (Vermunt

et al., 2000). Hence, there is no assurance that increasing ALA intake will increase tissue levels of EPA or DHA; in fact, some studies show that DHA remains unchanged or even decreases after ALA supplementation (Figures 7.2 and 7.3).

In human infants not consuming DHA or AA, the tendency in the 1980s was to raise the precursors in milk formula (first LA but more recently ALA) to "force" desaturation. However, if the results seen in adults in whom high intakes of ALA do not increase serum or blood cell DHA are applicable in infants, this strategy does not seem well-conceived. Indeed, recent data suggest that higher ALA intakes without DHA actually impair weight gain in healthy term infants (Jensen *et al.*, 1997).

Desaturase stimulation by ALA deficiency

Deficiency of ALA (or LA) increases hepatic microsomal desaturase activity and has often been used experimentally to maximize enzyme activity *in vitro* (Sprecher, 1968). It is important to remember, however, that this approach risks misrepresenting true desaturase activities *in vivo* because such experimental dietary deficiency conditions rarely exist outside the laboratory. This is another potential reason why the capacity to synthesize EPA and DHA from ALA could be mistakenly overestimated.

Cofactor nutrients

Desaturation of ALA or other polyunsaturates depends on electron transport from NADH via iron-containing cytochromes and activity of the final desaturase enzyme itself which also contains iron (Okayasu *et al.*, 1981). Iron deficiency is therefore one potential limitation on desaturation of ALA that is entirely independent of age, species, or dietary ALA concentration (Cunnane *et al.*, 1987). Several other nutrients are also cofactors in desaturation, notably zinc and vitamin B_6.

Zinc deficiency lowers DHA concentration more effectively than dietary deficiency of $\omega 3$ polyunsaturates, an effect that appears to occur through interruption of electron transport from NADH (Cunnane *et al.*, 1993b; Yang and Cunnane, 1994; Cunnane and Yang, 1995). Zinc deficiency also impairs food intake and usually leads to impaired weight gain and negative energy balance. Although uncommon as a solitary nutrient deficiency, there are many secondary causes of zinc deficiency, including alcoholism, and disorders of malabsorption. Thus, in addition to impairing ALA desaturation, dietary zinc deficiency and under-nutrition increase β-oxidation of ALA, thereby magnifying the problem of depending on ALA to synthesize DHA (Table 7.1).

Table 7.1 Example of increased β-oxidation of α-linolenate and linoleate with under nutrition induced by maternal zinc deficiency (Zn D) during pregnancy[1]

		Non-pregnant	*Pregnant*
α-linolenate	Control	11 ± 1	4 ± 2
	Zn D	$99 \pm 11*$	$49 \pm 26*$
Linoleate	Control	67 ± 1	41 ± 12
	Zn D	$80 \pm 5*$	$359 \pm 187*$

Notes
* $P < 0.05$ compared to respective contol.
[1] Data are expressed as mg/g body weight gained (Cunnane *et al.* 1993b).

In infants, the main implication of the susceptibility of desaturation/chain elonga-tion to insufficient amounts of the various cofactor nutrients is that, regardless of the maturity of the pathway and the potential to synthesize sufficient DHA, ALA con-version to DHA is simply not assured. Hence, pre-formed DHA in milk (breast or formula) bypasses these risks. In adults, malabsorption of these nutrients and diseases such as alcoholism similarly impair desaturation and chain elongation and reduce ALA conversion to DHA (Cunnane, 1995).

Interconvertibility

The shortest ω3 PUFA that can be desaturated and chain elongated is 14:3ω3 and, in principle, it can be converted to DHA. Retroconversion by two carbon chain short-ening is needed to make DHA from 24:6ω3 and also converts some DHA to EPA and even back to ALA. Hence, this interconvertibility within the ω3 PUFA family suggests that by starting with any ω3 PUFA between 14 and 24 carbons long, any other ω3 PUFA can be made – at least the enzymes to achieve this are present. This interconvertibility makes relying on tissue or plasma ω3 PUFA profiles a rather unreliable means of determining adequacy of intakes of individual ω3 PUFA.

Whole body fatty acid balance analysis

Whole body fatty acid balance analysis is a simple, direct method of determining ALA utilization and the capacity to synthesize DHA in experimental animals (Cunnane and Anderson, 1997). It can also be used to study LA metabolism but only works for LA and ALA because of the inability to synthesize these two fatty acids when the diet provides no other PUFA. This method involves measuring food and, hence, ALA intake for a designated study period, typically at least 20 days, and determining the differ-ence in fatty acid content (mg/whole body) between animals killed at the beginning and end of the balance period (Table 7.2). In growing animals, there is a clearly measurable accumulation of fatty acids, which is relatively simple to determine by

Table 7.2 Application of the whole body fatty acid balance method to determining the utilization of α-linolenate in young growing rats[1]

		α-linolenate	*Long chain ω3 polyunsaturates*
Intake	(mg)	6,731	0
Excretion	(mg)	149	31
	(% of intake)	2.2	–
Body Content (mg)	Start	592	96
	Finish	1,330	191
Accumulation	(mg)	738	95
	(% of intake)	10.9	1.4
Disappearance[2] (mg)		5,700	19[3]
	(% of intake)	84.9	–

Notes
[1] From Cunnane and Anderson (1997).
[2] Difference after subtracting Excretion and Accumulation from Intake.
[3] Estimated as 20% of accumulation.

quantitative lipid extraction and fatty acid analysis. Accumulation of ALA itself and of newly synthesized longer chain ω3 polyunsaturates can then be easily calculated and compared to the intake of ALA. The difference between the ALA consumed and that recovered in feces or in whole body ALA itself or longer chain ω3 polyunsaturates represents the ALA that was β-oxidized.

β-oxidation of α-linolenate

The whole body fatty acid balance method has shown that about 80 percent of dietary ALA is normally β-oxidized and less than 2 percent of ALA intake is converted to DHA by the young, rapidly growing rat consuming adequate but not excessive ALA (Table 7.2) (Cunnane and Anderson, 1997). Slightly different results are obtained depending on dietary total fat and ALA provided (Poumes-Balliaut *et al.*, 2001). Since the same methodology shows that about 65 percent of the DHA is also oxidized (Poumes-Balliaut *et al.*, 2001), estimates of ALA conversion to DHA done by balance methodology of changes in fatty acid profiles (Bryan *et al.*, 2001) need to be corrected for a loss by β-oxidation of more than half the DHA produced.

When looking at the parallel example of LA metabolism, the conversion of LA to AA calculated using the whole body fatty acid balanced method is in agreement with data on conversion of ^{13}C-LA to ^{13}C-AA in rats (Cunnane *et al.*, 1999b). Whole body fatty acid balance studies highlight the vulnerability of ALA accumulation and desaturation/chain elongation to even transient nutritional deficit (Chen and Cunnane, 1993; Chen *et al.*, 1995; Cunnane and Yang, 1995).

This technique provides a clear approach to understanding dietary controls and limitations on whole body capacity to synthesize DHA from ALA. In addition to providing the whole body perspective on DHA synthesis, this method allows determination of those tissues which, in a net sense, retain or produce more DHA. This is done by analyzing the change in fatty acid content of key organs individually. The implication is that one can determine organs that are net "producers" and those that are net "consumers" of DHA. The liver is well-known as a "producer" while the brain is probably a net "consumer" of DHA. Capacities and requirements of most other organs to produce and use DHA are mostly unknown. The neonatal gut may be able to contribute significantly to DHA synthesis (Menard *et al.*, 1998).

Useful as it is for estimating the capacity to synthesize DHA and the actual utilization of ALA in animal studies, direct whole body fatty acid balance can't be done in living humans. However, we are developing minimally invasive methods involving fat biopsies combined with magnetic resonance imaging to estimate whole body content of polyunsaturates in living humans and the degree to which the whole body content of these fatty acids changes with energy deficit (Cunnane *et al.*, 2001).

Accumulation of docosahexaenoate in infants

Estimates of the capacity to synthesize DHA in infants can also be derived from a combination of autopsy data on organ weights and tissue fatty acid profiles. This approach has recently shown that DHA accumulates in a term infant between birth and 6 months of age at about 10 mg/d (Table 7.3) (Cunnane *et al.*, 2000). To achieve this accumulation, about 20 mg/d would be required owing to obligate losses via β-oxidation (Poumes-Balliaut *et al.*, 2001) and carbon recycling (Sheaff-Greiner *et al.*, 1996) of DHA itself. A DHA requirement of 20 mg/d is easily met by the DHA content of most

Table 7.3 Accumulation of docosahexaenoate (DHA) in breast-fed (B) and formula-fed (F) infants over the first 6 months of life[1]

		Term	*6 months*	*Accumulation (mg/d)*
Brain	B	720	1,625	5.0
	F		1,170	2.5
Body Fat	B	1,050	1,200	0.8
	F		0	−5.8
Whole Body	B	3,800	5,685	10.3
	F		2,870	−5.1

Note
[1] From Cunnane *et al.* (2000). Because of obligatory losses to β-oxidation and carbon recycling (see text for details), requirement for DHA exceeds accumulation by about two-fold or 20 mg/d. Breast milk provides about a three-fold excess of DHA. Formulas not containing DHA provide about 400 mg/d as α-linolenate. Thus, a 5% conversion is required to obtain 20 mg/d as DHA and this seems unlikely from available data in animals.

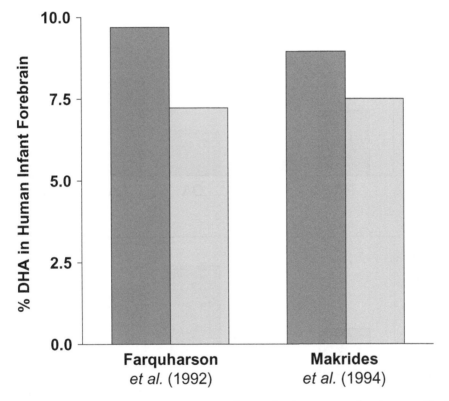

Figure 7.4 Docosahexaenoate levels in the brain of infants consuming breast milk (left, darker bar) or a milk formula providing only α-linolenate (right, lighter bar).

breast milk. However, to synthesize this amount of DHA from ALA would require a conversion rate of about 5 percent (20 mg/d from an ALA intake averaging 400 mg/d). No data from humans or animals support a capacity to make this amount of DHA in human infants. The lower accumulation of DHA in the brain when only ALA is given supports the need to supply pre-formed DHA (Figure 7.4).

   The point is that new approaches are needed to determine the capacity to synthesize DHA in humans and this requires creative and minimally-invasive ways of accessing the fatty acid content and composition of the body. Direct or indirect whole body fatty acid balance analysis offers one such approach.

β-oxidation and carbon recycling

All fatty acids undergo a greater or lesser degree of β-oxidation in the body and ALA is no exception. In fact, a variety of methods in rats and humans show that ALA is one of the most easily β-oxidized long chain fatty acids in the Western diet (Figure 7.5). There are two main implications of this: (a) under normal conditions, β-oxidation is probably by far the dominant pathway of ALA metabolism in mammals, and (b) β-oxidation gives rise to substantial "carbon recycling" of the ALA skeleton into *de novo* lipid synthesis.

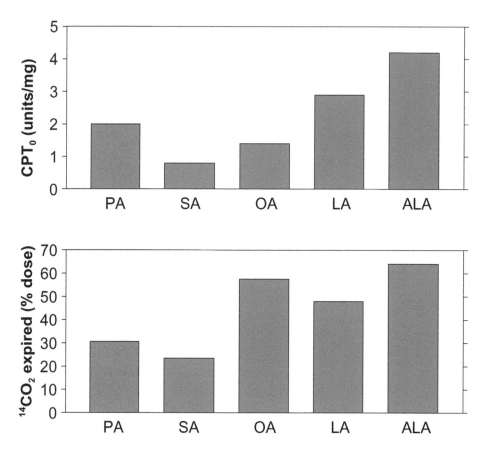

Figure 7.5 *In vitro* (top) and *in vivo* (bottom) comparisons of the β-oxidation of long chain fatty acids (PA – palmitate, SA – stearate, OA – oleate, LA – linoleate, ALA – α-linolenate). The top panel shows CPT_0 (carnitine palmitoyl transferase) measured in rat liver (Gavino and Gavino, 1991). The bottom panel shows the percent of the ^{14}C-labeled fatty acid that is expired as [14]CO_2 within 24 h (Leyton *et al.*, 1987).

Table 7.4 Comparison of the incorporation of carbon from α-linolenate into brain docosahexaenoate (DHA) compared to brain *de novo* lipogenesis (DNL) in suckling rats[1]

Time point (h)	DHA	DNL	DHA/DNL (%)
3	0.02	0.73	2.7
6	0.03	1.38	2.2
12	0.06	2.63	2.3
24	0.11	2.51	4.4
48	0.40	3.04	13.2
192	0.44	8.03	5.5

Note
[1] Data are shown in mg ^{13}C/brain after an oral dose of ^{13}C-α-linolenate. The data for *de novo* lipogenesis are derived from values obtained for cholesterol and palmitate only (from Menard *et al.* 1998).

Not only the whole body capacity to synthesize DHA but also whole body ALA β-oxidation can be accurately determined by doing whole body fatty acid balance analysis after feeding laboratory animals a diet devoid of ω3 polyunsaturates, except for ALA. This simple but unique method provides very useful information on the utilization or "disposal" of ALA, i.e. its loss from the body pool of ω3 PUFA. Although the whole body fatty acid balance method clearly indicates the "disappearance" or apparent β-oxidation of ALA (or LA), it cannot distinguish between carbon from ALA that is completely β-oxidized to CO_2 or carbon that is retained in newly-synthesized compounds, especially lipids.

Using ^{14}C-ALA, studies done more than 20 years ago qualitatively indicated the potential for carbon recycling of ALA into lipids synthesized *de novo* (Dhopeshwarker and Sunramanian, 1975; Sinclair, 1975). This earlier work has been largely overlooked until our own whole body fatty acid balance data indicated the large proportion of ALA that is normally β-oxidized. This led us to *quantitatively* assess carbon recycling from ALA in suckling rat pups (Table 7.4) (Cunnane *et al.*, 1994; Menard *et al.*, 1998). The results of this work demonstrated that carbon recycling in the suckling rat consumed about *30-fold more* of a physiological dose of ^{13}C-ALA than its conversion to DHA. Low levels of DHA partially reduce carbon recycling of ALA but do not eliminate it (Willard *et al.*, 2001).

A different model using specific LA deficiency was used to address this issue but ALA deficiency would work equally well. The rationale was that if carbon recycling still occurred while rats were deficient in dietary and body stores of LA (or ALA) over several weeks, recycling could not reasonably be attributed to overflow due to an excess availability of body stores of LA or longer chain ω6 polyunsaturates. Hence, the minimal or "obligatory" amount of carbon recycling from linoleate could be determined. LA-deficient rats were shown to recycle about 20 percent of a tracer dose of ^{14}C-LA into sterols and fatty acids synthesized *de novo*, compared to about 30 percent in controls (Cunnane *et al.*, 1998; Belza *et al.*, 1999).

Thus, β-oxidation and carbon recycling consume a quantitatively significant proportion of LA intake and, by inference, of ALA intake. Carbon recycling of ALA is as quantitatively important as desaturation and chain elongation to DHA, or perhaps more so. Its biological importance in ALA metabolism, however, is unclear. One possibility is that ALA is redundant if DHA is in the diet. This may be the case during breast feeding and in populations consuming seafood. Hence, rather than being "essential," ALA would become dispensable. As a readily β-oxidized fatty acid, ALA would therefore become available and be used for fatty acid and cholesterol synthesis. Cholesterol is extremely important for normal embryonic development and brain cholesterol must be synthesized *in situ*. Hence, ALA may contribute to normal brain development as a reserve for DHA synthesis but also for brain lipid synthesis (Cunnane *et al.*, 1999b).

Food sources and intake of ALA

The leaves of green plants contain up to 80 percent of their total fatty acids as ALA. However, the total lipid content of plants is generally very low, so the net yield of ALA from green plants even for vegetarians is also low (Tinoco, 1982; Periera *et al.*, 2001). Hexadecatrienoate ($16:3\omega3$) is also present in edible green leafy plants and can be converted to ALA (Cunnane, 1995; Periera *et al.*, 2001). ALA is present in commercial plant/vegetable oils (sunflower, safflower, corn, and olive) in very low amounts, but traditional flaxseed (50–60 percent ALA), soybean (about 7 percent ALA), and low erucic acid rapeseed (canola, 10 percent ALA) are notable exceptions that have been successfully commercialized. Wild purslane is an edible plant very rich in ALA. As long as canola and soybean oils remain non-hydrogenated, they continue to be good dietary sources of ALA. Hence, food products made with non-hydrogenated canola or soybean oil, including salad oils and dressings, margarines, and baked products, are important sources of dietary ALA (Table 7.5).

Table 7.5 Average daily intake of α-linolenic acid

Food	% intake ALA
Margarine	16[1]
Fats, oils, dressings, mayonnaise	15[2]
Meat (chicken, beef, pork, lamb)	12[3]
Vegetables	8[4]
Milk	7[3]
Cheese, yogurt, milk products	6[3]
Butter	6[4]
Breads (white, dark)	5[3]
Cookies	5[4]
Shortening	4[4]
Potatoes	4[4]
Sauces	3[4]
Fish and seafood	3[3]
Fruit	3[4]
Other	3[5]
Σ	100

Notes
[1] Mean of Hu *et al.* (1999) and Voskuil *et al.* (1996).
[2] Mean of Lewis *et al.* (1995) and Hu *et al.* (1999).
[3] Mean of Lewis *et al.* (1995) and Voskuil *et al.* (1996).
[4] Voskuil *et al.* (1996).
[5] Accounts for foods with low levels of α-linolenic acid (e.g. leafy green vegetables, certain fruits and eggs), Hu *et al.* (1999).

Legumes, pulses, starches, fruit and vegetables may have up to 30 percent of their fatty acids as ALA but all are relatively low in fat. Among foods of animal origin, ALA concentration is highest in fat tissue (1–4 percent) so its availability from meat is directly proportional to fattiness of meat cuts (Romans *et al.*, 1995). Therefore, in meats from domesticated animals, the lower the visible fat content, the lower the overall ALA content. Food sources and estimated ALA intake from common foods are listed in Table 7.5. The average dietary intake of ALA in North America has been estimated from food disappearance data to be 1–2 g/day or 0.5 percent of energy intake (Hunter, 1990; Kris-Etherton *et al.*, 2000; Sanders, 2000). Fatty acid intake data from diet questionnaires used in the Health Professionals Follow-Up Study gave a similar estimate of mean ALA intake as 1.05 g/day with a 95 percent confidence interval of 0.58–1.35 g/day (Giovannucci *et al.*, 1993; Hu *et al.*, 1999). Many factors influence ALA intake with vegetarians obtaining more ALA from green vegetables while non-vegetarians obtain a significant amount of ALA from processed meat and dairy products.

Dietary ALA always raises the spectre of lipid peroxidation which is a major reason why processed foods tend to be as low as possible in ALA. A consequence of the peroxidation process is isomerization of ALA from its all *cis* form to one of several *trans*-isomers that are possible. The *trans*-isomers of LA appear to be somewhat more easily β-oxidized which would help with their removal from the body, but this does not occur with *trans*-isomers of ALA (Bretillon *et al.*, 2001). Since *trans*-isomers of ALA are present in the diet and can, in large amounts, raise blood cholesterol (Vermunt *et al.*, 2001), continued caution about including ALA without adequate antioxidants in processed foods is warranted. Nevertheless, it is encouraging to see that various induces of lipid peroxidation in human plasma can remain within normal limits on diets enriched in ALA (Cunnane *et al.*, 1995a; Sodergren *et al.*, 2001), again indicating the value of appropriate antioxidant intake.

ALA requirement and deficiency

Studies in female rats provided the first estimate of ALA requirement in mammals. This value, 0.5 percent of energy intake, has remained the dominant value for ALA requirement and has been used for several species, including humans (Holman, 1971). Suggestions have also been made to set an absolute requirement (mg/d) as opposed to a relative one (Bjerve *et al.*, 1987). Nevertheless, the requirement for ALA in humans is still widely thought to be 0.5 percent of energy intake. In North America, ALA intake approximates 0.5 percent of energy or about 1 g/d. If this intake estimate is correct, it means that half the population is consuming less than the required intake. Contrary to recent assertions (Conner, 1999), there are *not* "ample food sources" of ALA in the North American diet because, on a population basis, these ALA sources clearly do not provide sufficient ALA for at least half the population. In addition, those foods that are good sources of ALA tend also to be rich in LA. Given the protective effects of consuming > 1 g/d of ALA for risk of cardiovascular disease and cancer (Dolecek, 1992; Guallar *et al.*, 1999; Hu *et al.*, 1999), increasing ALA intake in the population to 1.5 or even 2 g/d should be a public health policy goal. This position has recently been supported by a workshop (de Deckere *et al.*, 1998).

ALA deficiency symptoms seen in young rapidly growing laboratory animals have not frequently been observed in humans. In fact, only two case reports claiming specific ALA deficiency are in the literature (Holman *et al.*, 1982; Bjerve *et al.*, 1988). These two cases had few, if any, overlapping symptoms. They, and other reports of ALA deficiency in elderly hospitalized patients (Bjerve *et al.*, 1987), generated commentaries disputing their classification as cases of specific ALA deficiency (Koletzko, 1987; Koletzko and Cunnane, 1988). One of the issues was whether LA deficiency was also present as suggested by abnormal values for ω6 polyunsaturates in serum.

Since these well-known studies were conducted, it has remained difficult to reproducibly demonstrate the exact symptoms of ALA deficiency, especially in humans. This difficulty continues to exist for several reasons: (a) different LA-rich dietary oils exacerbate ALA deficiency to a greater or lesser extent, (b) only low amounts of ALA are required and can be synthesized by retroconversion from the larger body pool of DHA, and (c) so long as there is some DHA in tissues, ALA deficiency symptoms may remain absent or muted.

Many animal studies in the past three decades have established that when all other ω3 polyunsaturates are absent from the diet, the deficiency of ALA results in reduced levels of EPA, DPA, and DHA in all tissues investigated. In tissues that contain high amounts of DHA, particularly the brain and retina, the depletion of DHA has reproducible deleterious consequences in several non-human species, including primates, especially during early development. In some cases, the depletion of DHA can be prevented/reversed by dietary ALA but this seems to be unlikely in human infants (Farquharson *et al.*, 1992; Makrides *et al.*, 1994).

Function of ALA and ω3 polyunsaturates

Clearly, when EPA and DHA are absent from the diet, ALA becomes the limiting source of these fatty acids, and when all three are deficient in the diet, depletion of EPA and DHA is inevitable. The depletion of tissue levels of ω3 polyunsaturates depends on fat tissue stores and these stores can be reduced most easily by fasting or semi-starvation or by markedly increased LA intake.

EPA and DHA have specific metabolic and/or structural roles but the question remains as to whether ALA has a function of its own. EPA is an effective competitor of AA both for esterification into membrane phospholipids and for the cyclooxygenase used during eicosanoid synthesis. Eicosanoids derived from EPA may competitively block or inhibit those produced from AA. DHA has irreplaceable effects in complex excitable membranes of neurons and photoreceptors (Dratz and Deese, 1986; Salem *et al.*, 1996; Crawford *et al.*, 1999). ALA has recently been shown to inhibit cyclooxygenase-2 (Ringborn *et al.*, 2001), to inhibit leukotriene production (Okamoto *et al.*, 2000), and to be a precursor to plant isoprostanes (Imbusch and Mueller, 2000). Some of these effects of ALA appear to be important in the leaves of green plants, most of which concentrate ALA, but they may also be useful to human health as well.

Whether ALA has any specific biological functions that only it performs is unknown at the moment. Its levels in membrane phospholipids are too low to perform the functions typically associated with a membrane fatty acid. Nevertheless, higher ALA intake is associated with a greater reduction in risk of breast cancer metastasis (Bougnoux *et al.*, 1994) than the longer chain ω3 polyunsaturates. Higher levels of ALA in adipose tissue are associated with lower risk of heart attack but this is not observed for DHA (Guallar *et al.*, 1999). Hence, it is important to continue to search for and verify new functions and new routes of metabolism of ALA that might account for its potentially unique properties.

Early human development

Important developments in understanding the role of ALA in human health have come from studies in human infants and from animal models of early infant development. Breast milk accumulates fat (and ALA) more as a function of maternal stores rather than directly from the diet (Koletzko *et al.*, 2001). Prior to 1980, the fat in most milk formulas contained little or no ALA because the oils used were rich in LA and low in ALA, e.g. corn or sunflower (Cuthbertson, 1999). Studies of infant body composition and development then began to show that plasma and brain DHA accumulation, brain development, and IQ were potentially vulnerable to the lack of ALA (Clandinin *et al.*, 1980; Carlson *et al.*, 1987; Farquharson *et al.*, 1992; Leaf *et al.*, 1992; Lucas *et al.*, 1992; Martinez, 1992; Makrides *et al.*, 1995; Crawford *et al.*, 1999).

These studies also began to question whether ALA alone was sufficient to meet the DHA needs of infants; i.e. differences in "bioequivalence" between ALA and DHA started to become obvious (Crawford, 1993; Woods *et al.*, 1996). Tracer studies also show that pre-formed DHA accumulates seven-fold more rapidly in the neonatal monkey brain than DHA synthesized endogenously from ALA (Su *et al.*, 1999). DHA is transferred across the human term placenta about three-fold more easily than ALA (Haggarty *et al.*, 1997). These studies suggest that DHA is preferred over ALA for fetal and neonatal development, a preference that should be reflected in milk formulas.

Despite efforts to minimize confounding variables, it would have been difficult to argue on the strength of the few human studies done in the 1980s that ALA was essential for normal infant development. Rather, extensive animal studies at the same time (Bourre *et al.*, 1984; Neuringer *et al.*, 1984; Bourre *et al.*, 1989; Foote *et al.*, 1990; Arbuckle *et al.*, 1991) demonstrated that the deficits seen earlier in ALA deficient rats (Lamptey and Walker, 1976) would probably occur and have similar harmful effects in human infants. These and earlier studies also indicated that excess intake of LA was able to contribute to ALA deficiency in human infants. This led to acceptance of the essentiality of ALA for infant development and a gradual reduction in the LA:ALA ratio in milk formulas.

The issue for intake of polyunsaturated fatty acids from infant milk formulas has therefore shifted from neglect of the necessity for ALA to whether ALA is sufficient without longer chain ω3 polyunsaturates. Initial attempts to address this issue by including EPA were not successful because the negative impact of its competition with AA was not fully appreciated. Thus, EPA-supplemented infant formulas led, in some cases, to comparatively smaller head circumference than in infants not receiving EPA (Carlson *et al.*, 1993; Carlson *et al.*, 1996). This resulted in concern that longer chain ω 3 polyunsaturates were potentially harmful for infant development, a contentious issue that has not disappeared despite the abandonment of the supplementation of milk formulas with EPA.

Positive correlations between infant body growth and AA (Koletzko and Braun, 1991; Crawford, 1993; Xiang *et al.*, 1999) suggested that if longer chain ω3 polyunsaturates were to be included in infant formulas, AA should also be present. During the 1990s, infant formulas providing a dietary source of AA and DHA instead of EPA were developed and approved for marketing in many countries but have only just been approved for use in North America. The apparent risk of EPA supplementation without AA has not appeared when formulas with AA and DHA are given. Furthermore, these supplemented formulas appear in many (short-term) studies to be superior to those not providing DHA and AA (Agostoni *et al.*, 1995; Makrides *et al.*, 1995; Birch *et al.*, 1998; Willatts *et al.*, 1998).

Some studies show no difference in growth or indices of visual development in term infants not receiving DHA or AA compared to breast-fed infants (Auestad *et al.*, 1997; Lucas *et al.*, 1999). It has therefore been claimed that, with sufficient ALA and an LA:ALA ratio under 10:1, growth and development of formula-fed infants is equivalent to breast-fed infants. A recent study disputes this by showing that a formula with an LA:ALA ratio of 4.8:1 and ALA at 3.2 percent of fatty acids resulted in infants being 0.7 kg lighter at 4 months than infants receiving a formula with an LA:ALA ratio of 44:1 (Jensen *et al.*, 1997). One key issue in these studies comparing formulas to breast milk is the level of DHA in breast milk. Data from the USA show that DHA is frequently <0.2 percent of breast milk fatty acids, a value much lower than in many other countries (Hamosh and Salem, 1998). Hence, the breast milk control group may be receiving insufficient DHA itself and would therefore be an inappropriate control group.

Unambiguous studies of nutrient effects on infant development are difficult and expensive to conduct and need to be long-term to convincingly show significant developmental differences. They also need good controls with a sufficiently high DHA intake to distinguish potentially important differences in neurodevelopment. Many studies can be criticized for shortcomings in these areas (Wainwright and Ward, 1997; Morley, 1998; Carlson and Neuringer, 1999). Overcoming these shortcomings requires both scientific skill in study design and commitment of sufficient funds to conduct such studies. Whether this will be possible in the near future is unclear. In the meantime, many countries are marketing the DHA- and AA-supplemented formulas. A large 18-month trial evaluating AA- and DHA-supplemented formulas consumed by term infants refutes the safety fears that have been raised (Life Sciences Research Office Report, 1998).

Two definitive trials clarifying the role of DHA and AA in long-term "functional" measures of infant development have recently been reported and strongly suggest that the healthy term but not pre-term infant can grow and develop normally on formulas lacking arachidonate and DHA (Auestad *et al.*, 2001; O'Conner *et al.*, 2001). However, two other reports still show that provision of only ALA in milk formulas given to *term* infants leads to inadequate brain DHA accumulation in comparison to breast-fed infants (Figure 7.4). This suboptimal level of brain DHA accumulation occurs despite a demonstrated ability to synthesize DHA from ALA in term and pre-term infants (Carnielli *et al.*, 1996; Salem *et al.*, 1996; Sauerwald *et al.*, 1997) and the presence of about 1,000 mg of DHA in body fat at term (Cunnane *et al.*, 2000) (Table 7.3). These are important observations that strongly support the need for inclusion of DHA in milk formulas so as to guarantee brain DHA accumulation.

Human disease

The most promising research on ALA in human disease relates to the two major killers in Westernized societies – cardiovascular disease and cancer. Other interesting but less well-established progress concerning the potential preventative or therapeutic role of ALA in human diseases will be discussed under "Other new developments" (see p. 169).

Cardiovascular disease

Factors contributing to an elevated risk of heart attack are elevated serum cholesterol, obesity, hypertension, increased platelet aggregability and heart rhythm instability. In each case, there is controversy over whether supplemental dietary ALA is beneficial or ineffective in changing these risks. Some studies show that thrombosis tendency, platelet aggregation, and heart rhythm stability are responsive to ALA supplementation in the form of flaxseed oil (Renaud and Nordøy, 1983; Budowski *et al.*, 1984; Nordstrom *et al.*, 1995; Kang and Leaf, 1996; Freese and Mutanen, 1997) but others show no effect of ALA on these parameters despite significant changes in plasma ALA and EPA after ALA supplementation (Chan *et al.*, 1993; Kelley *et al.*, 1993; Allman-Farinelli *et al.*, 1999; Li *et al.*, 1999). Plasma ALA levels do not appear to correlate with heart rate stability (Christensen *et al.*, 2000).

Free ALA, EPA or DHA bound to albumin and infused intravenously all prevent sudden cardiac death in dogs undergoing coronary artery occlusion (Billman *et al.*, 1999). Similarly, ALA also prevents neuronal damage in an animal model of transient global ischemia (Lauritzen *et al.*, 2000). Both of these effects may relate to a vasodilatory effect of ALA (Sarabi *et al.*, 2001).

ALA may also have beneficial effects in hypertension (Berry and Hirsch, 1986; Salonen *et al.*, 1988; Singer *et al.*, 1990) but this is also disputed (Bursztyn, 1987). Most studies show that supplemental ALA does not have a significant effect on serum cholesterol levels (Sanders and Roshanai, 1983; Kestin *et al.*, 1990; Mantzioris *et al.*, 1994; Pang *et al.*, 1998) but some have reported that it does reduce serum cholesterol (Chan *et al.*, 1991). Higher adipose levels of ALA are associated with lower serum cholesterol (de Vries *et al.*, 1991). Central adiposity has been reported to be negatively associated with the content of ω3 polyunsaturates in adipose biopsies (Garaulet *et al.*, 2001).

Virtually all epidemiological studies suggest that ALA intake in the upper quintile of the American average intake (> 1 g/d) has a significant risk-lowering effect for heart disease in middle age (Figure 7.6) (Dolecek, 1992; Hu *et al.*, 1999; Djousse *et al.*, 2001) but this appears not to be the case in the elderly (Oomen *et al.*, 2001). Similar protective results were obtained in comparing quintiles of adipose tissue ALA to the risk of myocardial infarction (Guallar *et al.*, 1999). Collectively, these studies suggest ALA is one of the most "cardioprotective" nutrients available in the human diet. Clinical intervention trials support these observations, though blood pressure, platelet aggregability and serum cholesterol were unchanged in this study (de Lorgeril *et al.*, 1994). Hence, there seems to be unanimous and unambiguous support for a moderately high intake of ALA to reduce the risk of heart attack but the mechanism of this effect is far from clear. The heart rhythm-stabilizing effects are intriguing and may be important. Whether the apparent benefits of ALA depend on even a small amount of conversion to EPA and DHA is also unclear.

In addition to ALA, milled flaxseed contains soluble fiber which consistently lowers LDL cholesterol in humans with normal or moderately elevated serum lipids (Bierenbaum *et al.*, 1993; Cunnane *et al.*, 1993a; Cunnane *et al.*, 1995a; Jenkins *et al.*, 1999). Components in flaxseed other than ALA may actually be more important in reducing the risk of atherosclerosis, particularly in experimental animals (Prasad *et al.*, 1998). Whereas extracted flaxseed oil may protect against some of the major risks for heart attack (platelet aggregation and heart rhythm instability), milled flaxseed protects against the same risks but also helps in the reduction of serum LDL cholesterol. It

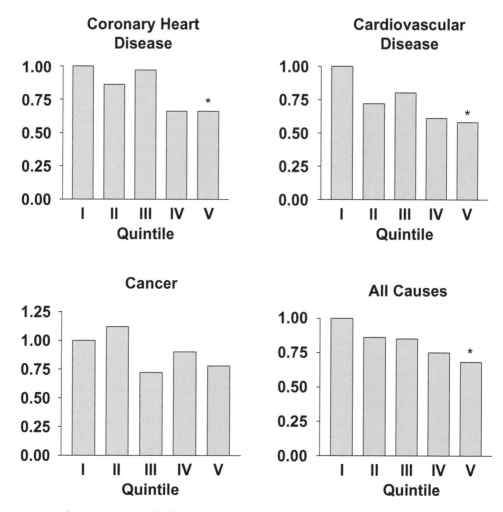

Figure 7.6 The relative risk of death from four major causes compared to α-linolenate intake (Quintile 1 – 0.42 en %, Quintile 2 – 0.54 en %, Quintile 3 – 0.63 en %, Quintile 4 – 0.73 en %, Quintile 5 – 0.98 en %). (From Dolecek, 1992).

also contains antioxidants to protect the ALA and other constituents like lignans and soluble fiber that probably act synergistically with ALA (Cunnane, 1995).

Given the apparent cardiovascular risk-reducing attributes of higher ALA intake and the marginally adequate intake of ALA in Westernized societies, it seems prudent to recommend increased ALA intake (de Deckere *et al.*, 1998; Harper and Jacobson, 2001)

Cancer

Cancer is the other major killer in Westernized societies. Studies in animals and epidemiological studies commonly show that higher intakes of ω3 polyunsaturates lower the risk of cancer (Dolecek, 1992; Johnston, 1995). Lower plasma levels of ALA have been found in patients with colorectal cancer (Baro *et al.*, 1998). Elevated ALA and, to a lesser extent, DHA levels in breast tissue are associated with significantly lower risk of breast cancer metastasis (Bougnoux *et al.*, 1994; Klein *et al.*, 2000). Several unsubstantiated mechanisms of this effect have been suggested, including reduction of the synthesis of "2 series" eicosanoids derived from AA.

Typically for ALA, however, the inverse relation between ALA intake and cancer risk just described is not a consistent finding. Indeed, increased breast and prostatic cancer risk have been observed as intake of ALA increases (Giovannucci *et al.*, 1993; Vatten *et al.*, 1993; De Stefani *et al.*, 1998; De Stefani *et al.*, 2000; Michaud *et al.*, 2001; Newcomer *et al.*, 2001). It is unclear why these opposite findings with respect to cancer risk occur but the consumption of red meat and processed meat may have an influence. Processed meat contributes significantly to ALA intake in omnivores (Hu *et al.*, 1999) and also increases prostatic cancer risk (Giovannucci *et al.*, 1993). Intake of neither vegetables (relatively high in ALA) nor fish (high in EPA and DHA) increases cancer risk. Thus, it seems intuitively unlikely that ALA *per se* is responsible for the increased cancer risk almost exclusively in the prostate; rather, it seems reasonable that it is fat from meat that is the predominant risk factor and that ALA is simply a marker of higher intake of fat from meat. One of the recent reports demonstrates that the risk of prostate cancer was partially attenuated when variation in ALA intake was corrected for (Michaud *et al.*, 2001), supporting the possibility that ALA is a marker but not a causative factor for prostate cancer risk. It follows that higher ALA intake in vegetarians would not be associated with increased cancer risk but this remains to be established.

It is not yet clear whether different ω3 polyunsaturates (ALA versus EPA or DHA) have different benefits for cancer risk reduction or how these fatty acids achieve this effect. As in the reduction of heart attack risk, milled flaxseed may be more beneficial for cancer risk reduction than flaxseed oil alone. This additional benefit of whole flaxseed comes in part from the soluble fiber (mucilage) in flaxseed, which, like other dietary soluble fiber sources, reduces serum lipids independently of ALA (Prasad *et al.*, 1998), and also colon cancer risk. In addition, lignans in flaxseed have been shown to reduce experimental tumor growth in laboratory animals (Thompson, 1995). Whether ALA acts independently or synergistically with these other components is unknown at the moment.

Other new developments

Adiposity and fatty acid synthesis

ω3 polyunsaturated fatty acids have been associated with reduction in adiposity in several animal models, and recently, in a human epidemiological study relating dietary fatty acids to central obesity (Garaulet *et al.*, 2001). This may partly be a function of more rapid β-oxidation of ω3 polyunsaturates compared to monounsaturates or saturates (Figure 7.5). However, recent studies suggest other complimentary effects on the balance between fatty acid β-oxidation and synthesis. Raised intake of ALA in rats down-regulates genes controlling adipocyte differentiation and adipose tissue growth, including peroxisome proliferator-activated receptor α, adipocyte P2, and adipsin, when the diet contains ALA-rich perilla oil compared to safflower oil (mainly LA), olive oil (mainly oleate) or beef tallow (mainly saturates) (Okuno *et al.*, 1997). Hepatic fatty acid synthase is decreased and carnitine palmitoyl transferase is increased by ω3 polyunsaturates compared to LA which may also contribute to minimizing adiposity (Ikeda *et al.*, 1998; Kumamoto and Ide, 1998). Given the increasing prevalence of obesity, especially in adolescents, a beneficial role of ALA in controlling adiposity and fatty acid β-oxidation warrants further investigation in humans.

Arthritis and immune function

ω3 polyunsaturates attenuate eicosanoid synthesis from AA. Although EPA and DHA are thought to be more effective than ALA in this respect (Kelley *et al.*, 1991; Whelan *et al.*, 1991), ALA clearly inhibits eicosanoid and leukotriene synthesis (Okamoto *et al.*, 2000; Ringborn *et al.*, 2001), effects which are likely to contribute to a beneficial effect on the risk of atopy in children (Duchen *et al.*, 2000). Fish oils and fish consumption have achieved some success in the treatment of autoimmune diseases such as arthritis in part because of their eicosanoid-suppressing effects (Whelan *et al.*, 1991; Kremer, 1996; Shapiro *et al.*, 1996). ALA-rich oils have also been tested in rheumatoid arthritis but, so far, without success (Nordstrom *et al.*, 1995). Probiotic bacteria in the gut help control the risk of inflammatory conditions of the gut. One such bacterium, *Lactobacillus casei* Shirota, binds more easily to intestinal Caco-2 cells in culture in the presence of ALA (Kankaanpaa *et al.*, 2001), suggesting the ALA may contribute to healthier gut microflora.

Increasing the food sources of ALA

ALA clearly has potential to modify the risk of both cancer and heart disease. It also appears that, as with fish, about half the population is consuming less than the recommended amount of ALA. Hence, there are good reasons to consume more ALA but Western diets commonly provide either insufficient amounts of ALA or too much LA (Lands, 2001), or both.

Flaxseed and flaxseed oil are excellent, concentrated sources of ALA but, despite growing interest in the health food marketplace, flaxseed is still not widely consumed in North America. The stability of ALA in products baked with flaxseed is now reasonably well-established (Chen *et al.*, 1994). Plasma vitamin E levels are not decreased nor are measures of peroxidation significantly increased by consumption of up to 50 g/d traditional flaxseed provided in muffins (Cunnane *et al.*, 1995b; Sodergren *et al.*, 2001).

Canola and soybean oils are moderately good sources of ALA but they are widely hydrogenated, negating their value. Processed meats have moderately high amounts of ALA but are contraindicated due to their content of saturated fatty acids. Hence, other, novel food sources of ALA have been developed and successfully marketed, including baked products, ω3-polyunsaturated fatty acid-enriched eggs (Ferrier *et al.*, 1995; Sim and Qi, 1995) and pork (Romans *et al.*, 1995). Because of their popularity in the food supply, eggs and pork are viable means of moderately raising ALA intake of large numbers of people.

Conclusion

The majority of evidence points to ALA as an important nutrient for the reduction of the risk of the major killer diseases in Westernized societies – cancer and cardiovascular disease. Much of the evidence is still controversial and it is unclear how ALA achieves the observed benefits because so little of it seems to be converted to the equally or more beneficial EPA and DHA. Increased ALA intake may also be beneficial for the chronic, non-killer diseases like those involving increased inflammation. ALA is necessary but, by itself, is barely sufficient for normal infant development. The intake of ALA is inadequate in about half the North American population, especially because of the low consumption of fish, non-hydrogenated oils and vegetables.

If the health benefits of ALA are to be realized, its intake will need to increase. This will require alternative dietary sources, including ALA-enriched eggs and pork, which have been developed on a small scale, to become much more widely available. The uncertainty of how to interpret the apparently increased risk of prostate cancer when ALA intake is high casts a shadow over these recommendations but it is my view that this correlation is an artifact of ALA being relatively high in processed meat and animal fat compared to other foods. New developments in understanding the metabolism of ALA point to β-oxidation and carbon recycling as quantitatively significant routes in its metabolism.

Acknowledgments

The support and collaboration of several colleagues, particularly Dennis McIntosh, Lilian Thompson and David Jenkins, is greatly appreciated. Research on ALA that was performed by the author's group was supported by NSERC, MRC, The Flax Council of Canada, Flax Growers–Western Canada, Unilever-Lipton, Martek Biosciences, and Milupa AG. Valerie Francescutti provided valuable technical assistance.

References

Ackman, R.G., and Cunnane, S.C. (1992). Long-chain polyunsaturated fatty acids. In Padley, F.B. (ed.) *Advances in Applied Lipid Research*, Vol. 1, JAI Press, London, pp. 161–215.

Adam, O., Wolfram, G., and Zöllner, N. (1986). Effect of α-linolenic acid in the human diet on linoleic acid metabolism and prostaglandin metabolism. *J. Lipid Res.* 27, 421–426.

Agostoni, C., Trojan, S., Bellu, R., Riva, E., and Giovannini, M. (1995). Neurodevelopmental quotient of healthy term infants at 4 months and feeding practice: The role of long-chain polyunsaturated fatty acids. *Pediatr. Res.* 38, 262–266.

Allman-Farinelli, M.A., Hall, D., Kingham, K., Pang, D., Petacz, P., and Favaloro, E.J. (1999). Comparison of the effects of two low fat diets with different alpha-linolenic:Linoleic acid ratios on coagulation and fibrinolysis. *Atherosclerosis* 142, 159–168.

Arbuckle, L.D., Rious, F.M., Mackinnon, M.J., Hrboticky, N., and Innis, S.M. (1991). Response of n-3 and n-6 fatty acids in piglet brain, liver, and plasma to increasing, but low, fish oil supplementation of formula. *J. Nutr.* 121, 1536–1547.

Auestad, N., Halter, R., Hall, R.T., Blatter, M., Bogle, M., Burks, W., Erickson, J.R., Fitzgerald, K.M., Dobson, V., Innis, S.M., Singer, L.T., Montalto, M.B., Jacobs, J.R., Qui, W., and Bornstein, M.H. (2001). Growth and development in term infants fed long chain polyunsaturated fatty acids: A double-masked, randomized, parallel, prospective multivariate trial. *Pediatr. Res.* 108, 372–381.

Auestad, N., Montalto, M.B., Hall, R.T., Fitzgerald, K.M., Wheeler, R.E., Connor, W.E., Taylor, J.A., and Hartmann, E.E. (1997). Visual acuity, erythrocyte fatty acid composition, and growth in term infants fed formulas with long chain polyunsaturated fatty acids for one year. *Pediatr. Res.* 41, 1–10.

Baro, L., Hermoso, J.C., Nunez, M.C., Jimenez-Rios, J.A., and Gil, A. (1998). Abnormalities in plasma and red blood cell fatty acid profiles of patients with colorectal cancer. *Br. J. Cancer* 77, 1978–1983.

Beitz, J., Mest, H.-J., and Forster, W. (1981). Influence of linseed oil diet on the pattern of serum phospholipids in man. *Acta Biol. Med. Germ.* 40, K31–K35.

Belza, K., Anderson, M.J., Ryan, M.A., and Cunnane, S.C. (1999). Carbon recycling from linoleate during severe dietary linoleate deficiency. *Lipids* 34, S129–S130.

Berry, E.M., and Hirsch, J. (1986). Does dietary linolenic acid influence blood pressure? *Am. J. Clin. Nutr.* 44, 336–340.

Bierenbaum, M.L., Reichstein, R., and Watkins, T.R. (1993). Reducing atherogenic risk in hyperlipiemic humans with flaxseed supplementation: A preliminary report. *J. Am. Coll. Nutr.* 12, 501–504.

Billman, G.E., Kang, J.X., and Leaf, A. (1999). Prevention of sudden cardiac death by dietary pure omega-3 polyunsaturated fatty acids in dogs. *Circulation* 99, 2452–2457.

Birch, E.E., Hoffman, D.R., Uauy, R., Birch, D.G., and Prestidge, F. (1998). Visual acuity and the essentiality of docosahexaenoic acid and arachidonic acid in the diet of term infants. *Pediatr. Res.* 44, 201–209.

Bjerve, K.S., Lovold Mostad, I., and Thorensen, L. (1987). Alpha-linolenic acid deficiency in patients on long term gastric tube feeding: Estimation of linolenic acid and long chain unsaturated n-3 fatty acid requirement in man. *Am. J. Clin. Nutr.* 45, 66–77.

Bjerve, K.S., Thoresen, L., and Borsting, S. (1988). Linseed and cod liver oil induce rapid growth in a 7-year-old girl with n-3 fatty acid deficiency. *J. Parent. Enter. Nutr* 12, 521–525.

Bougnoux, P., Koscielny, S., Chajes, V., Descamps, P., Couet, C., and Calais, G. (1994). Alpha-linolenic acid content of adipose breast tissue: A host determinant of the risk of early metastasis in breast cancer. *Br. J. Cancer* 70, 330–334.

Bourre, J.-M., Francois, M., Youyou, A., Dumont, O., Piciotti, M., Pascal, G., and Durand, G. (1989). The effects of dietary alpha-linolenic acid on the composition of nerve membranes, enzymatic activity, amplitude of electrophysiological parameters, resistance to poisons, and performance of learning tasks in rats. *J. Nutr.* 119, 1880–1892.

Bourre, J.M., Pascal, G., Durand, G., Masson, M., Dumont, O., and Piciotti, M. (1984). Alterations in the fatty acid composition of rat brain cells (neurons, astrocytes, and oligodendrocytes) and of subcellular fractions (myelin, synaptosomes) induced by a diet devoid of n-3 fatty acids. *J. Neurochem.* 43, 342–348.

Bretillon, L., Chardigny, J.M., Sebedio, J.L., Noel, J.P., Scrimgeour, C.M., Fernie, C.L., Loreau, O., Gachon, P., and Beaufrere, B. (2001). Isomerization increases the postprandial oxidation of linoleic acid but not α-linolenic acid in men. *J. Lipid Red.* 42, 995–997.

Bryan, D.L., Hart, P., Forsyth, K., and Gibson, R. (2001). Incorporation of α-linolenic acid and linoleic acid into human respiratory epithelial cell lines. *Lipids* 36, 713–716.

Budowski, P., Trostler, N., Lupo, M., Vaisman, N., and Eldor, A. (1984). Effect of linseed oil ingestion on plasma lipid fatty acid composition and platelet aggregability in healthy volunteers. *Nutr. Res.* 4, 343–346.

Burr, G.O., and Burr, M.M. (1930). The nature and role of fatty acids essential in nutrition. *J. Biol. Chem.* 86, 587–621.

Bursztyn, P. (1987). Does dietary linolenic acid influence blood pressure? *Am. J. Clin. Nutr.* 45, 1541–1548.

Carlson, S.E., and Neuringer, M. (1999). Polyunsaturated fatty acid status and neurodevelopment: A summary and critical analysis of the Literature. *Lipids* 34, 171–178.

Carlson, S.E., Rhodes, P.G., Rao, V.S., and Goldgar, D.E. (1987). Effect of fish oil supplementation on the n-3 faty acid content of red blood cell membranes in preterm infants. *Pediatr. Res.* 21, 507–510.

Carlson, S.E., Werkman, S.H., Rhodes, P.G., and Tolley, E.A. (1993). Visual acuity development in healthy preterm infants: Effect of marine-oil supplementation. *Am. J. Clin. Nutr.*
58, 35–42.

Carlson, S.E., Werkman, S., and Tolley, E. (1996). Effect of long chain n-3 supplementation
on visual acuity and growth in preterm infants with and without bronchopulmonary
dysplasia. *Am. J. Clin. Nutr.* **63**, 687–697.

Carnielli, V.P., Wattimena, D.J.L., Luijendijk, I.H.T., Boerlage, A., Degenhart, H.J., and Sauer,
P.J.J. (1996). The very low birth weight premature infant is capable of synthesizing
arachidonic and docosahexaenoic acid from linoleic and α-linolenic acids. *Pediatr. Res.* **40**,
169–171.

Chan, J.K., Bruce, V.M., and McDonald, B.E. (1991). Dietary α-linolenic acid is as effective
as oleic acid and linoleic acid in lowering blood cholesterol in normolipemic men. *Am. J.
Clin. Nutr.* **53**, 1230–1234.

Chan, J.K., McDonald, B.E., Gerrard, J.M., Bruce, V.M., Weaver, B.J., and Holub, B.J. (1993).
Effect of dietary α-linolenic acid and its ratio to linoleic acid on platelet and plasma fatty
acids and thrombogenesis. *Lipids* **28**, 811–817.

Chen, Z.-Y., and Cunnane, S.C. (1993). Refeeding after fasting increases apparent oxidation of
n-6 and n-3 fatty acids in pregnant rats. *Metabolism* **42**, 1206–1211.

Chen, Z.-Y., Menard, C.R., and Cunnane, S.C. (1995). Moderate, selective depletion of
linoleate and alpha-linolenate in weight-cycled rats. *Am. J. Physiol.* **268**, 498–505.

Chen, Z.-Y., Ratnayake, W.M.N., and Cunnane, S.C. (1994). Oxidative stability of flaxseed
lipids during baking. *J. Am. Oil Chem. Soc.* **71**, 629–632.

Christensen, J.H., Christensen, M.S., Toft, E., Dyerberg, J., and Schmidt, E.B. (2000). α-linolenic
acid and heart rate variability. *Nutr. Metab. Cardiovasc. Dis.* **10**, 57–61.

Clandinin, M.T., Chappell, J.E., Leong, S., Heim, T., Swyer, P.R., and Chance, G.W. (1980).
Extrauterine fatty acid accretion in infant brain: Implications for fatty acid requirements.
Early Hum. Dev. **4**, 131–138.

Conner, W.E. (1999). α-linolenic acid in health and disease. *Am. J. Clin. Nutr.* **69**, 827–828.

Crawford, M.A. (1993). The role of essential fatty acids in neural development: Implications
for perinatal nutrition. *Am. J. Clin. Nutr.* **57 (Suppl.)**, 703S–709S.

Crawford, M.A., Bloom, M., Broadhurst, C.L., Schmidt, W.F., Cunnane, S.C., Galli, C.,
Gehbremeskel, K., Linseisen, F., Lloyd-Smith, J., and Parkington, J. (1999). Evidence for
the unique function of docosahexaenoic acid during the evolution of the modern hominid
brain. *Lipids* **34**, S1–9.

Crawford, M.A., and Sinclair, A.J. (1972). The limitations of whole tissue analysis to define
linolenic acid deficiency. *J. Nutr.* **102**, 1315–1322.

Cunnane, S.C. (1991). Third Toronto Essential Fatty Acid Workshop on α-linolenic acid in
human nutrition and disease. *Nutrition* **7**, 435–446.

Cunnane, S.C. (1995). Metabolism and function of α-linolenic acid in humans. In Cunnane,
S.C., and Thompson, L.U. (eds), *Flaxseed in Human Nutrition*, AOCS Press, Champaign, Il,
pp. 99–127.

Cunnane, S.C., and Anderson, M.J. (1997). Majority of dietary linoleate in growing rats is
β-oxidized or stored in visceral fat. *J. Nutr.* **127**, 146–152.

Cunnane, S.C., Belza, K., Anderson, M.J., and Ryan, M.A. (1998). Substantial carbon recycling
from linoleate into products of de novo lipogenesis occurs in rat liver even under conditions
of extreme dietary linoleate deficiency. *J. Lipid Res.* **39**, 2271–2276.

Cunnane, S.C., Francescutti, V., Brenna, J.T., and Crawford, M.A. (2000). Whole body fatty
acid accumulation data suggest that without preformed dietary docosahexaenoate, the formula fed infant cannot meet its docosahexaenoate requirement. *Lipids* **35**, 105–111.

Cunnane, S.C., Ganguli, S., Menard, C., Liede, A.C., Hamadeh, M.J., Chen, Z.-Y., Wolever,
T.M.S., and Jenkins, D.J.A. (1993a). High α-linolenic acid flaxseed (*Linum usitatissimum*):
Some nutritional properties in humans. *Br. J. Nutr.* **49**, 443–453.

Cunnane, S.C., Hamadeh, M.J., Liede, A.C., Thompson, L.U., Wolever, T.M.S., and Jenkins, D.J.A. (1995a). Nutritional attributes of traditional flaxseed in healthy young adults. *Am. J. Clin. Nutr.* 61, 62–68.

Cunnane, S.C., McAdoo, K.R., and Horrobin, D.F. (1987). Iron intake influences essential fatty acids and lipid composition of rat plasma and erythrocytes. *J. Nutr.* 117, 1514–1519.

Cunnane, S.C., Menard, C.R., Likhodii, S.S., Brenna, J.T., and Crawford, M.A. (1999a). Carbon recycling into de novo lipogenesis is a major pathway in neonatal metabolism of linoleate and α-linolenate. *Prost. Leuk. Essent. Fatty Acids* 60, 387–392.

Cunnane, S.C., Ross, R., Bannister, J.L., and Jenkins, D.J.A. (1999b). Magnetic resonance imaging-based balance analysis of linoleate utilization during weight loss in obese humans. *Lipids* 34, S89–S90.

Cunnane, S.C., Ross, R., Bannister, J.L., and Jenkins, D.J.A. (2001). β-oxidation of linolate in obese men undergoing weight loss. *Am. J. Clin. Nutr.* 73, 713–716.

Cunnane, S.C., Ryan, M.A., Craig, K.A., Brookes, S., Koletzko, B., Demmelmair, H., Sunger, J., and Kyle, D.J. (1995b). Synthesis of linoleate and α-linolenate by chain elongation in the rat. *Lipids* 20, 781–783.

Cunnane, S.C., and Thompson, L.U. (1995). *Flaxseed in Human Nutrition*. AOCS Press. Champaign, IL. 384 p.

Cunnane, S.C., Williams, S.C.R., Bell, J.D., Brookes, S., Craig, K., Iles, R.A., and Crawford, M.A. (1994). Utilization of uniformly labeled ^{13}C-polyunsaturated fatty acids in the synthesis of long chain fatty acids and cholesterol accumulating in the neonatal rat brain. *J. Neurochem.* 62, 2429–2436.

Cunnane, S.C., and Yang, J. (1995). Zinc deficiency impairs whole-body accumulation of polyunsaturates and increases the utilization of [1-^{14}C] linoleate for de novo lipid synthesis in pregnant rats. *Can. J. Physiol. Pharmacol.* 73, 1246–1252.

Cunnane, S.C., Yang, J., and Chen, Z.-Y. (1993b). Low zinc intake increases apparent oxidation of linoleic and α-linolenic acids in the pregnant rat. *Can. J. Physiol. Pharmacol.* 71, 205–210.

Cuthbertson, W.J.F. (1999). Evolution of infant nutrition. *Br. J. Nutr.* 81, 359–371.

de Deckere, E.A.M., Korver, O., Verschuren, P.M., and Katan, M.B. (1998). Health aspects of fish and n-3 polyunsaturated fatty acids from plant and marine origin. *Eur. J. Clin. Nutr.* 52, 749–753.

de Lorgeril, M., Renaud, S., Mamelle, N., Salen, P., Martin, J.-L., Monjaud, I., Guidollet, J., Touboul, P., and Delaye, J. (1994). Mediterranean alpha-linolenic acid-rich diet in secondary prevention of coronary heart disease. *The Lancet* 343, 1454–1459.

De Stefani, E., Deneo-Pellegrini, H., Boffetta, P., Ronco, A., and Mendilaharsu, M. (2000). α-linolenic acid and risk of prostate cancer: A case-control study in Uruguay. *Cancer Epidemiol. Biomarkers Prev.* 9, 335–338.

De Stefani, E., Deneo-Pellegrini, H., Mandilaharsu, M., and Ronco, A. (1998). Essential fatty acids and breast cancer: A case-control study in Uruguay. *Br. J. Cancer* 76, 491–494.

de Vries, P.M.J.M., Folkers, H.B.M., de Fijter, C.W.H., van der Meulen, J., van der Veen, M.L., Popp-Snijders, C., and Oe, L.P. (1991). Adipose tissue fatty acid composition and its relation to diet plasma lipid concentrations in hemodialysis patients. *Am. J. Clin. Nutr.* 53, 469–473.

Dhopeshwarker, G.A., and Subramanian, C. (1975). Metabolism of linolenic acid in the developing brain. Incorporation of radioactivity from [1-^{14}C] linolenic acid into brain fatty acids. *Lipids* 10, 230–241.

Djousse, L., Pankow, J.S., and Eckfeldt, H.J. (2001). Relation between dietary linoleic acid and coronary heart disease in the National Heart, Lung and Blood Institute Family Heart Study. *Am. J. Clin. Nutr.* 74, 612–619.

Dolecek, T.A. (1992). Epidemiological evidence of relationships between dietary polyunsaturated fatty acids and mortality in the multiple risk factor intervention trial. *Proc. Soc. Exp. Biol. Med.* 200, 177–182.

Dratz, E.A., and Deese, A.J. (1986). The role of docisahexaenoic acid (22:6ω3) in biological membranes: Examples from photoreceptors and model membrane bilayers. In Simopoulos, A.P., Kifer, R.R., and Martin, R.E. (eds), *Health Effects of Polyunsaturated Fatty Acids in Seafoods.* Academic Press, Orlando, FL, pp. 319–351.

Duchen, K., Casas, R., Fageras-Bottcher, M., Yu, G., and Bjorksten, B. (2000). Human milk polyunsaturated long chain fatty acids and secretory immunoglobulin A antibodies and early childhood allergy. *Pediatr. Allergy Immunol.* 11, 29–39.

Dyerberg, J., Bang, H.O., and Aagaard, O. (1980). Alpha-linolenic acid and eicosapentaenoic acid. *Lancet* 26, 199.

Farquharson, J., Cockburn, R., Patrick, W.A., Jamieson, E.C., and Logan, R.W. (1992). Infant cerebral cortex phospholipid fatty acid composition and diet. *Lancet* 340, 810–813.

Ferrier, L.K., Caston, L.J., Leeson, S., Squires, J., Weaver, B.J., and Holub, B.J. (1995). α-linolenic acid- and docosahexaenoic acid-enriched eggs from hens fed flaxseed: Influence on blood lipids and platelet phospholipid fatty acids in humans. *Am. J. Clin. Nutr.* 62, 81–86.

Foote, K.D., Hrboticky, N., Mackinnon, M.J., and Innis, S.M. (1990). Brain synaptosomal, liver, plasma, and red blood cell lipids in piglets fed exclusively on a vegetable-oil-containing formula with and without fish-oil supplements. *Am. J. Clin. Nutr.* 51, 1001–1006.

Freese, R., and Mutanen, M. (1997). α-linolenic acid and marine long-chain n-3 fatty acids differ only slightly in their effects on hemostatic factors in healthy subjects. *Am. J. Clin. Nutr.* 66, 591–598.

Garaulet, M., Perez-Liamas, F., Perez-Ayala, M., Martinez, P., de Medina, F.S., Tebar, F.J., and Zamora, S. (2001). Site-specific differences in the fatty acid composition of abdominal adipose tissue in an obese population from a Mediterranean area: Relation with dietary fatty acids, plasma lipid profile, serum insulin, and central obesity. *Am. J. Clin. Nutr.* 74, 585–591.

Gavino, G.R., and Gavino, V.C. (1991). Rat liver outer mitochondrial carnitine palmitoyltransferase activity towards long-chain polyunsaturated fatty acids and the CoA esters. *Lipids* 26, 266–270.

Gerster, H. (1998). Can adults adequately convert α-linolenic acid (18:3n-3) to eicosapentaenoic acid (20:5n-3) and docosahexaenoic acid (22:6n-3)? *Int. J. Vitamin Nutr. Res.* 68, 159–173.

Giovannucci, E., Rimm, E.B., Colditz, G.A., Stampfer, M.J., Ascherio, A., Chute, C.C., and Willett, W.C. (1993). A prospective study of dietary fat and risk of prostate cancer. *J. Nat. Cancer Inst.* 85, 1571–1579.

Guallar, E., Aro, A., Jimenez, F.J., Martin-Moreno, J.M., Salminen, I., van't Veer, P., Kardinaal, A.F., Gomez-Aracena, J., Martin, B.C., Kohlmeier, L., Kark, J.D., Mazaev, V.P., Ringstad, J., Guillen, J., Rienersma, R.A., Huttunen, J.K., Thamm, M., and Kok, F.J. (1999). Omega-3 fatty acids in adipose tissue and risk of myocardial infarction: The EURAMIC study. *Arterioscler. Thromb. Vasc. Biol.* 19, 1111–1118.

Haggarty, P., Page, K., Abramovich, D.R., Ashton, J., and Brown, D. (1997). Long-chain polyunsaturated fatty acid transport across the perfused human placenta. *Placenta* 18, 635–642.

Hamosh, M., and Salem, N. (1998). Long chain polyunsaturated fatty acids. *Biol. Neonate* 74, 106–120.

Harper, C.R., and Jacobson, T.A. (2001). The fats of life: The role of omega 3 fatty acids in the prevention of coronary heart disease. *Arch. Intern. Med.* 161, 2185–2192.

Holman, R.T. (ed.) (1971). *Progress in the Chemistry of Fats and Other Lipids.* Pergamon Press, Oxford.

Holman, R.T., Johnson, S.B., and Hatch, T.F. (1982). A case of human linolenic acid deficiency involving neurological abnormalities. *Am. J. Clin. Nutr.* 35, 617–623.

Hu, F.B., Stampfer, M.J., Manson, J.E., Rimm, E.B., Wolk, A., Colditz, G.A., Hennekens, C.H., and Willett, W.C. (1999). Dietary intake of α-linolenic acid and risk of fatal ischemic heart disease among women. *Am. J. Clin. Nutr.* 69, 890–897.

Hunter, E.J. (1990). n-3 fatty acids from vegetable oils. *Am. J. Clin. Nutr.* 51, 809–814.

Ikeda, I., Cha, J.Y., Yanagita, T., Nakatani, N., Oogami, K., Imaizumi, K., and Yazawa, K. (1998). Effects of dietary alpha-linolenic, eicosapentaenoic and docosahexaenoic acids on hepatic lipogenesis and beta-oxidation in rats. *Biosci. Biotech. Biochem.* 62, 675–680.

Imbusch, R., and Mueller, M.J. (2000). Formation of isoprostane F(2)-like compounds (phytoprostanes F(1)) from α-linolenic acid in plants. *Free Radical Biol. Med.* 28, 720–726.

Jenkins, D.J., Kendall, C.W., Vidgen, E., Agarwal, S., Rao, A.V., Rosenberg, R.S., Diamandis, E.P., Novokmet, R., Mehling, C.C., Perera, T., Griffin, L.C., and Cunnane, S.C. (1999). Health aspects of partially defatted flaxseed, including effects on serum lipids, oxidative measures, and ex vivo androgen and progestin activity: A controlled crossover trial. *Am. J. Clin. Nutr.* 69, 395–402.

Jensen, C.L., Prager, T.C., Fraley, J.K., Chen, H., Anderson, R.E., and Heird, W.C. (1997). Effect of dietary linoleic/alpha-linolenic acid ratio on growth and visual function of term infants. *J. Pediatr.* 131, 200–209.

Johnston, P.V. (1986). Linolenate metabolism. *Nutr. Rev.* 44, 315–316.

Johnston, P.V. (1995). Flaxseed oil and cancer: α-linolenic acid and carcinogenesis. In Cunnane, S.C., and Thompson, L.U. (eds), *Flaxseed in Human Nutrition*, AOCS Press, Champaign, IL, pp. 207–218.

Kang, J.X., and Leaf, A.A. (1996). Antiarrhythmic effects of polyunsaturated fatty acids. *Circulation* 94, 1774–1780.

Kankaanpaa, P.E., Salminen, S.J., Isolauri, E., and Lee, Y.K. (2001). The influence of polyunsaturated fatty acids on probioltic growth and adhesion. *FEMS Microbiol. Lett.* 194, 149–153.

Kelley, D.S., Branch, L.B., Love, J.E., Taylor, P.C., Rivera, Y.M., and Iacono, J.M. (1991). Dietary α-linolenic acid and immunocompetence in humans. *Am. J. Clin. Nutr.* 53, 40–46.

Kelley, D.S., Nelson, G.J., Love, J.E., Branch, L.B., Taylor, P.C., Schmidt, P.C., Mackey, B.E., and Iacono, J.M. (1993). Dietary α-linolenic acid alters tissue fatty acid composition, but not blood lipids, lipoproteins or coagulation status in humans. *Lipids* 28, 533–537.

Kestin, M., Clifton, P., Belling, G.B., and Nestel, P.J. (1990). n-3 fatty acids of marine origin lower systolic blood pressure and triglycerides but raise LDL cholesterol compared with n-3 and n-6 fatty acids from plants. *Am. J. Clin. Nutr.* 51, 1028–1034.

Klein, V., Chajes, V., Germain, E., Schulgen, G., Pinault, M., Malvy, D., Lefrancq, T., Fignon, A., Le Floch, O., Lhuillery, C., and Bougnoux, P. (2000). Low α-linolenic acid content of adipose breast tissue is associated with an increased risk of breast cancer. *Eur. J. Cancer* 36, 335–340.

Koletzko, B. (1987). Omega-3 fatty acid requirement. *Am. J. Clin. Nutr.* 46, 374–377.

Koletzko, B., and Braun, B. (1991). Arachidonic acid and early human growth: Is there a relation? *Ann. Nutr. Metab.* 35, 128–131.

Koletzko, B., and Cunnane, S.C. (1988). Human alpha-linolenic acid deficiency. *Am. J. Clin. Nutr.* 47, 1084–1087.

Koletzko, B., Rodriquez-Palmero, M., Demmelmair, H., Fidler, M., Jensen, R., and Sauerwald, T. (2001). Physiological aspects of human milk lipids. *Early Hum. Dev.* 65, S3–S18.

Kremer, J.M. (1996). Effects of modulation of inflammatory and immune parameters in patients with rheumatic and inflammatory disease receiving dietary supplementation of n-3 and n-6 fatty acids. *Lipids* 31, S243–S247.

Kris-Etherton, P.M., Taylor, D.S., Yu-Poth, S., Huth, P., Moriarty, K., Fishell, V., Hargrove, R.L., Zaho, G., and Etherton, T.D. (2000). Polyunsaturated fatty acids in the food chain in the United States. *Am. J. Clin. Nutr.* 71, 179S–188S.

Kumamoto, T., and Ide, T. (1998). Comparative effects of alpha- and gamma-linolenic acids on rat liver fatty acid oxidation. *Lipids* 33, 647–654.

Lamptey, M.S., and Walker, B.L. (1976). A possible essential role for dietary linolenic acid in the development of the young rat. *J. Nutr.* 106, 86–93.

Lands, W.E.M. (2001). Impact of daily food choices on health promotion and disease prevention. In Hamazaki, T., and Okuyama, H. (eds), *Fatty Acids and Lipids – New Findings*, Karger SA, Basel, pp. 1–5.

Lauritzen, I., Blondeau, N., Heurteaux, C., Widmann, C., Romey, G., and Lazdunski, M. (2000). Polyunsaturated fatty acids are potent neuroprotectors. *EMBO J.* 19, 1784–1793.

Layne, K.S., Goh, Y.K., Jumpsen, J.A., Ryan, E.A., Chow, P., and Clandinin, M.T. (1996). Normal subjects consuming physiological levels of 18:3(n-3) and 20:5(n-3) from flaxseed or fish oils have characteristic differences in plasma lipid and lipoprotein fatty acid levels. *J. Nutr.* 126, 2130–2140.

Leaf, A.A., Leighfield, M.J., Costeloe, K.L., and Crawford, M.A. (1992). Factors affecting long-chain polyunsaturated fatty acid composition of plasma choline phosphoglycerides in preterm infants. *J. Pediatr. Gastroenterol. Nutr.* 14, 300–308.

Lewis, N.M., Widga, A.C., Buck, J.S., and Frederick, A.M. (1995). Survey of omega-3 fatty acids in diets of midwest low-income pregnant women. *J. Agromed.* 2, 49–57.

Leyton, J., Drury, P.J., and Crawford, M.A. (1987). Differential oxidation of saturated and unsaturated fatty acids *in vivo* in the rat. *Br. J. Nutr.* 57, 383–393.

Li, D., Sinclair, A., Wilson, A., Nakkote, S., Kelly, F., Abedin, L., Mann, N., and Turner, A. (1999). Effect of dietary alpha-linolenic acid on thrombotic risk factors in vegetarian men. *Am. J. Clin. Nutr.* 69, 872–882.

Life Sciences Research Office Report (1998). *Assessment of Nutrient Requirements for Infant Formulas.* 2059S–1078S p.

Lucas, A., Morley, R., Cole, T.J., Lister, G., and Leeson-Payne, C. (1992). Breast milk and subsequent intelligence quotient in children born preterm. *Lancet* 339, 261–264.

Lucas, A., Stafford, M., and Morley, R. (1999). Efficacy and safety of long-chain polyunsaturated fatty acid supplementation of infant-formula milk: A randomized trial. *Lancet* 354, 1948–1954.

Makrides, M., Neumann, M.A., Byard, R.W., Simmer, K., and Gibson, R.A. (1994). Fatty acid composition of brain, retina, and erythrocytes in breast- and formula-fed infants. *Am. J. Clin. Nutr.* 60, 189–194.

Makrides, M., Neumann, M., Simmer, K., Pater, J., and Gibson, R. (1995). Are long chain polyunsaturated fatty acids essential nutrients in infancy? *Lancet* 345, 1463–1468.

Mantzioris, E., James, M.J., Gibson, R.A., and Cleland, L.G. (1994). Dietary substitution with an α-linolenic acid-rich vegetable oil increases eicosapentaenoic acid concentrations in tissues. *Am. J. Clin. Nutr.* 59, 1304–1309.

Martinez, M. (1992). Tissue levels of polyunsaturated fatty acids during early human development. *J. Pediatr.* 120, 129S–138S.

Menard, C.R., Goodman, K., Corso, T., Brenna, J.T., and Cunnane, S.C. (1998). Recycling of carbon into lipids synthesized de novo is a quantitatively important pathway of [U-13C]-α-linolenate utilization in the developing rat brain. *J. Neurochem.* 71, 2151–2158.

Michaud, D.S., Augustsson, K., Rimm, E.B., Stampfer, M.J., Willett, W.C., and Giovannucci, E. (2001). A prospective study on intake of animal products and risk of prostate cancer. *Cancer Causes Control* 12, 557–567.

Morley, R. (1998). Nutrition and cognitive development. *Nutrition* 14, 752–754.

Neuringer, M., Connor, W.E., van Petten, C., and Barstad, L. (1984). Dietary omega-3 fatty acid deficiency and visual loss in infant rhesus monkeys. *J. Clin. Invest.* 73, 272–276.

Newcomer, L.M., King, I.B., Wicklund, K.G., and Stanford, J.L. (2001). The association of fatty acids with prostate cancer risk. *Prostate* 47, 262–268b.

Nordstrom, D.C.E., Honkanen, V.E.A., Nasu, Y., Antila, E., Friman, C., and Konttinen, Y.T. (1995). Alpha-linolenic acid in the treatment of rheumatoid arthritis. A double-blind, placebo-controlled and randomized study: Flaxseed vs. safflower seed. *Rheumatol. Int.* 14, 231–234.

O'Conner, D.L., Hall, R., Adamkin, D., Auestad, N., Castillo, M., Conner, W.E., Conner, S.L., Fitzgerald, K., Groh-Wargo, S., Hartmann, E.E., Jacobs, J., Janowsky, J., Lucas, A.,

Margeson, D., Mena, P., Neuringer, M., Neisen, M., Singer, L., Stephenson, T., Szabo, J., and Zemon, V. (2001). Growth and development in preterm infants fed long chain polyunsaturated fatty acids: A prospective, randomized controlled clinical trial. *Pediatrics* 108, 359–371.

Okamoto, M., Mitsunobu, F., Ashida, K., Mifune, T., Hosaki, Y., Tsugeno, H., Harada, S., and Tanizaki, Y. (2000). Effects of dietary supplementation with n-3 fatty acids compared with n-6 fatty acids on bronchial asthma. *Intern. Med.* 39, 107–111.

Okayasu, T., Nagao, M., Ishibashi, I., and Imai, Y. (1981). Purification and partial characterization of linoleoyl-CoA desaturase from rat liver microsomes. *Arch. Biochem. Biophys.* 206, 21–28.

Okuno, M., Kajiwara, K., Imai, S., Kobayashi, T., Honma, N., Maki, T., Suruga, K., Goda, T., Takase, S., Muto, Y., and Moriwaki, H. (1997). Perilla oil prevents the excessive growth of visceral adipose tissue in rats by down-regulating adipocyte differentiation. *J. Nutr.* 127, 1752–1757.

Oomen, C.M., Ocke, M.C., Feskens, E.J., Kok, F.J., and Kromhout, D. (2001). α-linolenic acid intake is not beneficially associated with 10 year risk of coronary artery disease incidence: The Zutphen Elderly Study. *Am. J. Clin. Nutr.* 74, 457–463.

Owren, P.A., Hellem, A.J., and Ödegaard, A. (1964). Linolenic acid for the prevention of thrombosis and myocardial infarction. *Lancet* 13, 975–978.

Palowsky, R.J., Hibbeln, J.R., Novotny, J.A., and Salem, N. (2001). Physiological compartmental analysis of α-linolenic acid metabolism in adult humans. *J. Lipid Red.* 42, 1257–1265.

Pang, D., Allman-Farinelli, M.A., Wong, T., Barnes, R., and Kingham, K.M. (1998). Replacement of linoleic acid with alpha-linolenic acid does not alter blood lipids in normolipidaemic men. *Br. J. Nutr.* 80, 163–167.

Periera, C., Li, D., and Sinclair, A.J. (2001). The α-linolenic acid content of green vegetables commonly available in Australia. *Int. J. Vitamin Nutr. Res.* 71, 223–228.

Poumes-Balliaut, C., Langelier, B., Houlier, F., Alessandri, J.M., Durand, G., Latge, C., and Guesnet, P. (2001). Comparative bioavailability of dietary α-linolenic acid and docosahexaenoic acids in the growing rat. *Lipids* 36, 793–800.

Prasad, K., Mantha, S.V., Kalra, J., Muir, A.D., and Westcott, N.D. (1998). Reduction of hypercholerterolemic atherosclerosis by CDC-flaxseed with very low α-linolenic acid. *Atherosclerosis* 136, 367–375.

Renaud, S., and Nordøy, A. (1983). "Small is beautiful": α-linolenic acid and eicosapentaenoic acid in man. *Lancet* 1 (8834), 1169.

Ringborn, T., Huss, U., Stenholm, A., Flock, S., Skattebol, L., Perera, P., and Bohlin, L. (2001). Cox-2 inhibitory effects of naturally occurring and modified fatty acids. *J. Natural Prod.* 64, 745–749.

Romans, J., Wulf, D.M., Johnson, R.C., Libal, G.W., and Costello, W.J. (1995). Flaxseed and the composition and quality of pork. In Cunnane, S.C., and Thompson, L.U. (eds), *Flaxseed in Human Nutrition*, AOCS Press, Champaign, Illinois, pp. 348–362.

Salem, N., Wegher, B., Mena, P., and Uauy, R. (1996). Arachidonic and docosahexaenoic acids are biosynthesized from their 18 carbon precursors in human infants. *Proc. Natl. Acad. Sci. USA* 93, 49–54.

Salonen, J.T., Salonen, R., Ihanainen, M., Parviainen, M., Seppanen, R., Kantola, M., Seppanen, K., and Rauramaa, R. (1988). Blood pressure, dietary fats, and antioxidants. *Am. J. Clin. Nutr.* 48, 1226–1232.

Sanders, T.A.B. (2000). Polyunsaturated fatty acids in the food chain in Europe. *Am. J. Clin. Nutr.* 71, 176S–178S.

Sanders, T.A.B., and Roshanai, F. (1983). The influence of different types of ω3 unsaturated fatty acids on blood lipids and platelet function in healthy volunteers. *Clin. Sci.* 64, 91–99.

Sarabi, M., Vessby, B., Millgard, J., and Lind, L. (2001). Endothelium-dependent vasodilation is related to the fatty acid composition of serum lipids in healthy subjects. *Atherosclerosis* 156, 349–355.

Sauerwald, T.V., Hachey, D.L., and Jensen, C. (1997). Intermediates in endogenous synthesis of 22:6n-3 by term and preterm infants. *Pediatr. Res.* 41, 183–186.

Shapiro, J.A., Koepsell, T.D., Voigt, L.F., Dugowson, C.E., Kestin, M., and Nelson, J.L. (1996). Diet and rheumatoid arthritis in women: A possible protective effect of fish consumption. *Epidemiology* 7, 256–263.

Sheaff-Greiner, R.C., Zhang, Q., Goodman, K.J., Guissani, D.A., Nathanielsz, P.W., and Brenna, J.T. (1996). Linoleate, α-linolenate and docosahexaenoate recycling into saturated and monounsaturated fatty acids is a major pathway in pregnant or lactating adults and fetal or infant rhesus monkeys. *J. Lipid Red.* 137, 243–254.

Sim, J.S., and Qi, G.-H. (1995). Designing poultry products using flaxseed. In Cunnane, S.C., and Thompson, L.U. (eds), *Flaxseed in Human Nutrition*, AOCS Press, Champaign, IL, pp. 315–333.

Sinclair, A.J. (1975). Incorporation of radioactive polyunsaturated fatty acids into liver and brain of the developing rat. *Lipids* 10, 175–184.

Sinclair, A.J., and Crawford, M.A. (1972). Accumulation of arachidonate and docosahexaenoate in the developing rat brain. *J. Neurochem.* 19, 1753–1758.

Singer, P., Jaeger, W., Berger, I., Barleben, H., Wirth, M., Richter-Heinrich, E., Voigt, S., and Godicke, W. (1990). Effects of dietary oleic, linoleic, and alpha-linolenic acids on blood pressure, serum lipids, lipoproteins, and the formation of eicosanoid precursors in patients with mild essential hypertension. *J. Hum. Hypertens* 4, 227–233.

Sodergren, E., Gustafsson, I.B., Basu, I., Nourooz-Zadeh, J., Nalsen, C., Turpeinen, A., Berglund, L., and Vessby, B. (2001). A diet containing rapeseed oil-based fats does not increase lipid peroxidation in humans when compared to a diet rich in saturated fatty acids. *Eur. J. Clin. Nutr.* 55, 922–931.

Sprecher, H. (1968). The synthesis and metabolism of hexadeca-4,7,10-trienoate, eicosa-8,11,14-trienoate, docosa-10,13,16-trienoate, and docosa-6,9,12,15-tetraenoate in the rat. *Biochim. Biophys. Acta* 1, 519–530.

Su, H.M., Bernardo, L., Mirmiran, M., Ma, X.H., Corso, T.N., Nathanielsz, P.W., and Brenna, J.T. (1999). Bioequivalence of dietary α-linolenic and docosahexaenoic acids as sources of docosahexaenoate accretion in brain and associated organs of neonatal baboons. *Pediatr. Res.* 45, 1–7.

Thompson, L.U. (1995). Flaxseed, lignans and cancer. In Cunnane, S.C., and Thompson, L.U. (eds), *Flaxseed in Human Nutrition*, AOCS Press, Champaign, IL, pp. 219–236.

Tinoco, J. (1982). Dietary requirements and functions of α-linolenic acid in animals. *Prog. Lipid Res.* 21, 1–38.

Vatten, L.J., Bjerve, K.S., Andersen, A., and Jellum, E. (1993). Polyunsaturated fatty acids in serum phospholipids and risk of breast cancer: Case control study from the Janus serum bank in Norway. *Eur. J. Cancer* 29A, 532–538.

Vermunt, S.H., Beaufrere, B., Riemersma, R.A., Sebedio, J.L., Chardigny, J.M., Mensink, R.P., and TransLinE Investigators (2001). Dietary trans-α-linolenic acid from deodorized rapeseed oil and plasma lipids and lipoproteins in health men: The TransLinE Study. *Br. J. Nutr.* 85, 387–392.

Vermunt, S.H., Mensink, R.P., Simonis, M.M., and Hornstra, G. (2000). Effects of dietary α-linolenic acid on the conversion and oxidation of ^{13}C-α-linolenic acid. *Lipids* 35, 137–142.

Voskuil, D.W., Feskens, E.J.M., Katan, M.B., and Kromhout, D. (1996). Intake and sources of α-linolenic acid in Dutch elderly men. *Eur. J. Clin. Nutr.* 50, 784–787.

Wainwright, P.E., and Ward, G.R. (1997). Early nutrition and behaviour: A conceptual framework for critical analysis of research. In Dobbing, J. (ed.) *Developing Brain and Behaviour: The Role of Lipids in Infant Formula*. Academic Press, London, pp. 387–425.

Whelan, J., Broughton, K.S., and Kinsella, J.E. (1991). The comparative effects of dietary α-linolenic acid and fish oil on 4- and 5-series leukotriene formation *in vivo*. *Lipids* 26, 119–126.

Willard, D.E., Harman, S.D., Kaduce, T.L., Presuss, M., Moore, S.A., Robbins, M.E.C., and Spector, A.A. (2001). Docosahexaenoic acid synthesis from n-3 polyunsaturated fatty acids in differentiated rat brain astrocytes. *J. Lipid Red.* 42, 1368–1376.

Willatts, P., Forsyth, J.S., DiModugno, M.K., Varma, S., and Colvin, M. (1998). Influence of long-chain polyunsaturated fatty acids on infant cognitive function. *Lipids* 33, 973–980.

Woods, J., Ward, J., and Salem, N. (1996). Is DHA necessary in infant formula? Evaluation of high α-linolenic acid diets in the neonatal rat. *Pediatr. Res.* 40, 687–694.

Xiang, M., Lei, T., and Zetterstrom, R. (1999). Composition of long chain polyunsaturated fatty acids in human milk and growth of young infants in rural areas of northern China. *Acta Paediatr.* 88, 126–131.

Yang, J., and Cunnane, S.C. (1994). Quantitative measurements of dietary and [1-^{14}C] linoleate metabolism in pregnant rats: Specific influence of moderate zinc depletion independent of food intake. *Can. J. Physiol. Pharmacol.* 72, 1180–1185.

8 The role of flaxseed lignans in hormone-dependent and independent cancer

Sharon E. Rickard-Bon and Lilian U. Thompson

Introduction

Over the last 20 years, there has been a growing interest in the role of phytoestrogens – plant-derived compounds with estrogenic activity – in health and disease. The major classes of phytoestrogens examined are the isoflavones, coumestans, and lignans found at high levels in soybean, clover, and flaxseed, respectively (Rickard and Thompson, 1997). Lignans, the focus of this chapter, are diphenolic compounds that generally have a 2,3-dibenzylbutane skeleton (Ayers and Loike, 1990). They are believed to be by-products of the pathway for lignin synthesis (Setchell, 1995), and many have exhibited antibacterial, antifungal, and antimitotic activity (Ayers and Loike, 1990). These activities suggest that plant lignans are phytoalexins produced under stress which may play a role in the plant host-defense systems (Adlercreutz, 1998).

Much of the heightened interest in lignans came about with the discovery of mammalian lignans in the urine of humans, monkeys and rats (Setchell *et al.*, 1980a, 1980b; Stitch *et al.*, 1980; Setchell *et al.*, 1981a). Mammalian lignans differ from plant lignans in that the hydroxy substituents on the aromatic rings are in the *meta* position and not in the *para* position. The major mammalian lignans were identified as enterolactone (EL, MW = 298 g/mol) and enterodiol (ED, MW = 302 g/mol) by Setchell and colleagues in the early 1980s (Setchell *et al.*, 1981c). At first, EL and ED were believed to be of ovarian origin. This belief arose due to their cyclical pattern of urinary excretion during the menstrual cycle and increased urinary excretion during early pregnancy (Setchell *et al.*, 1980a, 1980b, 1981a). However, it has now become apparent that ED and EL are formed by colonic bacterial action (Axelson and Setchell, 1981; Setchell *et al.*, 1981b; Borriello *et al.*, 1985) on plant precursors present in the diet (Axelson *et al.*, 1982; Coert *et al.*, 1982; Thompson *et al.*, 1991). As shown in Figure 8.1, ED seems to be formed directly from the plant lignan secoisolariciresinol (SECO, MW = 362), whereas the plant lignan precursor for EL is matairesinol (MAT, MW = 358) (Borriello *et al.*, 1985). Enterolactone can also be formed by oxidation of ED. Both ED and EL undergo enterohepatic circulation and are excreted in the urine as glucuronide or sulphate conjugates (Axelson and Setchell, 1980; Adlercreutz *et al.*, 1995). Urinary lignan excretion has been used as an indicator of colonic lignan production and hence dietary lignan intake.

Flaxseed is the richest known source of ED and EL and is unique among plants in that the predominant plant lignan is the diglycoside of SECO (SDG, MW = 686) as opposed to MAT (Axelson *et al.*, 1982; Thompson *et al.*, 1991). Using extraction with enzymatic hydrolysis, the SDG content of full-fat flaxseed was determined to be 1.5 mg/g (0.8 mg SECO/g) (Obermeyer *et al.*, 1995), 2 mg/g (1 mg SECO/g) (Rickard *et al.*, 1996), or 3.8 mg/g (2 mg SECO/g) (Haggans *et al.*, 1999). In contrast, an SDG

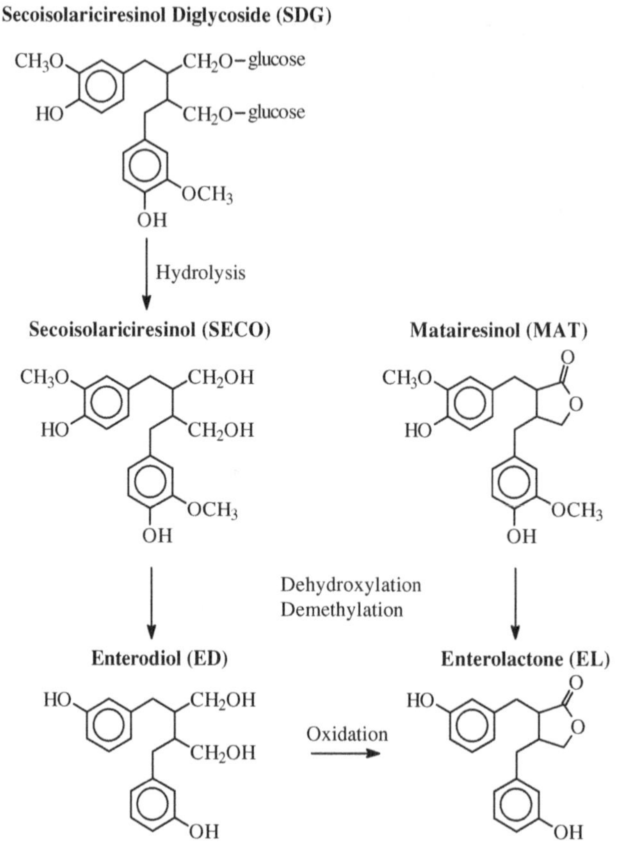

Secoisolariciresinol Diglycoside (SDG)

Hydrolysis

Secoisolariciresinol (SECO)   **Matairesinol (MAT)**

Dehydroxylation
Demethylation

Enterodiol (ED)   **Enterolactone (EL)**

Oxidation

Figure 8.1 Conversion of the plant lignans secoisolariciresinol diglycoside (SDG) and matairesinol (MAT) found in flaxseed to the mammalian lignans enterodiol (ED) and enterolactone (EL) by colonic bacteria.

level of 7.0 mg/g (3.7 mg SECO/g) (Mazur *et al.*, 1996; Westcott and Muir, 1996) has been determined in full-fat flaxseed using extraction with chemical hydrolysis. Although results involving enzymatic hydrolysis appear to be lower, they may be more representative of what may be available physiologically. The MAT content of full-fat flaxseed is estimated to be 11 μg/g (Mazur *et al.*, 1996).

In addition to being rich in mammalian lignans, flaxseed is one of the richest edible plant sources of omega-3 fatty acids with an α-linolenic acid (ALA) content of about 57 percent. Both lignans and ALA have exhibited anticarcinogenic effects *in vitro* (Hirano *et al.*, 1990; Mousavi and Adlercreutz, 1992; Grammatikos *et al.*, 1994; Chajes *et al.*, 1995; Wang and Kurzer, 1998).

The idea that flaxseed could have chemopreventive properties evolved from a small study where urinary lignan excretion, an indicator of dietary lignan intake, was found to be lower in women with breast cancer (Adlercreutz *et al.*, 1982). This observation then led to a series of experiments conducted *in vitro*, in animals, and in humans to gain a better understanding of the role of flaxseed and its lignans on different types of cancer. These studies (summarized in Table 8.1) will be reviewed in this chapter and the hormonal and non-hormonal mechanisms of lignans (summarized in Table 8.2)

Table 8.1 Studies examining the effect of flaxseed and its lignans on carcinogenesis*

Reference	Subjects/experimental design	Results
Breast cancer: *in vitro* studies		
Welshons et al., 1987	MCF-7 and T47D human breast cancer cells	↑ DNA synthesis with 1–10 μM EL ↓ DNA synthesis with 100 μM EL
Wang and Kurzer, 1997	MCF-7 and MDA-MB-231 human breast cancer cells	↑↓ DNA synthesis in MCF-7 cells with 10–50 μM EL ↓ DNA synthesis in MCF-7 cells with > 50 μM EL ↓ DNA synthesis in MDA-MB-231 cells with > 100 μM EL
Wang and Kurzer, 1998	MCF-7 human breast cancer cells	↑ DNA synthesis with 10 μM EL in presence of 0.01 nM estradiol
Mousavi and Adlercreutz, 1992	MCF-7 human breast cancer cells	↑ cell proliferation with 0.5–10 μM EL
Sathyamoorthy et al., 1994	MCF-7 human breast cancer cells	↓ cell proliferation with > 10 μM EL ↑ cell proliferation with 1 μM EL ED had no effect
Hirano et al., 1990	ZR-75-1 breast cancer cells	↓ cell proliferation with ED, EL, and various mammalian lignan derivatives at 30–33 μM
Chen and Thompson, 1999	MDA-MB-435 and MDA-MB-231 human breast cancer cells	↓ cell invasion at 1–5 μM ED or EL ↓ cell adhesion to ECM with ED (5 μM) or EL (1 μM) ↓ invasion and adhesion enhanced with tamoxifen with ED and EL
Breast cancer: animal studies		
Tou and Thompson, 1999	Female, Sprague-Dawley rats; 5% or 10% flaxseed, 1.82% flaxseed oil, or 1.5 mg SDG/d during pregnancy and lactation	→ number of mammary gland terminal end buds with flaxseed and SDG
Serraino and Thompson, 1991	Female Sprague-Dawley rats; 5% and 10% full-fat and defatted flaxseed from 21–50 days of age	→ mammary epithelial cell proliferation and nuclear aberration in terminal end buds and alveolar buds with full-fat flaxseed
Serraino and Thompson, 1992a	Female Sprague-Dawley rats; DMBA (5 mg/rat); 5% flaxseed exposure during initiation (21–50 days of age) and/or promotion stages of carcinogenesis (1–21 weeks post-DMBA)	→ tumor incidence and number of tumors/group with flaxseed at initiation stage → tumor size and tendency for ↑ tumor multiplicity with flaxseed during promotion stage → tumor multiplicity with flaxseed feeding throughout experimental period

Table 8.1 (Cont'd)

Reference	Subjects/experimental design	Results
Thompson et al., 1996a	Female Sprague-Dawley rats; DMBA (5 mg/rat); 2.5% or 5% SDG/d, and 1.5 mg SDG/d at late promotion/early progression stage of carcinogenesis (13–20 weeks post DMBA)	↓ established tumor volume with flaxseed, flaxseed oil, and SDG ↓ new tumor volume, average number of new tumors/group, and new tumor incidence with SDG
Thompson et al., 1996b	Female Sprague-Dawley rats; DMBA (5 mg/rat); 1.5 mg SDG/d at the promotion stage of carcinogenesis (1–20 weeks post-DMBA)	↓ number of tumors/group and tumor multiplicity with SDG trend to ↓ tumor volume with 5% flaxseed
Rickard et al., 1999	Female Sprague-Dawley rats; MNU (50 mg/kg BW); 2.5% and 5% flaxseed and equivalent doses of SDG (0.7 mg/d = LSDG and 1.4 mg/d = HSDG) at promotion stage of carcinogenesis (1–22 weeks post-MNU)	↓ tumor multiplicity with HSDG but ↑ tumor multiplicity with LSDG ↓ tumor invasiveness and grade with flaxseed or SDG

Breast cancer: clinical studies

Reference	Subjects/experimental design	Results
Adlercreutz et al., 1982	Observational study; postmenopausal women; omnivores, vegetarians, and breast cancer patients	↓ urinary EL in breast cancer patients compared to other two groups
Adlercreutz et al., 1986	Observational study; premenopausal women; omnivores, lactovegetarian, and macrobiotics	↑ urinary ED and EL in lactovegetarians and macrobiotics
Ingram et al., 1997	Case-control study; pre- and post-menopausal women	OR = 0.36 for breast cancer risk in highest quartile of urinary EL excretion

Colon cancer: in vitro studies

Reference	Subjects/experimental design	Results
Sung et al., 1998	LS174T, T84, Caco-2, and HCT-15 human colon cancer cells	↓ cell proliferation with 100 μM of ED or EL

Colon cancer: animal studies

Reference	Subjects/experimental design	Results
Serraino and Thompson, 1992b	Male Sprague-Dawley rats; AOM (15 mg/kg BW); 5% or 10% full-fat and defatted flaxseed for 4 weeks	↓ number of aberrant crypts and aberrant crypt foci with full-fat or defatted flaxseed ↓ cell proliferation in all groups except 5% defatted flaxseed

Reference	Model/treatment	Results
Jenab and Thompson, 1996	Male Sprague-Dawley rats; AOM (15 mg/kg BW); 2.5% or 5% full-fat and defatted flaxseed and 1.5 mg SDG/d for 100 days	↓ aberrant crypt multiplicity with full-fat or defatted flaxseed ↓ number of aberrant crypts and aberrant crypt multiplicity with SDG
Other cancers: *in vitro* studies Hirano et al., 1994	HL-60 human promyelocytic leukemia cells	↓ cell proliferation with 0.03–3 μM MAT
Other cancers: animal studies Yan et al., 1998	Male C57BL/6 mice, transplantable mouse melanoma cell line B16BL6; flaxseed supplementation (2.5%, 5%, or 10%)	↓ number and cross-sectional area of tumors that metastasized to the lung
Other cancers: animal studies Li et al., 1999	Male C57BL/6 mice, transplantable mouse melanoma cell line B16BL6; SDG supplementation equivalent to 2.5%, 5%, or 10% flaxseed (73, 147, and 293 μmol/kg diet, respectively)	↓ number and cross-sectional area of tumors that metastasized to the lung
Landstrom et al., 1998	Rats, transplantable Dunning R3327 PAP human prostate cancer cells; 33% rye bran and endosperm diets	delayed tumor development
Tou et al., 1998	Male Sprague-Dawley rats; 5% or 10% flaxseed, 1.82% flaxseed oil, and 1.5 mg SDG/d	↑ cell proliferation in prostate with 10% flaxseed ↓ cell proliferation in prostate with 5% flaxseed
Other cancers: clinical studies Morton et al., 1997	Observational study; British, Chinese, and Portuguese men	↑ prostatic fluid EL levels in Portuguese men with intermediate prostate cancer risk

Note

* Abbreviations: AOM = azoxymethane, DMBA = dimethylbenz(a)anthracene, ED = enterodiol, EL = enterolactone, ECM = extracellular matrix, OR = odds ratio, MAT = matairesinol, MNU = N-methyl-N-nitrosourea, and SDG = secoisolariciresinol diglycoside.

Table 8.2 Hormonal and non-hormonal activities of plant and mammalian lignans*

Reference	Biological effect	Effective concentration or level tested
Sathyamoorthy et al., 1994	Stimulation of pS2 mRNA expression in MCF-7 breast cancer cells in vitro	1 µM EL
Welshons et al., 1987	Stimulation of the progesterone receptor in MCF-7 breast cancer cells and immature rat uterine cells in vitro	EL, EC_{50} = 10 µM (MCF-7), 1–100 µM (uterine cells)
Welshons et al., 1987	Stimulation of prolactin synthesis in normal pituitary cells in vitro	ED, EC_{50} = 100 µM EL, EC_{50} = 100 µM ED, EC_{50} > 100 µM
Adlercreutz et al., 1993a	Inhibition of aromatase in placental microsomes in vitro	EL, IC_{50} = 14 µM didemethoxymatairesinol, IC_{50} = 6 µM
Adlercreutz et al., 1993a	Decreased estrone production via aromatase inhibition in JEG-3 human choriocarcinoma cells in vitro	1–100 µM EL
Wang et al., 1994	Inhibition of aromatase in preadipocytes in vitro	EL, IC_{50} = 74 µM 3'-demethoxy-3O-demethylmatairesinol, IC_{50} = 84 µM didemethoxymatairesinol, IC_{50} = 60 µM
Garreau et al., 1991	Inhibition of estrone and estradiol binding to rat alpha-fetoprotein (data for estradiol not given) in vitro	ED or EL, 0.5–50 µM
Adlercreutz et al., 1992a	Stimulation of SHBG synthesis in HepG2 human liver cancer cells in vitro	0.5–10 µM EL
Ganßer and Spiteller, 1995	Reduction in binding activity of dihydrotestosterone to SHBG in vitro	40% inhibition at 100 µM SECO
Martin et al., 1995	Inhibition of estradiol and testosterone binding to SHBG in vitro	EL, IC_{50} = 10 µM for estradiol displacement EL, IC_{50} = 40 µM for testosterone displacement
Shultz et al., 1991	No significant changes in plasma testosterone, free testosterone, or SHBG in males	13.5 g flaxseed/day
Phipps et al., 1993	No significant changes in plasma estradiol, estrone, DHEAS, SHBG, testosterone, or progesterone in premenopausal women; lengthened luteal phase of the menstrual cycle; increased luteal phase progesterone to estradiol ratio	10 g flaxseed/day
Hutchins et al., 1999	Decreased serum estradiol in postmenopausal women	5 or 10 g flaxseed/day
Tou et al., 1999	Increased serum estradiol levels in female rat offspring; increased serum testosterone and estradiol in male rat offspring	Lifetime exposure to 10% flaxseed
Rickard and Thompson, unpublished	Decreased serum estradiol in female rats	After weaning: 1.5 or 3.0 mg SDG/day; 5% or 10% flaxseed

Reference	Effect	Dose
Haggans et al., 1999	Increased urinary excretion of 2-hydroxy estradiol, 2-hydroxyestrone, and the 2/16α-hydroxyestrone ratio	10 g flaxseed/day
Evans et al., 1995	Reduction of 5α-reductase activity in genital skin fibroblasts and prostate tissue homogenates	At 100 μM, ED = 22% decrease and EL = 74% decrease (fibroblasts); IC_{50} for EL in prostate = 14 μM (70–80% inhibition at 40 μM, type I isozyme)
Evans et al., 1995	Reduction in 17-beta-hydroxysteroid dehydrogenase activity in genital skin fibroblasts	At 100 μM, ED = 79% decrease and EL = 98% decrease
Jenab and Thompson, 1996	Stimulation of specific and/or total activity of cecal beta-glucuronidase in AOM-treated male rats	2.5% defatted flaxseed and 5% defatted or full-fat flaxseed
Jenab et al., 1999	Stimulation of specific and total activity of cecal β-glucuronidase activity in non-carcinogen treated female rats	5% and 10% flaxseed, 1.5 and 3.0 mg SDG/day
Orcheson et al., 1998	Increased estrous cycle length and trend for higher number of irregular cycling/acyclic rats in persistant diestrus	5% flaxseed (cycle length); 1.5 mg or 3.0 mg SDG/d, 10% flaxseed (irregular cycling/acyclic rats)
Tou et al., 1998; Tou et al., 1999	Increased ovarian and/or uterine weight, early onset of puberty, increased estrous cycle length, and increased persistent estrus in female rat offspring; decreased sex gland and prostate weight in male rat offspring	Lifetime exposure or exposure during pregnancy and lactation to 10% flaxseed
Tou et al., 1998; Tou et al., 1999	Decreased ovarian weight, delayed puberty onset, and increased persistent diestrus in female rat offspring	Lifetime exposure to 5% flaxseed or exposure during pregnancy and lactation to 5% flaxseed or 1.5 mg SDG/day
Waters and Knowler, 1982	Depression of the stimulation of uterine RNA synthesis by estradiol	0.3, 3.0, or 30 μg EL/rat
Adlercreutz et al., 1992a	Binding to the nuclear type II estrogen receptor (bioflavonoid receptor) in rat uterine cells	10–100 μM MAT or EL; 100 μM ED
Kitts et al., 1999	Hydroxy radical scavenging activity; inhibition of DNA scission	10–100 μM ED, EL, and SDG; SDG at 36–2912 μM
Prasad, 1997	Hydroxy radical scavenging activity	
Yuan et al., 1999	Reduction in hepatic glutathione reductase activity	10% flaxseed and 3 mg SDG/day
Fotsis et al., 1993	Inhibition of bovine and human vascular endothelial cell proliferation	10–100 μM EL

Note

* Abbreviations: DHEAS = dehydroepiandrosterone sulphate, ED = enterodiol, EL = enterolactone, EC_{50} = concentration that was 50% effective, ECM = extracellular matrix, IC_{50} = concentration at which 50% inhibition was observed, MAT = matairesinol, SDG = secoisolariciresinol diglycoside, SECO = secoisolariciresinol, and SHBG = sex hormone binding globulin.

discussed. Although the focus of the discussion will be on the lignan component of flaxseed, the effects of its oil will also be included where appropriate.

Breast cancer

In vitro *studies*

The mammalian lignans have been shown to have stimulatory as well as inhibitory effects on indices of cell growth in breast cancer cells *in vitro*, depending on the concentrations used. Using DNA synthesis as a marker of cell growth, EL (1–10 µM) was found to be stimulatory in the estrogen-dependent breast cancer cell lines MCF-7 and T47D (Welshons *et al.*, 1987; Wang and Kurzer, 1997, 1998), but higher levels (> 50 µM) were inhibitory (Welshons *et al.*, 1987; Wang and Kurzer, 1997). In terms of cell proliferation, 0.5 to 10 µM EL was found to stimulate MCF-7 cells (Mousavi and Adlercreutz, 1992; Sathyamoorthy *et al.*, 1994), whereas concentrations above 10 µM were inhibitory (Mousavi and Adlercreutz, 1992). At a concentration of 10 µg/mL (approximately 30–33 µM), ED, EL, and various mammalian lignan derivatives were shown to inhibit cell proliferation of the estrogen-dependent cell line ZR-75-1 by 18–68 percent (Hirano *et al.*, 1990). Interestingly, Hirano and colleagues (1990) found that although the ED derivative called hattalin (2,3-dibenzylbutane-1,4-diol) inhibited cell proliferation, DNA synthesis was unaffected.

The stimulatory or inhibitory effects of the mammalian lignans may also be dependent on the presence of a stronger estrogen or antiestrogen. The stimulatory effects of the mammalian lignans were found to be 10^3–10^6 times weaker than the endogenous steroidal estrogen estradiol (Welshons *et al.*, 1987; Sathyamoorthy *et al.*, 1994). Mousavi and Adlercreutz (1992) observed that the combination of 1 nM estradiol and 1 µM EL, both having the same stimulatory effect on MCF-7 cell proliferation, resulted in a reduction in cell growth back to the levels in control cells. They suggested that in the presence of a stronger estrogen, the mammalian lignans inhibit rather than stimulate breast cancer cell growth. In contrast, a recent study by Wang and Kurzer (1998) showed that 10 µM EL stimulated DNA synthesis in MCF-7 cells despite the presence of 0.01 nM estradiol; other concentrations of EL (0.5–5 µM) or estradiol (0.1–1.0 nM) had no effect. This increased DNA synthesis by EL was inhibited by tamoxifen, the antiestrogenic drug used to treat breast cancer. Welshons *et al.* (1987) had observed the inhibitory effect of tamoxifen on the induction of DNA synthesis by mammalian lignans over a decade earlier, but in the absence of estrogen. EL has been shown to inhibit DNA synthesis in the estrogen-independent breast cancer cell line MDA-MB-231 but only at very high concentrations (> 100 µM) (Wang and Kurzer, 1997). Because plasma concentrations of ED and EL have been found to be up to 0.3 µM in vegetarians (Adlercreutz *et al.*, 1993b) and to range from 100 nM to 1.7 µM after 25 g/day flaxseed supplementation (Morton *et al.*, 1994; Nesbitt *et al.*, 1999), the *in vitro* study results suggest that mammalian lignans would be stimulatory at the concentrations to which breast cancer cells are exposed *in vivo*. Although it is possible that lignans may accumulate at higher concentrations in breast tissue, a recent study in rats suggested that this was not the case (Rickard and Thompson, 1998).

In addition to effects on cell growth, mammalian lignans have been examined for their potential to inhibit the invasion and adhesion of metastatic breast cancer cells. Using Matrigel, an extract of the basement membrane, Chen and Thompson (1999) observed that ED and EL inhibited the invasion of the estrogen-independent cell lines MDA-MB-435 and MDA-MB-231 at concentrations as low as 1 µM and 0.5 µM, respectively. Both lignans also inhibited adhesion of the breast cancer cells to the extracellular matrix, but EL was more potent (1 µM) than ED (5 µM) (Chen and Thompson, 1999). The reduction in invasion and adhesion by tamoxifen was enhanced by the addition of ED and EL (Chen and Thompson, 1999), suggesting that mammalian lignans may have a synergistic effect with certain breast cancer drugs in inhibiting metastasis.

Animal studies

Because flaxseed is the richest known plant source of precursors to ED and EL, flaxseed has been used to assess the role of mammalian lignans in reducing breast cancer risk using rodents as a model. In contrast to the studies done *in vitro*, mammalian lignan exposure *in vivo* has been shown to be protective at the initiation (before carcinogen), promotion (after carcinogen), and progression (visible tumors) stages of carcinogenesis at relatively low levels.

In the rat mammary gland, the hypothesized target of carcinogens is the terminal end bud (TEB) because of its high proliferative index (Russo and Russo, 1996). The higher number of TEBs at the time of carcinogen exposure (usually at 50 days of age), the greater the number of malignant tumors found during adulthood (Russo and Russo, 1996). In our laboratory, flaxseed feeding at the 5 percent level (5%F) to rat dams during pregnancy and lactation resulted in atrophy of TEBs in their female offspring at postnatal day 50 (Tou and Thompson, 1999). A similar effect on TEB number was observed when pure SDG, but not flax oil, was fed at levels found in the 5%F diet, suggesting that lignans were the component of flaxseed responsible for the observed effect. In contrast, 10 percent flaxseed (10%F) feeding during the same period reduced TEB number by stimulating differentiation of TEBs to alveolar buds (ABs) (Tou and Thompson, 1999). Thus, flaxseed feeding during pregnancy and lactation may reduce breast cancer risk in the offspring as indicated by lower TEB number, but the mechanism by which the TEBs are reduced is dependent on the flaxseed dose.

In addition to reducing the number of TEBs in the rat mammary gland, flaxseed feeding for four weeks from 21 days of age to 50 days of age at the 5 or 10 percent levels was found to significantly reduce mammary epithelial cell proliferation and nuclear aberration (Serraino and Thompson, 1991). Nuclear aberrations, which were induced by an intragastric dose of 100 mg DMBA/kg body weight in corn oil, have been correlated with carcinogen exposure and may indicate susceptibility of cells to carcinogens (Sharley and Bruce, 1986). Mammary epithelial cell proliferation was assessed using mitotic index (number of epithelial cells arrested in metaphase per 100 cells) and labeling index (number of tritium-labeled epithelial cells per 100 cells). The 5%F diet resulted in significantly lower mitotic index in the TEBs, lower labeling index in the TEBs and ABs, and lower number of nuclear aberrations in the TEBs. When 10%F was fed, the mitotic index was reduced in the TEBs and the number of nuclear aberrations decreased in the ABs (Serraino and Thompson, 1991). Lignan excretion was inversely related to nuclear aberration ($r = 0.940$, $p < 0.025$), and nuclear aberrations were positively correlated to the labeling and mitotic indices determined in the mammary gland tissue (Serraino and Thompson, 1991). Feeding defatted flaxseed at the same level as the full-fat flaxseed was less effective in reducing cell proliferation and nuclear aberration, suggesting that the oil component of flaxseed may also be playing a role.

Results from mammary tumorigenesis studies with flaxseed suggested that the effect on different tumor parameters (incidence, number, and size) was dependent on the timing of flaxseed feeding in relation to carcinogen. Rats given 5%F for four weeks prior to DMBA administration and then switched to high fat (20% corn oil) basal diet for 21 weeks had a 21 percent lower tumor incidence and the lowest number of tumors per group throughout the experimental period (42 percent lower by the end of treatment) compared to control (Serraino and Thompson, 1992a). When the feeding regimen was reversed (high fat diet for four weeks, followed by DMBA, and then 5%F for 21 weeks), the rats had the smallest tumors (67 percent smaller than control), but there was a non-significant tendency for higher number of tumors per tumor-bearing rat (tumor multiplicity) (Serraino and Thompson, 1992a). In contrast, rats fed 5%F from four weeks before DMBA to 21 weeks after DMBA had the lowest tumor multiplicity, which was significant compared to the group fed 5%F after DMBA but not compared to 5%F before DMBA or control (Serraino and Thompson, 1992a). Tumor size in the group fed flaxseed throughout the experimental period was intermediate between the other two groups. Even when flaxseed feeding (at the 2.5 and 5 percent levels) was started 13 weeks after DMBA treatment and continued for seven weeks, significant reductions in established tumor (present at start of treatment) volume by nearly 75 percent were observed with no effect on established tumor number (Thompson *et al.*, 1996a). Thus, flaxseed diet introduced during the initiation stage had a larger effect on tumor incidence and multiplicity, whereas giving the flaxseed diet during the promotion or progression stages of carcinogenesis affected tumor size to a greater extent.

However, because of the different constituents present in flaxseed, it could not be concluded that the protective effects seen with flaxseed were due to its lignan content alone. As mentioned in the introduction, flax oil is a rich source of the omega-3 fatty acid ALA, which has exhibited anticarcinogenic activity *in vitro* (Grammatikos *et al.*, 1994; Chajes *et al.*, 1995). In addition, feeding mice a 10 percent flax oil diet was found to reduce the growth of mouse mammary tumor cells injected into BALB/c mice (Fritsche and Johnston, 1990). Feeding 5%F to rats, which contains only 1.82 percent flax oil, was sufficient to significantly increase the ALA content of the mammary gland and tumors (Serraino *et al.*, 1992; Thompson *et al.*, 1996a; Rickard *et al.*, 1999).

To determine the role of lignans alone on mammary tumorigenesis, the major flaxseed lignan precursor SDG was isolated using a modified form of the method developed by Klosterman and his colleagues (Klosterman and Smith, 1954; Bakke and Klosterman, 1956). Two separate studies were then conducted to determine the effect of SDG at the early promotion stage (Thompson *et al.*, 1996b) and at the late promotion/early progression stages of carcinogenesis (Thompson *et al.*, 1996a). In the first study, rats fed 1.5 mg SDG/d (equivalent to the amount consumed in a 5%F diet) starting one week after DMBA until 20 weeks after DMBA had 46 percent fewer tumors per group and 37 percent less tumor multiplicity (Thompson *et al.*, 1996b). No effect on tumor volume was observed. In the second study, a daily gavage of 1.5 mg SDG started 13 weeks after DMBA treatment and continued for seven weeks reduced established tumor (present at start of treatment) volume by 54 percent, the new tumor (appearing during treatment) volume by 75 percent, the average new tumor number per group by 50 percent, and the new tumor incidence by 27 percent (Thompson *et al.*, 1996a). Feeding 5%F or the level of oil equivalent to the 5%F diet (1.82%) significantly reduced established tumor volume by 75 percent and 54 percent, respectively, but had no effect on new tumor size or established or new tumor number (Thompson *et al.*, 1996a). Thus, although both the lignan and oil components of flaxseed appeared to play a role in its anticarcinogenic activity, the SDG had a stronger inhibitory effect on tumor number, particularly during the early and late promotion stages, whereas the flaxseed oil was more effective at reducing the size of visible tumors.

All of the chemically induced carcinogenesis studies described above have used the indirect-acting carcinogen DMBA. It has been suggested that the N-methyl-N-nitrosourea (MNU) model is more related to the human breast cancer histologically, in endocrine responsiveness, and in metastatic behavior (Gullino *et al.*, 1975). So the objective of our most recent study was to examine the effect of flaxseed (2.5%F and 5%F) and equivalent levels of SDG (LSDG = 0.7 mg/d and HSDG = 1.4 mg/d, respectively) on MNU-induced tumor promotion (Rickard *et al.*, 1999). As seen in previous tumor promotion studies with flaxseed (Serraino and Thompson, 1992a), the 5%F group had the smallest tumors throughout the experimental period which ended at 22 weeks post-MNU, but this did not reach significance. Throughout treatment, the HSDG group had the lowest tumor multiplicity, whereas the LSDG group had the highest, suggesting that HSDG inhibited and LSDG promoted MNU-induced tumor development. All the treatment groups significantly decreased tumor invasiveness and grade, determined histologically, in comparison to control, suggesting that flaxseed and its SDG delayed the progression of MNU-induced mammary tumorigenesis.

Two possible reasons for the lack of a significant inhibitory effect of flaxseed on MNU tumor parameters (e.g. size, number, incidence) may be the carcinogen dose and the type of control diet used. First, a larger dose of carcinogen was used (9 mg MNU/animal versus 5 mg DMBA/animal), which may have been too high to observe the subtle protective effects of dietary flaxseed. Second, the use of a 20 percent soybean oil-based control diet may have conferred a protective effect in comparison to the 20 percent corn oil diet used in previous studies because of its higher omega-3 fatty acid content. It was calculated that the 20 percent soybean oil diet contained 1.6 percent ALA, whereas the 20 percent corn oil diet contained 0.24 percent ALA (Rickard *et al.*, 1999). Interestingly, supplementation of the 20 percent corn oil diet with 5 percent flaxseed only increased dietary ALA levels to 1.22 percent (Rickard *et al.*, 1999), which was lower than the ALA content of the control diet used in the MNU tumor study. The tumor ALA levels were higher with soybean oil used as a fat source (control = 0.99 ± 0.21%, 5%F = 1.58 ± 0.30%) than with corn oil as the fat source (control = 0.23 ± 0.06%, 5%F = 0.67 ± 0.08%) (Rickard *et al.*, 1999). The inverse correlation observed between log-transformed tumor volume and tumor ALA content ($y = -0.204 x + 0.573$, $r = -0.419$, $p = 0.014$, $n = 34$) suggested that the ALA may have played a role in reducing tumor growth.

Clinical studies

There have been few clinical studies examining the relationship between mammalian lignans and breast cancer risk. In a small epidemiological study, Adlercreutz and colleagues (1982) found lower urinary EL levels in women with breast cancer compared to women without breast cancer consuming an omnivorous or vegetarian diet. The lower lignan excretion in the breast cancer group compared to the omnivorous group did not appear to be due to diet since dietary fiber levels were similar. However, the authors hypothesized that because there were data to suggest that urinary lignan excretion may be influenced by hormones (Setchell *et al.*, 1980a, 1980b, 1981a), differences in hormonal conditions may have played a role (Adlercreutz *et al.*, 1982). Higher urinary levels of ED and EL were also found in women considered to have lower breast cancer risk (lactovegetarians and macrobiotics) in another study (Adlercreutz *et al.*, 1986). A more recent and larger case-control study of pre- and post-menopausal women with newly diagnosed breast cancer determined that the odds ratio of breast cancer decreased with increasing urinary EL, with the highest quartile of EL excretion associated with a 64 percent reduction in breast cancer risk (Ingram *et al.*, 1997). At present, there are clinical trials in progress investigating the effect of short-term flaxseed feeding on cancer biomarkers in women newly diagnosed with breast cancer.

Colon cancer

In vitro *studies*

In contrast to breast cancer, few studies have investigated the effect of flax lignans on the growth of colon tumor cells *in vitro*. Sung *et al.* (1998) tested the effect of ED and EL on four different colon tumor cell lines (LS174T, T84, Caco-2, and HCT-15). These cell lines were found to be estrogen-independent since the addition of estradiol at 0.5, 1, and 10 nM were not able to stimulate cell proliferation. Cell proliferation was inhibited by ED (LS174T, T84, Caco-2) and EL (HCT-15, Caco-2) at a concentration of 100 μM, with EL being twice as effective as ED. In humans consuming 50 g flaxseed, the mammalian lignan concentration could reach 665 μM in the colon based on a colonic content of 220 g and 2.93 μmoles SDG/g flaxseed (Sung *et al.*, 1998). Since the concentrations used could be easily achieved *in vivo*, the inhibitory effects observed *in vitro* for ED and EL in the colon were physiologically possible.

Animal studies

All of the animal studies conducted to examine the role of flaxseed and its lignan on colon carcinogenesis have used the aberrant crypt, a putative precursor lesion, as a marker because studies taken to the colon tumor stage typically take one year to complete. The aberrant crypts differ from normal colon crypts in that they are larger in size, have a thicker epithelial lining, and have an increased pericryptal zone. Multiple aberrant crypts can occur in localized regions of the colon and are called aberrant crypt foci (ACF). The term "aberrant crypt multiplicity" is used to describe the number of aberrant crypts per focus. For example, a multiplicity of two would indicate that there are two aberrant crypts in one focus. Because the aberrant crypts are precursor lesions, we should be cautious in applying the results observed in the studies below to what may happen at the tumor stage.

As in the breast cancer studies, early investigations into the role of lignans on colon cancer risk markers have used flaxseed as the dietary source. Serraino and Thompson (1992b) treated rats with the colon carcinogen azoxymethane (AOM, 15 mg/kg body weight) and one week later treated the animals to a 20 percent corn oil diet alone or supplemented with 5% or 10% full-fat or defatted flaxseed for four weeks. The number of aberrant crypts and aberrant crypt foci in the rats fed the flaxseed diets was reduced by 50 percent. Except for the 5 percent defatted flaxseed group, flaxseed feeding also reduced the cell proliferation activity (labeling index) with the 5 percent full-fat flaxseed group having the greatest effect. Because both the defatted and full-fat flaxseed groups had similar effects on aberrant crypt formation, this suggested that the lignan component of flaxseed was more important than the oil component for its inhibitory effect in this model. Higher flaxseed levels (10%) did not appear to confer any additional benefit in comparison to lower doses (5%).

Because 10 percent flaxseed was not any more protective than 5 percent flaxseed, lower doses of full-fat and defatted flaxseed (2.5% and 5%) were used in a subsequent longer term study of 100 days (Jenab and Thompson, 1996). SDG at the level of the 5 percent full-fat flaxseed group was included as an additional treatment group. All the flaxseed treatment groups (full-fat and defatted) significantly reduced the aberrant crypt multiplicity in the distal colon of AOM treated rats. Both flaxseed doses had similar effects, suggesting that even lower doses of flaxseed could be protective. In corroboration with the results of the short-term study, the anticarcinogenic effect of the flaxseed appeared to be due mainly to its lignan component. The SDG group had significantly less aberrant crypts and lower aberrant crypt multiplicity. In addition, a significant negative relationship was observed between urinary lignan excretion and aberrant crypt multiplicity. An interesting observation was that the control group also had four microadenomas and two polyps, suggesting that this group was further along in the carcinogenesis process.

Other cancers

In vitro *studies*

Besides breast and colon cancer cells, the only other cancer cell line in which flax lignans have been tested and shown to have an effect *in vitro* was the human promyelocytic leukemia cell line HL-60 (Hirano *et al.*, 1994). At levels of 0.03 to 3 μM, MAT was found to reduce the growth of HL-60 cells by 45–50 percent (IC_{50} = 0.11 μM) without any cytotoxicity. In contrast, routine anticancer agents (e.g. etopside, methotrexate) exhibited severe cytotoxicity (0% cell survival) at similar concentrations (Hirano *et al.*, 1994). The only mammalian lignan tested in this study was ED, but it was ineffective in this cell line.

Animal studies

Yan and colleagues (1998) examined the effect of flaxseed on lung metastasis of the mouse melanoma cell line B16BL6. After two weeks on a basal diet supplemented with 2.5, 5, or 10 percent flaxseed, the number of lung tumors was 32, 54 and 63 percent lower than controls. Tumor cross-sectional area was also decreased in a dose-dependent manner. When the effects of SDG at levels found in the flaxseed diets were tested (i.e. 73, 147 and 293 μmol SDG/kg diet), similar reductions in the number and size of lung metastasises were found (Li *et al.*, 1999), indicating that the lignans were partly responsible for the effect observed with flaxseed. This was the first study to show that flaxseed and its lignans could inhibit cancer cell metastasis *in vivo*. It should be noted, however, that the melanoma cells were injected directly into the bloodstream in this model, bypassing the first steps of the metastatic process. These steps are angiogenesis (the generation of new capillaries) of the primary tumor, detachment of metastatic cells from the tumor, attachment and degradation of the surrounding basement membrane, and intravasation by the metastatic cells into the blood circulation (Duffy, 1996). Although there is evidence *in vitro* that lignans can inhibit angiogenesis (Fotsis *et al.*, 1993), as well as the invasion and adhesion of cancer cells (Chen and Thompson, 1999), the effect of lignans on these stages of metastasis *in vivo* remains to be determined.

There is some evidence to suggest that lignans may have an effect on prostate cancer development. Mammalian lignans produced from 33 percent rye bran diets were believed to be partly responsible for the delay in the development of the Dunning R3327 PAP prostate tumor transplanted in rats because the 33 percent rye endosperm diets, which did not contain lignans, had no effect (Landstrom *et al.*, 1998). However, the lower energy intake in the rye bran group could have contributed the effects observed. Cell proliferation in the prostate was found to be inhibited in the male offspring of rat dams fed 5 percent flaxseed during pregnancy and lactation (Tou *et al.*, 1998) or throughout their lifetime until 132 days of age (Tou *et al.*, 1999). In contrast, flaxseed at the 10 percent dose appeared to be stimulatory in both cases, suggesting that flaxseed could have adverse or beneficial effects in the prostate depending on the level of exposure during this critical period of reproductive development.

Clinical studies

Morton and colleagues (1997) have examined the potential relationship between prostate cancer risk and phytoestrogen levels in the plasma and prostatic fluid in three ethnic groups: British, Portuguese and Chinese. The Portuguese men were found to have much higher EL levels in prostatic fluid but similar plasma EL levels in comparison to the other two groups. However, international comparisons indicate that the prostate cancer risk in Portugal is only slightly lower than in England, and Hong Kong has a much lower incidence of prostate cancer than both (Parkin *et al.*, 1997). It should be noted the Chinese men in this study had the highest levels of the isoflavone phytoestrogens such as daidzein and equol in their plasma and prostatic fluid due to their high soybean intake, suggesting that soybean isoflavones may play a role in their reduced prostate cancer risk. There was little correlation (r = 0.18) between the plasma and prostatic fluid EL concentration, and the prostatic fluid EL concentrations were much higher (4- to 40-fold higher) than plasma levels (Morton *et al.*, 1997). A similar result was observed in an earlier study where human seminal plasma concentrations of EL were 2.5 to 25 times higher than blood plasma levels (Dehennin *et al.*, 1982). This result suggests that lignans may accumulate in certain tissues to levels where they would potentially exert an anticancer effect.

Potential mechanisms

Because of structural similarities to synthetic estrogens such as diethylstilbestrol and the antiestrogen tamoxifen, many of the mechanisms explored to explain the biological effects of lignans have been related to hormonal activity (Table 8.2). Mammalian lignans have increased the level or expression of estrogen-dependent products such as the progesterone receptor and pS2 mRNA expression in breast cancer cells (Welshons *et al.*, 1987; Sathyamoorthy *et al.*, 1994) and the prolactin receptor in pituitary cells *in vitro* (Welshons *et al.*, 1987), all considered to be estrogenic effects. However, a majority of the evidence suggests that lignans can affect different aspects of estrogen and androgen metabolism and hence bioavailability:

1. EL and theoretical intermediates between MAT and EL have inhibited the activity of aromatase, an enzyme which converts androgens to estrogens, in placental microsomes, the JEG-3 human choriocarcinoma cell line (Adlercreutz *et al.*, 1993a) and preadipocytes (Wang *et al.*, 1994) *in vitro*. Because aromatization in breast fat has been shown to be higher in breast cancer patients as compared to women with benign breast disease (Miller and O'Neill, 1989), antagonism of aromatase, and hence reduction of endogenous estrogen synthesis, by lignans may be one mechanism whereby breast cancer risk is reduced.

2. Lignans have competitively inhibited estrogen binding to the rat carrier protein α-fetoprotein (Garreau *et al.*, 1991). Alpha-fetoprotein has a high affinity for steroid estrogens and is thought to be an immunomodulator and to regulate the growth of estrogen-sensitive cells (Garreau *et al.*, 1991).

3. Lignans have stimulated sex hormone binding globulin synthesis (SHBG) *in vitro* and were positively related to plasma SHBG in postmenopausal women (Adlercreutz *et al.*, 1992a). Because EL was only ten times weaker than estradiol at stimulating SHBG and its *in vivo* concentrations were found to be 100 to 10,000 times greater than estradiol, EL might be a more physiological regulator of SHBG than estradiol (Adlercreutz *et al.*, 1992b). Although an increase in SHBG would theoretically reduce the levels of free estradiol and testosterone and hence their biological activity in peripheral tissues, plant and mammalian lignans appear to compete with hormones for binding to SHBG (Ganßer and Spiteller, 1995; Martin *et al.*, 1995). On the other hand, increased binding of lignans to SHBG may facilitate the transport of lignans to target tissues. Nevertheless, studies in human males and premenopausal women have found no significant changes in SHBG levels with 10–13.5 g/day flaxseed feeding (Shultz *et al.*, 1991; Phipps *et al.*, 1993). This may be related to the low level of flaxseed used and/or due to lack of control for the intake of other phytoestrogens in the case of the study by Shultz *et al.* (1991).

4. There is some evidence that lignans may reduce circulating hormonal levels. Urinary lignan levels have been negatively correlated with percentage of free plasma estradiol and testosterone in postmenopausal Finnish women (Adlercreutz *et al.*, 1992b). Dietary supplementation with flaxseed (5 g or 10 g) for seven weeks was found to significantly decrease serum estradiol in postmenopausal women (Hutchins *et al.*, 1999), an effect associated with reduced breast cancer risk (Toniolo, 1997). In contrast, other human studies found no significant changes in plasma hormone levels in males (Shultz *et al.*, 1991) or premenopausal females (Phipps *et al.*, 1993). However, studies in rats have indicated that flaxseed or SDG consumption might increase or decrease plasma hormone levels depending on the dose used and the timing of flaxseed exposure (Rickard and Thompson, unpublished data; Tou *et al.*, 1999).

5. Flaxseed lignans may also affect the oxidative metabolism of estradiol. Haggans and colleagues (1999) found that flaxseed supplementation at 10 g/day in postmenopausal women increased the urinary excretion of 2-hydroxyestrogens (2-hydroxyestradiol and 2-hydroxyestrone) and the 2/16α-hydroxyestrone ratio. The 16α-hydroxyestrone is a more potent estrogen than the 2-hydroxyestrogen metabolites and has been associated with increased breast cancer risk (Fishman *et al.*, 1995).

6. Both ED and EL have reduced the activity of 5α-reductase, which converts testosterone to dihydrotestosterone, and of 17β-hydroxysteroid dehydrogenase, which

converts estrone to the more active estrogen estradiol, in genital skin fibroblasts and prostate tissue (Evans *et al.*, 1995). Decreases in dihydrotestosterone levels might slow the growth of prostate tumors (Adlercreutz and Mazur, 1997). An interesting observation was that a cocktail of seven phytoestrogens, which included ED and EL, had much lower inhibitory concentrations on 5α-reductase activity in comparison to the compound alone (Evans *et al.*, 1995). Because this latter scenario would be more representative of *in vivo* situations, ED and EL might be more biologically active at lower concentrations than previously thought due to the presence of other phytoestrogens in the diet.

7. The mammalian lignans produced from flaxseed and feeding in rats have increased in β-glucuronidase activity (Jenab and Thompson, 1996, Jenab *et al.*, 1999). This increased activity was believed to be due to the increased amount of ED- and EL-glucuronide in the colon because of the enterohepatic circulation of the lignans (Axelson and Setchell, 1981; Setchell *et al.*, 1981b). Thus, the much higher levels of lignan conjugates might saturate the enzyme and prevent it from acting on estrogen glucuronide (Jenab and Thompson, 1996), thereby preventing its absorption and increasing its excretion in the feces (Adlercreutz and Martin, 1980). However, flaxseed feeding on estrogen excretion has not been examined.

8. Lignans might potentially compete for sulphatase activity in cells. EL has been shown to be rapidly conjugated in the monosulphate form in MCF-7 breast cancer cells (Mousavi and Adlercreutz, 1992) and HepG2 liver cancer cells (Adlercreutz *et al.*, 1992a). Because estrone-sulphate is the major circulating form of estrogen and is contained in high amounts in breast cancer cells, the presence of EL-sulphate might reduce the hydrolysis of estrone-sulphate and its subsequent conversion to estradiol via 17-β-hydroxysteroid dehydrogenase (see point 6).

In addition to altering hormonal levels and metabolism, lignans appear to affect estrogen action. Daily supplementation of the diets of 18 premenopausal women with 10 g raw flaxseed over three menstrual cycles was found to increase the mean luteal phase length of the menstrual cycle and to increase the luteal progesterone to estradiol ratio (Phipps *et al.*, 1993). These effects were believed to be due to the antiestrogenic activity of the lignans produced with flaxseed consumption because similar increases in luteal phase length in premenopausal women were observed with the antiestrogenic drug tamoxifen (Lumsden *et al.*, 1989). Studies in rats suggest that alterations in estrogen-dependent processes such as estrous cycling, puberty onset, and weights of sex organs are dependent on the dose and timing of exposure to flaxseed or SDG (Orcheson *et al.*, 1998; Tou *et al.*, 1998, 1999). For example, 10 percent flaxseed exposure during pregnancy and lactation or throughout the lifetime of the rat resulted in increased ovarian and/or uterine weight, early onset of puberty, increased estrous cycle length, increased persistent estrus, and increased serum estradiol levels in females (Tou *et al.*, 1998, 1999), all considered to be estrogenic effects. On the other hand, 5 percent flaxseed exposure during these periods resulted in decreased ovarian weight, delayed puberty onset, and increased persistent diestrus (Tou *et al.*, 1998, 1999), considered to be anti-estrogenic effects. Exposure to flaxseed or SDG after weaning has resulted in no effect on these parameters (Tou *et al.*, 1999) or in antiestrogenic effects such as increased estrous cycle length and increased persistent diestrus (Orcheson *et al.*, 1998).

The estrogenic or antiestrogenic action of lignans may be mediated via the estrogen receptor (ER). Estrogen controls the growth, differentiation, and function of tissues via the ER. Recently, a new isoform of the ER, called ERβ, was discovered in the following tissues: prostate, ovary, uterus, epidiymis, testis, bladder, lung, thymus, colon, small intestine, vessel wall, hypothalamus, cerebellum, and brain cortex (Kuiper *et al.*, 1998). Although not tested with lignans, structurally similar phytoestrogens were found to preferentially bind to ERβ versus ERα (the original ER) and to have agonistic rather than antagonistic activity (Kuiper *et al.*, 1998). Because the relative levels of these two receptors differ depending on the tissue (Kuiper *et al.*, 1998), the biological activity of lignans and other phytoestrogens may change from tissue to tissue. Both Mat and EL have been shown to competitively bind to the nuclear type II ER ("bioflavonoid receptor") (Adlercreutz *et al.*, 1992a), believed to control uterine growth (Markaverich and Clark, 1979; Markaverich *et al.*, 1988). Earlier studies with EL showed that it decreased the stimulation of uterine RNA synthesis by estradiol in immature rats four-fold when given 22 hours before estradiol (Waters and Knowler, 1982), which may be mediated by the type II ER. Although not generally considered to be a hormone-dependent cancer, colon tissues and tumors have been shown to contain nuclear type II ER (Piantelli *et al.*, 1990). Thus, one mechanism for the inhibitory effect of lignans on colon cancer risk may be estrogen-dependent.

There are also non-hormone-dependent mechanisms of actions of lignans. One such mechanism is antioxidant activity. Both plant (SDG) and mammalian (ED and EL) lignans have exhibited hydroxy radical scavenging activity *in vitro* (Prasad, 1997; Kitts *et al.*, 1999) and have inhibited DNA scission (Kitts *et al.*, 1999). Free radicals formed endogenously through cellular respiration or obtained externally from food, polluted air or cigarette smoke can attack DNA, protein, and lipids in cellular membranes causing cell and tissue damage (Thompson, 1994). Flaxseed and SDG feeding in rats has also resulted in reduced hepatic glutathione reductase activity without affecting glutathione levels, suggesting an antioxidant sparing effect of flaxseed lignans (Yuan *et al.*, 1999). Another hormone-independent mechanism mentioned previously is the moderate inhibition of angiogenesis, an important step in the metastatic cascade, *in vitro* by EL (Fotsis *et al.*, 1993).

Implications and conclusions

The *in vitro* data, particularly in the case of breast cancer, suggest that the level of lignan exposure to produce anticancer effects would need to be much higher than physiological levels and that at physiological concentrations, lignans would be stimulatory. It should be noted that studies *in vitro* generally examine one compound at a time to discern its effect in a particular system. This, however, is not the case *in vivo* where phytoestrogens would be interacting with similar compounds as well as other food components. The study by Evans *et al.* (1995) illustrated this point nicely by finding lower effective concentrations of phytoestrogens when used together rather than individually. Animal data appear to be consistent in the anticarcinogenic effects of flaxseed or its mammalian lignan precursor SDG when fed at easily achievable doses in humans. The most effective doses of flaxseed (5%) and SDG (1.5 mg/day) tested in rats are roughly equivalent to daily doses of 25 g flaxseed or 50 mg SDG in humans. Although optimal doses in humans still need to be determined, flaxseed feeding at 50 g/day for four weeks was shown to have beneficial rather than adverse effects (Cunnane *et al.*, 1993, 1995). Results from clinical studies on lignans, albeit circumstantial, support the notion that lignans reduce cancer risk.

To determine the efficacy and safety of flaxseed and its lignans for human applications, there are certain things that must be taken into consideration. First, there could be species differences in the metabolism and activity of lignans so optimal doses for humans may not be the same as effective concentrations determined in rat studies. Second, there is large inter-individual variability in biological levels of lignans that could be attributed to the base diet consumed (Kirkman *et al.*, 1995) and activity of the gut microflora (Adlercreutz, 1998). Gender differences in the relative levels of ED and EL have been found in humans (Kirkman *et al.*, 1995) and in rats (unpublished data), which may be attributed to gender differences in the colonic environment as well as the hormonal milieu (Kirkman *et al.*, 1995). Third, the biological levels of lignans after flaxseed supplementation could also be affected by variability in plant lignan levels in flaxseed (Thompson *et al.*, 1997), variability in flaxseed levels in commercial and homemade products (Nesbitt and Thompson, 1997), and continuous versus occasional dietary intake of lignan precursors (Rickard and Thompson, 1998; Nesbitt *et al.*, 1999). Fourth, the effect of processing on lignan availability needs to be considered since flaxseed would likely be consumed in the form of baked goods rather than in the ground, raw form as in the animal studies. To answer this question, Nesbitt *et al.* (1999) determined that urinary lignan excretion, a marker of colonic lignan production, was the same for 25 g flaxseed consumed raw or in bread or muffin forms, indicating that this type of processing does not affect lignan availability. Finally, the timing and dose of flaxseed exposure during reproductive development may be an important safety consideration. Flaxseed exposure during pregnancy and lactation in rats was found to have reproductive effects (Tou *et al.*, 1998, 1999), whereas no reproductive changes were observed with flaxseed exposure after weaning. Depending on the dose used, the reproductive changes could be estrogenic or antiestrogenic and may be adverse or beneficial. The consequences of these effects on disease outcome in the long term have not been assessed but do suggest that caution should be used in terms of phytoestrogen exposure during critical stages of reproductive development.

In conclusion, the majority of the evidence supports the hypothesis that flaxseed and its lignans may have beneficial effects in reducing the risk of hormone and non-hormone-dependent cancers. Potential mechanisms are the antagonism of estrogen and androgen metabolism and action as well as antiangiogenic and antioxidant activity. Factors affecting lignan bioavailability need to be assessed to determine optimal doses for humans. Until more research is done, lignan-rich food such as flaxseed should be used with caution during pregnancy or in susceptible populations such as infants or young children due to potential adverse effects during critical stages of reproductive development.

References

Adlercreutz, H. (1998). Evolution, nutrition, intestinal microflora, and prevention of cancer: A hypothesis. *Proc. Soc. Exp. Biol. Med.* 217, 241–246.

Adlercreutz, H., Bannwart, C., Wähälä, K., Mäkelä, T., Brunow, G., Hase, T., Arosemena, P.J., Kellis, J.T., and Vickery, L.E. (1993a). Inhibition of human aromatase by mammalian lignans and isoflavonoid phytoestrogens. *J. Steroid Biochem. Mol. Biol.* 44, 147–153.

Adlercreutz, H., Fotsis, T., Bannwart, C., Wähälä, K., Mäkelä, T., Brunow, G., and Hase, T. (1986). Determination of urinary lignans and phytoestrogens metabolites, potential antiestrogens and anticarcinogens, in urine of women on various habitual diets. *J. Steroid Biochem.* 25, 791–797.

Adlercreutz, H., Fotsis, T., Heikkinen, R., Dwyer, J.T., Woods, M., Goldin, B.R., and Gorbach, S.L. (1982). Excretion of the lignans enterolactone and enterodiol and of equol in omnivorous and vegetarian postmenopausal women and in women with breast cancer. *Lancet* ii, 1295–1299.

Adlercreutz, H., Fotsis, T., Lampe, J., Wähälä, K., Mäkelä, T., Brunow, G., and Hase, T. (1993b). Quantitative determination of lignans and isoflavonoids in plasma of omnivorous and vegetarian women by isotope dilution gas chromatography-mass spectrometry. *Scand. J. Clin. Lab. Invest.* 53, 5–18.

Adlercreutz, H., and Martin, F. (1980). Biliary excretion and intestinal metabolism of progesterone and estrogens in man. *J. Steroid Biochem.* 13, 231–244.

Adlercreutz, H., and Mazur, W. (1997). Phyto-oestrogens and Western diseases. *Ann Med* 29, 95–120.

Adlercreutz, H., Mousavi, Y., Clark, J., Höckerstedt, K., Hämäläinen, E., Wähälä, K., Mäkelä, T., and Hase, T. (1992a). Dietary phytoestrogens and cancer: *In vitro* and *in vivo* studies. *J. Steroid Biochem.* 41, 331–337.

Adlercreutz, H., Mousavi, Y., and Hockerstedt, K. (1992b). Diet and breast cancer. *Acta Oncol.* 31, 175–181.

Adlercreutz, H., van der Widlt, J., Kinzel, J., Attalla, H., Wähälä, K., Mäkelä, T., Hase, T., and Fotsis, T. (1995). Lignan and isoflavonoid conjugates in human urine. *J. Steroid Biochem. Mol. Biol.* 52, 97–103.

Axelson, M., and Setchell, K.D.R. (1980). Conjugation of lignans in human urine. *FEBS Lett.* 122, 49–53.

Axelson, M., and Setchell, K.D.R. (1981). The excretion of lignans in rats – evidence for an intestinal bacterial source for this new group of compounds. *FEBS Lett.* 123, 337–342.

Axelson, M., Sjövall, J., Gustaisson, B.E., and Setchell, K.D.R. (1982). Origin of lignans in mammals and identification of a precursor from plants. *Nature* 298, 659–670.

Ayers, D.C., and Loike, J.D. (1990). *Lignans: Chemical, Biological and Clinical Properties.* Cambridge University Press, Cambridge.

Bakke, J.E., and Klosterman, H.J. (1956). A new diglucoside from flaxseed. *Proc. N. Dakota Acad. Sci.* 10, 18–22.

Borriello, S.P., Setchell, K.D.R., Axelson, M., and Lawson, A.M. (1985). Production and metabolism of lignans by the human fecal flora. *J. Appl. Bacteriol.* 58, 37–43.

Chajes, V., Sattler, W., Stranzl, A., and Kostner, G.M. (1995). Influence of n-3 fatty acids on the growth of human breast cancer cells in vitro: Relationship to peroxides and vitamin E. *Breast Cancer Res. Treat.* 34, 199–212.

Chen, J., and Thompson, L.U. (1999). Lignans and tamoxifen, alone and in combination, inhibit the invasion and adhesion of metastatic human breast cancer cells *in vitro*. *Fed. Am. Soc. Exp. Biol. Journal* 13, A582.

Coert, A., Vouk Noordegraaf, C.A., Grown, M.B., and van der Vies, J. (1982). The dietary origin of the urinary lignan HPMF. *Experientia* 38, 904–905.

Cunnane, S.C., Ganguli, S., Menard, C., Liede, A.C., Hamadeh, M.J., Chen, Z.-Y., Wolever, T.M.S., and Jenkins, D.J.A. (1993). High α-linolenic acid flaxseed (*Linum usitatissimum*): Some nutritional properties in humans. *Br. J. Nutr.* 49, 443–453.

Cunnane, S.C., Hamadeh, M.J., Liede, A.C., Thompson, L.U., Wolever, T.M.S., and Jenkins, D.J.A. (1995). Nutritional attributes of traditional flaxseed in healthy young adults. *Am. J. Clin. Nutr.* 61, 62–68.

Dehennin, L., Reiffsteck, A., Jondet, M., and Thibier, M. (1982). Identification and quantitative estimation of a lignan in human and bovine semen. *J. Reprod. Fert.* 66, 305–309.

Duffy, M. (1996). The biochemistry of metastasis. *Adv. Clin. Chem.* 32, 135–165.

Evans, B.A.J., Griffiths, K., and Morton, M.S. (1995). Inhibition of 5 α-reductase in genital skin fibroblasts and prostate tissue by dietary lignans and isoflavonoids. *J. Endocrinol.* 147, 295–302.

Fishman, J., Osborne, M.P., and Telang, N.T. (1995). The role of estrogen in mammary carcinogenesis. *Ann. N.Y. Acad. Sci.* **768**, 91–100.

Fotsis, T., Pepper, M., Adlercreutz, H., Fleischmann, G., Hase, T., Montesano, R., and Schweigerer, L. (1993). Genistein, a dietary-derived inhibitor of *in vitro* angiogenesis. *Proc. Natl. Acad. Sci. USA.* **90**, 2690–2694.

Fritsche, K.L., and Johnston, P.V. (1990). Effect of dietary alpha-linolenic acid on growth, metastasis, fatty acid profile and prostaglandin production of two murine mammary adenocarcinomas. *J. Nutr.* **120**, 1601–1609.

Ganßer, D., and Spiteller, G. (1995). Plant constituents interfering with human sex hormone-binding globulin. Evaluation of a test method and its application to *Urtica dioica* root extracts. *Z. Naturforsch.* **50c**, 98–104.

Garreau, B., Vallette, G., Adlercreutz, H., Wähälä, K., Mäkelä, T., Benassayag, C., and Nunez, E.A. (1991). Phytoestrogens: New ligands for rat and human alpha-fetoprotein. *Biochem. Biophys. Acta* **1094**, 339.

Grammatikos, S.I., Subbaiah, P.V., Victor, T.A., and Miller, W.M. (1994). n-3 and n-6 fatty acid processing and growth effects in neoplastic and non-cancerous human mammary epithelial cell lines. *Br. J. Cancer* **70**, 219–227.

Gullino, P.M., Pettigrew, H.M., and Grantham, F.H. (1975). N-nitrosomethylurea as mammary gland carcinogen in rats. *J. Nat. Cancer Inst.* **54**, 401–409.

Haggans, C.J., Hutchins, A.M., Olson, A.M., Thomas, W., Martini, M.C., and Slavin, J.L. (1999). Effect of flaxseed consumption on urinary estrogen metabolites in postmenopausal women. *Nutr. Cancer* **33**, 188–195.

Hirano, T., Fukuoka, K., Oka, K., Naito, T., Hosaka, K., Mitsuhashi, H., and Matsumoto, Y. (1990). Antiproliferative activity of mammalian lignan derivatives against the human breast carcinoma cell line, ZR-75-1. *Cancer Invest.* **8**, 595–602.

Hirano, T., Gotoh, M., and Oka, K. (1994). Natural flavonoids and lignans are potent cytostatic agents against human leukemic HL-60 cells. *Life Sci.* **55**, 1061–1069.

Hutchins, A.M., Martini, M.C., Olson, B.A., Thomas, W., and Slavin, L. (1999). Dietary estrogens influence endogenous estrogen concentrations in post-menopausal women. *Fed. Am. Soc. Exp. Biol. Journal* **13**, A583.

Ingram, D., Sanders, K., Kolybaba, M., and Lopez, D. (1997). Case-control study of phyto-oestrogens and breast cancer. *The Lancet* **350**, 990–994.

Jenab, M., Richard, S.E., Orcheson, L.J., and Thompson, L.U. (1999). Flaxseed and lignans increase cecal β-glucuronidase activity in rats. *Nutr. Cancer* **33**, 154–158.

Jenab, M., and Thompson, L.U. (1996). The influence of flaxseed and lignans on colon carcinogenesis and β-glucuronidase activity. *Carcinogenesis* **17**, 1343–1348.

Kirkman, L.M., Lampe, J.W., Campbell, D.R., Martini, M.C., and Slavin, J.L. (1995). Urinary lignan and isoflavonoid excretion in men and women consuming vegetable and soy diets. *Nutr. Cancer* **24**, 1–12.

Kitts, D.D., Yuan, Y.V., Wijeckremene, A.N., and Thompson, L.U. (1999). Antioxidant activity of the flaxseed lignan secoisolariciresinol diglycoside and its mammalian lignan metabolites enterodiol and enterolactone. *Mol. Cell. Biochem.* **202**, 91–100.

Klosterman, H.J., and Smith, F. (1954). The isolation of β-hydroxy-β-methylglutaric acid from the seed of flax (*Linum usitatissimum*). *J. Am. Chem. Soc.* **76**, 1229–1230.

Kuiper, G.G., Lemmen, J.G., Carlsson, B., Corton, J.C., Safe, S.H., van der Saag, P.T., van der Burg, B., and Gustafsson, J.A. (1998). Interaction of estrogenic chemicals and phytoestrogens with estrogen receptor beta. *Endocrinol.* **139**, 4252–4263.

Landstrom, M., Zhang, J.X., Hallmans, G., Aman, P., Bergh, P., Damber, J.E., Mazur, W., Wähälä, K., and Adlercreutz, H. (1998). Inhibitory effects of soy and rye diets on the development of Dunning R3327 prostate adenocarcinoma in rats. *Prostate* **36**, 151–161.

Li, D., Yee, J.A., Thompson, L.U., and Yan, L. (1999). Dietary supplementation with secoisolariciresinol diglycoside reduces experimental metastasis of melanoma cells in mice. *Cancer Lett.* **142**, 91–96.

Lumsden, M.A., West, C.P., and Baird, D.T. (1989). Tamoxifen prolongs luteal phase in pre-menopausal women but has no effect on the size of uterine fibroids. *Clin. Endocrinol. (Oxford)* 31, 335–343.

Markaverich, B.M., and Clark, J.H. (1979). Two binding sites for estradiol in rat uterine nuclei: relationship to uterotrophic response. *Endocrinol.* 105, 1458–1462.

Markaverich, B.M., Roberts, R.R., Alejandro, M.A., Johnson, G.A., Middleditch, B.S., and Clark, J.H. (1988). Bioflavonoid interaction with rat uterine type II binding sites and cell growth inhibition. *J. Steroid Biochem.* 30, 71–78.

Martin, M., Haourigui, M., Pelissero, C., Benassayag, C., and Nunez, E. (1995). Interactions between phytoestrogens and human sex steroid binding protein. *Life Science* 58, 429–436.

Mazur, W., Fotsis, T., Wähälä, K., Ojala, S., Salakka, A., and Adlercreutz, H. (1996). Isotope dilution gas chromatographic-mass spectrometric method for the determination of isoflavonoids, coumestrol, and lignans in food samples. *Anal. Biochem.* 233, 169–180.

Miller, W.R., and O'Neill, J.S. (1989). The significance of steroid metabolism in human cancer. *J. Steroid Biochem. Mol. Biol.* 37, 317–325.

Morton, M.S., Chan, P.S., Cheng, C., Blacklock, N., Matos-Ferreira, A., Abranches-Monteiro, L., Correia, R., Lloyd, S., and Griffiths, K. (1997). Lignans and isoflavonoids in plasma and prostatic fluid in men: Samples from Portugal, Hong Kong, and the United Kingdom. *Prostate* 32, 122–128.

Morton, M.S., Wilcox, G., Wahlqvist, M.L., and Griffiths, K. (1994). Determination of lignans and isoflavonoids in human female plasma following dietary supplementation. *J. Endocrinol.* 142, 251–259.

Mousavi, Y., and Adlercreutz, H. (1992). Enterolactone and estradiol inhibit each other's proliferative effect on MCF-7 breast cancer cells in culture. *J. Steroid Biochem. Mol. Biol.* 41, 615–619.

Nesbitt, P.D., Lam, Y., and Thompson, L.U. (1999). Human metabolism of mammalian lignan precursors in raw and processed flaxseed. *Am. J. Clin. Nutr.* 69, 549–555.

Nesbitt, P.D., and Thompson, L.U. (1997). Lignans in homemade and commercial products containing flaxseed. *Nutr. Cancer* 29, 222–227.

Obermeyer, W.R., Musser, S.M., Betz, J.M., Casey, R.E., Pohland, A.E., and Page, S.W. (1995). Chemical studies of phytoestrogens and related compounds in dietary supplements: Flax and chapparral. *Proc. Soc. Expt. Biol. Med.* 208, 6–12.

Orcheson, L.J., Rickard, S.E., Seidl, M.M., and Thompson, L.U. (1998). Flaxseed and its mammalian lignan precursor cause a lengthening or cessation of estrous cycling in rats. *Cancer Lett.* 125, 69–76.

Parkin, D.M., Muir, C.S., Whelan, S.L., Gao, Y., Ferlay, J., and Powell, J. (1997). *Cancer Incidence in Five Continents.* Scientific Publication No. 143, International Agency for Research in Cancer, Lyon, France.

Phipps, W.R., Martini, M.C., Lampe, J.W., Slavin, J.L., and Kurzer, M.S. (1993). Effect of flaxseed ingestion on the menstrual cycle. *J. Clinical Endocrinol. Metab.* 77, 1215–1219.

Piantelli, M., Ricci, R., Larocca, M., Rinelli, A., Capelli, A., Rizzo, S., Scambia, G., and Ranelleti, F.O. (1990). Type II estrogen binding sites in human colorectal carcinoma. *J. Clin. Pathol.* 43, 1004–1006.

Prasad, K. (1997). Hydroxy radical-scavenging property of secoisolariciresinol diglucoside (SDG) isolated from flax-seed. *Mol. Cell. Biochem.* 168, 117–123.

Rickard, S.E., Orcheson, L.J., Seidl, M.M., Luyengi, L., Fong, H.H.S., and Thompson, L.U. (1996). Dose-dependent production of mammalian lignans in rats and *in vitro* from the purified precursor secoisolariciresinol diglycoside in flaxseed. *J. Nutr.* 126, 2012–2019.

Rickard, S.E., and Thompson, L.U. (1997). Phytoestrogens and lignans: Effects on reproduction and chronic disease. In Shahidi, F. (ed.) *Antinutrients and Phytochemicals in Foods*, ACS, Washington, DC, pp. 273–293.

Rickard, S.E., and Thompson, L.U. (1998). Chronic exposure to secoisolariciresinol diglycoside alters lignan disposition in rats. *J. Nutr.* 128, 615–623.

Rickard, S.E., Yuan, Y.V., Chen, J., and Thompson, L.U. (1999). Dose effects of flaxseed and its lignan on N-methyl-N-nitrosourea (MNU)-induced mammary tumorigenesis in rats. *Nutr. Cancer* **35**, 50–57.

Russo, I.H., and Russo, J. (1996). Mammary gland neoplasia in long-term rodent studies. *Environ. Health Perspect.* **104**, 938–967.

Sathyamoorthy, N., Wang, T.T., and Phang, J.M. (1994). Stimulation of pS2 expression by diet-derived compounds. *Cancer Res.* **54**, 957–961.

Serraino, M., and Thompson, L.U. (1991). The effect of flaxseed supplementation on early risk markers for mammary carcinogenesis. *Cancer Lett.* **60**, 135–142.

Serraino, M., and Thompson, L.U. (1992a). The effect of flaxseed supplementation on the initiation and promotional stages of mammary tumorigenesis. *Nutr. Cancer* **17**, 153–159.

Serraino, M., and Thompson, L.U. (1992b). Flaxseed supplementation and early markers of colon carcinogenesis. *Cancer Lett.* **63**, 159–165.

Serraino, M., Thompson, L.U., and Cunnane, S.C. (1992). Effect of low level flaxseed supplementation on the fatty acid composition of mammary glands and tumors in rats. *Nutr. Res.* **12**, 767–772.

Setchell, K.D.R. (1995). Discovery and potential clinical importance of mammalian lignans. In Cunnane, S.C., and Thompson, L.U. (eds), *Flaxseed in Human Nutrition*, AOCS Press, Champaign, Ill, pp. 82–98.

Setchell, K.D.R., Bull, R., and Adlercreutz, H. (1980a). Steroid excretion during the reproductive cycle and pregnancy of the velvet monkey (*Ceropithecus aethiopus pygerythrus*). *J. Steroid Biochem.* **12**, 375–384.

Setchell, K.D.R., Lawson, A.M., Axelson, M., and Adlercreutz, H. (1981a). The excretion of two new phenolic compounds during the human menstrual cycle and in pregnancy. In Adlercreutz, H., Bulbrook, R.D., van der Molen, H.J., Vermeulen, A., and Sciarra, F. (eds), *Endocrinological Cancer, Ovarian Function and Disease. Proc IX Meeting Int. Study Group for Steroid Hormones, Rome, Dec 5–7, 1979*, Vol. 515, Excerpta Medica, Amsterdam, pp. 207–215.

Setchell, K.D.R., Lawson, A.M., Borriello, S.P., Harkness, R., Gordon, H., Morgan, D.M.L., Kirk, D.N., Adlercreutz, H., Anderson, L.C., and Axelson, M. (1981b). Lignan formation in man–microbial involvement and possible roles in relation to cancer. *Lancet* ii, 4–7.

Setchell, K.D.R., Lawson, A.M., Conway, E., Taylor, N.F., Kirk, D.N., Cooley, G., Farrant, R.D., Wynn, S., and Axelson, M. (1981c). The definite identification of the lignans *trans*-2,3-bis(3-hydroxybenzyl-γ-butyrolactone and 2,3-bis(3-hydroxybenzyl)butane-1,4-diol in human and animal urine. *Biochem. J.* **197**, 447–458.

Setchell, K.D.R., Lawson, A.M., Mitchell, F.L., Adlercreutz, H., Kirk, D.N., and Axelson, M. (1980b). Lignans in man and animal species. *Nature* **287**, 740–742.

Sharkey, M., and Bruce, R. (1986). Quantitation of nuclear aberrations as a screen for agents damaging to mammary epithelium. *Carcinogenesis* **7**, 1991–1995.

Shultz, T.D., Bonorden, W.R., and Seaman, W.R. (1991). Effect of short-term flaxseed consumption of lignan and sex hormone metabolism in men. *Nutr. Res.* **11**, 1089–1100.

Stitch, S.R., Toumba, J.K., Groen, M.B., Funke, C.W., Leemhuis, J., Vink, G.F., and Woods, G.F. (1980). Excretion, isolation and structure of a new phenolic constituent in female urine. *Nature* **287**, 738–740.

Sung, M.-K., Lautens, M., and Thompson, L.U. (1998). Mammalian lignans inhibit the growth of estrogen-independent human colon tumor cells. *Anticancer Res.* **18**, 1405–1408.

Thompson, L.U. (1994). Antioxidants and hormone-mediated health benefits of whole grains. *Crit. Rev. Food Sci. Nutr.* **34**, 473–497.

Thompson, L.U., Rickard, S.E., Cheung, F., Kenaschuk, E.O., and Obermeyer, W.R. (1997). Variability in anticancer lignan levels in flaxseed. *Nutr. Cancer* **27**, 26–30.

Thompson, L.U., Rickard, S.E., Orcheson, L.J., and Seidl, M.M. (1996a). Flaxseed and its lignan and oil components reduce mammary tumor growth at a late stage of carcinogenesis. *Carcinogenesis* **17**, 1373–1376.

Thompson, L.U., Robb, P., Serraino, M., and Cheung, F. (1991). Mammalian lignan production from various foods. *Nutr. Cancer* 16, 43–52.

Thompson, L.U., Seidl, M.M., Rickard, S.E., Orcheson, L.J., and Fong, H.H.S. (1996b). Antitumorigenic effect of a mammalian lignan precursor from flaxseed. *Nutr. Cancer* 26, 159–165.

Toniolo, P.G. (1997). Endogenous estrogens and breast cancer risk: The case for prospective cohort studies. *Environ. Health Perspect.* 105 (Suppl.), 587–592.

Tou, J.C., Chen, J., and Thompson, L.U. (1998). Flaxseed and its lignan precursor, secoisolariciresinol diglucoside, affect pregnancy outcome and reproductive development in rats. *J. Nutr.* 128, 1861–1868.

Tou, J.C., Chen, J., and Thompson, L.U. (1999). Dose, timing, and duration of flaxseed exposure affect reproductive indices and sex hormone levels in rats. *J. Toxicol. Environ. Health* 56, 555–570.

Tou, J.C.L., and Thompson, L.U. (1999). Exposure to flaxseed or its lignan component during different developmental stages influences rat mammary gland structures. *Carcinogenesis* 20, 1831–1835.

Wang, C., and Kurzer, M.S. (1997). Phytoestrogen concentration determines effects on DNA synthesis in human breast cancer cells. *Nutr. Cancer* 28, 236–247.

Wang, C., and Kurzer, M.S. (1998). Effects of phytoestrogens on DNA synthesis in MCF-7 cells in the presence of estradiol or growth factors. *Nutr. Cancer* 31, 90–100.

Wang, C., Mäkelä, T., Hase, T., Adlercreutz, H., and Kurzer, M.S. (1994). Lignans and flavanoids inhibit aromatase enzyme in human preadipocytes. *J. Steroid Biochem. Molec. Biol.* 50, 205–212.

Waters, A.P., and Knowler, J.T. (1982). Effect of a lignan (HPMF) on RNA synthesis in the rat uterus. *J. Reprod. Fertil.* 66, 379–381.

Welshons, W.V., Murphy, C.S., Koch, R., Calaf, G., and Jordan, V.C. (1987). Stimulation of breast cancer cells *in vitro* by the environmental estrogen enterolactone and the phytoestrogen equol. *Breast Cancer Res. Treat.* 10, 169–175.

Westcott, N.D., and Muir, A.D. (1996). Variation in the concentration of the flaxseed lignan concentration with variety, location and year. In *Proc. of the Flax Institute of the United States*, Vol. 56, Flax Institute of the United States, Fargo, ND, pp. 77–80.

Yan, L., Lee, J.A., Li, D., McGuire, M.H., and Thompson, L.U. (1998). Dietary flaxseed supplementation and experimental metastasis of melanoma cells in mice. *Cancer Lett.* 124, 181–186.

Yuan, Y.V., Rickard, S.E., and Thompson, L.U. (1999). Short-term feeding of flaxseed or its lignan has minor influence on *in vivo* hepatic antioxidant status in young rats. *Nutr. Res.* 19, 1233–1243.

9 Flaxseed in the prevention of cardiovascular diseases

Kailash Prasad

Introduction

Cardiovascular disease (ischemic heart disease, stroke, peripheral vascular disease) is the greatest killer of people in the industrialized nations. Hypercholesterolemia is a major risk factor for endothelial dysfunction, atherosclerosis and related vascular diseases. The role of hypercholesterolemia in the genesis of atherosclerosis and its clinical sequela, particularly ischemic heart disease and stroke, is now well established.

This chapter looks at the effectiveness of flaxseed in lowering serum lipids and preventing hypercholesterolemic atherosclerosis. It will also delineate the component(s) of flaxseed which is(are) effective in retarding the development of hypercholesterolemic atherosclerosis. Special attention has been given to the role of oxidative stress in hypercholesterolemic atherosclerosis.

Flaxseed components and their pharmacological properties

Flaxseed has been used as food in many countries, especially in Asia, for centuries. Flax oil has been used as a cooking oil in many countries. The overall composition of flaxseed is: fat (36%), protein (24%), carbohydrate (24%), fiber (6%), water (5.5%) and ash (3.4%) (Kritchevsky *et al.*, 1991). The major fatty acids of flax oil are: α-linolenic (53.3%), linoleic (12.7%), oleic (20.2%), palmitic (5.3%) and stearic (4.1%). In general 32–45 percent of the mass of flaxseed is oil, of which 51–55 percent is α-linolenic acid (α-3 fatty acid) (Hettiarachchy *et al.*, 1990; Oomah and Mazza, 1993). Flaxseed is the richest source of α-3 fatty acid as α-linolenic acid (Hunter, 1990; Kelley *et al.*, 1991) and the plant lignan known as secoisolariciresinol diglucoside (Bakke and Klosterman, 1956). The α-linolenic acid content of vegetable oils such as canola, soybean and corn is 11.1, 6.8 and 1.0 percent of total fat, respectively (Exler and Weihrauch, 1985; Hunter, 1990). Fish oil usually contains 20–25 percent of the fat as ω-3 fatty acids such as eicosapentaenoic, docosahexaenoic, and docosapentaenoic acid (Exler and Weihrauch, 1985; Hunter, 1990). Consumption of α-linolenic acid from flaxseed increases levels of serum ω-3 polyunsaturates, including eicosapentaenoic acid (20:5n-3) and docosapentaenoic acid (22:5n-3) and can raise docosahexaenoic acid (22:6n-3), depending upon the amount of dietary α-linolenic acid (Sanders and Roshanai, 1983; Mantzioris *et al.*, 1994).

Flaxseed also contains elements which are non-nutritional (Poulton, 1989) and include cyanogenic glycosides, vitamin B_6 antagonist and phytic acid which inhibits absorption of zinc and calcium. The cyanogenic glycosides are linustatin, neolinustatin and linamarin. Cyanogenic glucosides are hydrolysed to hydrogen cyanide which is a potent inhibitor of cell respiration. Flaxseed contains a vitamin B_6 antagonist called linatine.

Since vitamin B_6 is involved in numerous reactions, its deficiency might lead to vitamin B_6 deficiency related disorders. Phytic acid might reduce the levels of zinc and calcium in the tissue and blood, leading to calcium and zinc deficiency related disorders. However, the quantity of these compounds (cyanogenic glycosides, vitamin B_6 antagonist and phytate) is very small when flaxseed is used in moderate amounts and is expected to be eliminated in baked products.

Atherosclerosis causes ischemic heart disease, stroke and other peripheral vascular diseases. Oxygen free radicals (OFRs) have been implicated in the development of hypercholesterolemic atherosclerosis (Steinberg, 1992; Prasad and Kalra, 1993; Prasad *et al.*, 1994; Prasad *et al.*, 1997b). OFRs produce endothelial dysfunction which is a requirement for the development of atherosclerosis (Ross, 1986). Possible sources of OFRs in hypercholesterolemia have been described in detail in other studies (Prasad and Kalra, 1993; Prasad, 2000). Hypercholesterolemic atherosclerosis is associated with oxidative stress (Steinberg, 1992; Prasad and Kalra, 1993; Prasad *et al.*, 1994; Prasad *et al.*, 1997b; Prasad, 2000). Antioxidants (vitamin E, purpurogallin, garlic, probucol) reduce the development of hypercholesterolemic atherosclerosis and decrease the oxidative stress (Prasad and Kalra, 1993; Prasad *et al.*, 1994; Prasad *et al.*, 1997a; Prasad *et al.*, 1997b).

Various components of flaxseed have pharmacological activities that could reduce oxidative stress and hence would retard the development of hypercholesterolemic atherosclerosis. α-linolenic acid suppresses production of interleukin-1 (IL-1) and tumor necrosis factor TNF (Endres *et al.*, 1989; Chandrasekar and Fernandes, 1994). Dietary ω-3 fatty acids reduce the production of inflammatory mediators and leukotrienes from monocytes (Lee *et al.*, 1985) and suppress the respiratory bursts of polymorphonuclear leukocytes (PMNLs) (Fisher *et al.*, 1990). Dietary ω-3 fatty acid also reduces the production of superoxide anions (oxygen free radicals) by monocytes (Fisher *et al.*, 1986). Lignans have anti-PAF (platelet activating factor) activity (Cox and Wood, 1987). The lignan secoisolariciresinol diglucoside (SDG) isolated from flaxseed also has antioxidant activity (Prasad, 1997b).

Flaxseed and hypercholesterolemic atherosclerosis

Oxidative stress from hypercholesterolemia is believed to be due to increased production of OFRs by PMNLs and through increased arachidonic acid metabolism (Prasad and Kalra, 1993). PAF, IL-1 and TNF stimulate PMNLs to produce OFRs (Shaw *et al.*, 1981; Paubert-Braquet *et al.*, 1988; Braquet *et al.*, 1989). Since flaxseed contains the lignan SDG (an antioxidant and PAF-receptor antagonist), and α-linolenic acid (suppressant of IL-1 and TNF), it should be able to reduce oxidative stress and the development of atherosclerosis due to hypercholesterolemia. In one study (Prasad, 1997a) rabbits were divided into four groups: Group I, control diet; Group II, control diet + flaxseed (7.5 g/kg body weight, orally); Group III, 1 percent cholesterol diet; Group IV, 1 percent cholesterol diet + flaxseed (7.5 g/kg body weight, orally). These rabbits were on their respective diets for eight weeks. Blood samples for serum triglycerides (TG), total cholesterol (TC) and OFR-producing activity of PMNLs were collected before and after various intervals on the respective diets. The aorta was removed at the end of eight weeks to assess the extent of atherosclerosis. Flaxseed reduced the development of hypercholesterolemic atherosclerosis by 46 percent. Figure 9.1 is a representative graph of the atherosclerotic changes on the endothelial surface of the aorta from the four experimental groups. The reduction in atherosclerosis was associated with a

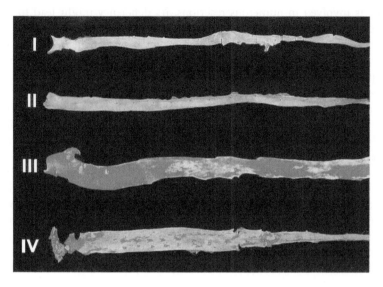

Figure 9.1 Intimal surface of aorta from four experimental groups showing lipid deposits (atherosclerotic changes) as bright red in color. (Reproduced from Prasad, 1997a with permission).

reduction in the levels of OFRs production by PMNLs. Flaxseed increased the serum total cholesterol (TC) slightly in the hypercholesterolemic group. Serum triglycerides (TG) remained unchanged. There are some studies where flax diets have been shown to affect serum lipids. Ratnayake *et al.* (1992) reported that although 10 percent flaxseed in diet did not affect the serum lipids, 20 percent and 40 percent flaxseed in diet lowered the serum TG, TC and low-density lipoprotein cholesterol (LDL-C) by 23 and 23 percent, 21 and 33 percent, and 33.7 and 67 percent respectively in rats. Kritchevsky *et al.* (1991) reported that 27.8, 41.7 and 55.6 percent flaxseed in diet did not change the serum TG levels in rats. However, the diets did lower the serum TC and LDL-C. Flaxseed and flax oil (15 percent in the diet) have been reported to prevent a rise in serum TC and TG in 5/6 renal ablation rat model (Ingram *et al.*, 1995). Ranhotra *et al.* (1993) also reported a decrease in the serum TC and serum non-high-density cholesterol in rats fed on flaxseed. The composition of flaxseed in the diet was 42.4 percent.

In humans, a dose of 50 g of flaxseed daily for four weeks lowered the serum TC by 9 percent and LDL-C by 18 percent (Cunnane *et al.*, 1993); 15 g of flaxseed daily for three months given to a hyperlipidemic human lowered the serum TC (18 mg/dl) and LDL-C (19 mg/dl) (Bierenbaum *et al.*, 1993). High-density lipoprotein-cholesterol (HDL-C) remained unchanged and TG was lowered slightly. A dose of 50 g of flaxseed daily for four weeks in humans lowered the plasma TC (6 percent), and LDL-C (9 percent) (Cunnane *et al.*, 1995). However, the serum HDL-C, TG and ratio of HDL-C/LDL-C remained unchanged. In another study flaxseed (30 g/daily for four weeks) given to humans reduced TC by 11 percent and LDL-C by 12 percent (Clark *et al.*, 1995). Flax oil given for 56 days to healthy humans had no effect on serum TG, TC, LDL-C and HDL-C (Kelley *et al.*, 1993). There are no other studies on the effects of flaxseed or flax oil on hypercholesterolemic atherosclerosis.

Flaxseed with very low α-linolenic acid and hypercholesterolemic atherosclerosis

The antiatherogenic effect observed with flaxseed could be due to α-linolenic acid, lignans (SDG) or both. α-linolenic acid in fish oil has been shown to reduce hypercholesterolemic atherosclerosis (Davis *et al.*, 1987; Zhu *et al.*, 1988). If the antiatherogenic effect of flaxseed is due to the α-linolenic acid content, flaxseed without α-linolenic acid would have no antiatherogenic effect. To ascertain if the antiatherogenic effect of flaxseed is due to α-linolenic acid, a low α-linolenic acid variety of flax developed by the Crop Development Center of the University of Saskatchewan (CDC-flaxseed) was used in hypercholesterolemic rabbits. CDC-flaxseed has an oil content (35 percent of total mass) and concentration of lignan SDG (16.4 mg/g versus 15.4 mg/g of defatted meal) similar to that of ordinary flaxseed but has only 2–3 percent of α-linolenic acid. The rabbits were assigned to four groups: Group I, control diet; Group II, control diet + CDC-flaxseed (7.5 g/kg body weight, orally); Group III, 1 percent cholesterol diet; and Group IV, 1 percent cholesterol diet + CDC-flaxseed (7.5 g/kg, body weight, orally). These rabbits were on their respective diets for eight weeks, at the end of which the aorta was removed to assess the extent of atherosclerosis. Blood samples were collected before and after various intervals for measurement of serum lipids. CDC-flaxseed at a dose of 7.5 g/kg body weight (similar to the dose of ordinary flaxseed) reduced the development of atherosclerosis by 69 percent (Figure 9.2) (Prasad *et al.*, 1998). The reduction of atherosclerosis was associated with a decrease in serum TC and LDL-C (Figure 9.3) and the ratio of TC/HDL-C and LDL-C/HDL-C by 14, 17, 28, and 24 percent, respectively (Prasad *et al.*, 1998). However, the reductions in serum TC, LDL-C and the ratio of LDL-C/HDL-C were not significant. Serum TG and VLDL-C were markedly elevated but serum HDL-C remained unchanged.

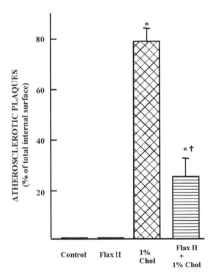

Figure 9.2 The extent of atherosclerotic changes in the four groups. Results are expressed as mean ± S.E. of % of total aortic intimal surface covered with atherosclerotic plaques. *p < 0.05, control versus other groups. †p < 0.05, 1% cholesterol vs 1% cholesterol + CDC-flaxseed (Flax II). (Reproduced from Prasad *et al.*, 1998 with permission).

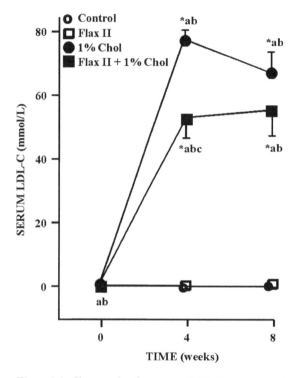

Figure 9.3 Changes in the serum LDL-C concentration of the four groups. Results are expressed as mean ± S.E. *p < 0.05, comparison of values at various times with respect to "0" time in the respective groups. ^{a}p < 0.05, control versus other groups. ^{b}p < 0.05, Flax II versus 1% cholesterol with or without Flax II. ^{c}p < 0.05, 1% cholesterol versus 1% cholesterol + Flax II. (Reproduced from Prasad *et al.*, 1998 with permission).

These results suggest that reduction in atherosclerosis by CDC-flaxseed is not due to α-linolenic acid but may be due to the lignan component of the flaxseed which might have a lipid lowering effect in addition to its antioxidant activity. Reduction in hypercholesterolemic atherosclerosis by antioxidants (vitamin E, probucol, garlic) has been reported earlier (Prasad and Kalra, 1993; Prasad *et al.*, 1994; Prasad *et al.*, 1997b). Decreases in serum TC and LDL-C, and increases in TG and VLDL-C, with CDC-flaxseed could be due to low levels of α-linolenic acid. α-linolenic acid decreases TG and VLDL-C (Phillipson *et al.*, 1985; Sanders *et al.*, 1985; Kestin *et al.*, 1990), and HDL-C (Harris *et al.*, 1993), and increases LDL-C (Nestel *et al.*, 1984; Harris *et al.*, 1993) and TC (Nestel *et al.*, 1984). The absence of α-linolenic acid may have the opposite effect to that seen with CDC-flaxseed.

The results of this study suggest that the antiatherogenic effect of flaxseed is not due to α-linolenic acid and may be due to SDG.

Secoisolariciresinol diglucoside (SDG) and hypercholesterolemic atherosclerosis

The studies of flaxseed with a high content of α-linolenic acid and low α-linolenic acid (CDC-flaxseed) suggest that the antiatherogenic effect of flaxseed could be due to its SDG content which has antioxidant activity (Prasad, 1997b). Because of the anti-PAF activity of lignans (Cox and Wood, 1987) SDG could inhibit the PAF-induced release of OFRs by PMNLs and because of the antioxidant activity of SDG, it could scavenge OFRs produced by the body. Antioxidants have been reported to retard hypercholesterolemic atherosclerosis (Prasad and Kalra, 1993; Prasad *et al.*, 1994; Prasad *et al.*, 1997a; Prasad *et al.*, 1997b). SDG has been isolated in pure form from flaxseed meal by Westcott and Muir (1998). The effects of SDG on the development of hypercholesterolemic atherosclerosis, serum lipid profile (TC, TG, LDL-C, HDL-C, VLDL-C), aortic malondialdehyde (MDA), lipid peroxidation (a measure of level of OFRs) and aortic chemiluminescence (AO-CL) (a measure of antioxidant reserve) have been investigated in rabbits by Prasad (1999). The rabbits were assigned to four groups: Group I, control diet; Group II, control diet + SDG (15 mg/kg body weight, orally); Group III, 1 percent cholesterol diet; and Group IV, 1 percent cholesterol diet + SDG (15 mg/kg body weight, orally). Blood samples were collected before and after four and eight weeks on these experimental diets for measurement of serum lipids. The aorta was removed at the end of eight weeks of the experimental diet to assess the extent of atherosclerosis, MDA and AO-CL. SDG reduced the development of hypercholesterolemic atherosclerosis by 73 percent. The reduction of atherosclerosis was associated with a decrease in TC by 33 percent, LDL-C by 35 percent, and risk ratio of TC/HDL-C (Figure 9.4) and LDL-C/HDL-C by approximately 64 percent.

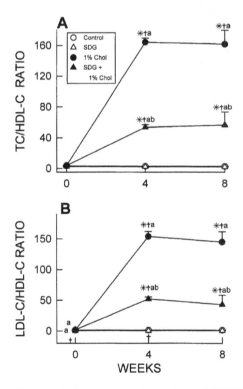

Figure 9.4 Changes in risk ratios of TC/HDL-C (A) and LDL-C/HDL-C (B) in four groups. Results are expressed as mean ± S.E. *p < 0.05, comparison of values at various times with respect to time "0" in respective groups. †p < 0.05, Group I versus other groups. ªp < 0.05, Group II versus Groups III and IV. ᵇp < 0.05, Group III versus Group IV. (Reproduced from Prasad, 1999 with permission).

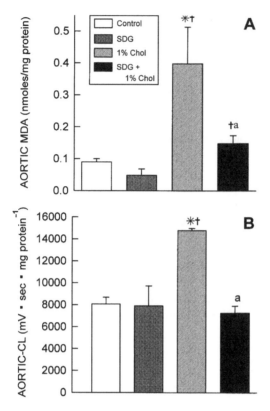

Figure 9.5 Aortic tissue MDA (A) and aortic-CL (B) in four groups. Results are expressed as means ± S.E. *p < 0.05, Group I versus other groups. †p < 0.05, Group II versus Groups III and IV. ªp < 0.05, Group III versus Group IV. (Reproduced from Prasad, 1999 with permission).

SDG initially raised serum HDL-C in high cholesterol-fed rabbits, but it remained unchanged at the end of eight weeks. Serum TG and VLDL-C remained unchanged in SDG-treated cholesterol-fed rabbits. Hypercholesterolemic atherosclerosis was associated with an increase in aortic tissue MDA and a decrease in the antioxidant reserve. The SDG-treated hypercholesterolemic group had lower MDA and higher antioxidant reserve (Figure 9.5).

These results suggest SDG is effective in reducing hypercholesterolemic atherosclerosis and this effect was associated with a decrease in serum TC, LDL-C, risk ratios and oxidative stress. SDG, therefore, may be effective in retarding the development of hypercholesterolemic atherosclerosis and lowering the relative risk of coronary artery disease, stroke and other peripheral vascular diseases.

Comments

The SDG compound is obtained from natural food products and hence it is expected to have no or minimal side-effects. This compound does not contain cyanogenic glycosides, phytate or vitamin B$_6$ antagonist elements. Hence it would be far superior to flaxseed. The dose required to accomplish the antiatherogenic effect is very small compared to the dose of flaxseed. The use of the SDG compound may prevent the development of hypercholesterolemic atherosclerosis and hence reduce the morbidity and mortality associated with heart attack, stroke and other peripheral vascular diseases.

Acknowledgment

The excellent secretarial assistance of Ms. Gloria Schneider is gratefully acknowledged.

References

Bakke, J.E., and Klosterman, H.J. (1956). A new diglucoside from flaxseed. *Proc. N. Dakota Acad. Sci.* 10, 18–22.

Bierenbaum, M.L., Reichstein, R., and Watkins, T.R. (1993). Reducing atherogenic risk in hyperlipemic humans with flaxseed supplementation: A preliminary report. *J. Am. Coll. Nutr.* 12, 501–504.

Braquet, P., Hosford, D., Braquet, M., Bourgain, R., and Bussolino, F. (1989). Role of cytokines and platelet-activating factor in microvascular immune injury. *Int. Arch. Allergy Appl. Immunol.* 88, 88–100.

Chandrasekar, B., and Fernandes, G. (1994). Decreased pro-inflammatory cytokines and increased antioxidant enzyme gene expression by omega-3 lipids in murine lupus nephritis. *Biochem. Biophys. Res. Comm.* 200, 893–898.

Clark, W.F., Parbtani, A., Huff, M.W., Spanner, E., De Salis, H., Chin-Yee, I., Philbrick, D.J., and Holub, B.J. (1995). Flaxseed: A potential treatment for lupus nephritis. *Kidney Int.* 48, 475–480.

Cox, C.P., and Wood, K.L. (1987). Selective antagonism of platelet-activating factor (PAF)-induced aggregation and secretion of washed rabbit platelets by CV-3988, L-652731, triazolam and alprazolam. *Thromb. Res.* 47, 249–257.

Cunnane, S.C., Ganguli, S., Menard, C., Liede, A.C., Hamadeh, M.J., Chen, Z.-Y., Wolever, T.M.S., and Jenkins, D.J.A. (1993). High α-linolenic acid flaxseed (*Linum usitatissimum*): Some nutritional properties in humans. *Br. J. Nutr.* 49, 443–453.

Cunnane, S.C., Hamadeh, M.J., Liede, A.C., Thompson, L.U., Wolever, T.M.S., and Jenkins, D.J.A. (1994). Nutritional attributes of traditional flaxseed in healthy young adults. *Am. J. Clin. Nutr.* 61, 62–68.

Davis, H.R., Bridenstine, R.T., Vesselinovitch, D., and Wissler, R.W. (1987). Fish oil inhibits development of atherosclerosis in rhesus monkeys. *Arteriosclerosis* 7, 441–449.

Endres, S., Ghorbani, R., Kelley, V.E., Georgilis, K., Lonnemann, G., van der Meer, J.W., Cannon, J.G., Rogers, T.S., Klemoner, M.S., Weber, P.C., *et al.* (1989). The effect of dietary supplementation with n-3 polyunsaturated fatty acids on the synthesis of interleukin-1 and tumor necrosis factor by mononuclear cells. *New England J. Med.* 320, 265–271.

Exler, J., and Weihrauch, J.L. (1985). *Provisional Table of Content of n-3 Fatty Acids and Other Fat Components in Selected Foods*. US Department of Agriculture. Washington, DC (Publication HNIS/PT-103).

Fisher, M., Levine, P.H., Weiner, B.H., Johnson, M.H., Doyle, E.M., Ellis, P.A., and Hoogasian, J.J. (1990). Dietary n-3 fatty acid supplementation reduces superoxide production and chemiluminescence in a monocyte-enriched preparation of leukocytes. *Am. J. Clin. Nutr.* 51, 804–808.

Fisher, M., Upchurch, K.S., and Levine, P.H. (1986). Effects of dietary fish oil supplementation on polymorphonuclear leukocyte inflammatory potential. *Inflammation* 10, 387–392.

Harris, W.S., Windsor, S.L., and Caspermeyer, J.J. (1993). Modification of lipid related atherosclerosis risk factors by omega-3 fatty acid ethyl esters in hypertriglyceridemic patients. *J. Nutr. Biochem.* 4, 706–712.

Hettiarachchy, N.S., Hareland, G.A., Ostenson, A., and Baldner-Shank, G. (1990). Chemical composition of eleven flaxseed varieties grown in North Dakota. *Proc. of the Flax Institute of the United States*, Flax Institute of the United States, pp. 36–40.

Hunter, J.E. (1990). n-3 fatty acids from vegetable oils. *Am. J. Clin. Nutr.* 51, 809–814.

Ingram, A.J., Parbtani, A., Clark, W.F., Spanner, E., Huff, M.W., Philbrick, D.J., and Holub, B.J. (1995). Effects of flaxseed and flax oil diets in a Rat-5/6 Renal Ablation Model. *Am. J. Kidney Dis.* 25, 320–329.

Kelley, D.S., Branch, L.B., Love, J.E., Taylor, P.C., Rivera, Y.M., and Iacono, J.M. (1991). Dietary α-linolenic acid and immunocompetence in humans. *Am. J. Clin. Nutr.* 53, 40–46.

Kelley, D.S., Nelson, G.J., Love, J.E., Branch, L.B., Taylor, P.C., Schmidt, P.C., Mackey, B.E., and Iacono, J.M. (1993). Dietary α-linolenic acid alters tissue fatty acid composition, but not blood lipids, lipoproteins or coagulation status in humans. *Lipids* 28, 533–537.

Kestin, M., Clifton, P., Belling, G.B., and Nestel, P.J. (1990). n-3 fatty acids of marine origin lower systolic blood pressure and triglycerides but raise LDL cholesterol compared with n-3 and n-6 fatty acids from plants. *Am. J. Clin. Nutr.* 51, 1028–1034.

Kritchevsky, D., Shirley, A.T., and Klurfeld, D.M. (1991). Influence of flaxseed on serum and liver lipids in rats. *J. Nutr. Biochem.* 2, 133–134.

Lee, T.H., Hoover, R.L., Williams, J.D., Sperling, R.I., Ravalese, J., III, Spur, B.W., Robinson, D.R., Corey, E.J., Lewis, R.A., and Austen, K.F. (1985). Effect of dietary enrichment with eicosapentaenoic acid and docosahexaenoic acid on *in vitro* neutrophil and monocyte leukotriene generation and neutrophil function. *New England J. Med.* 312, 1217–1224.

Mantzioris, E., James, M.J., Gibson, R.A., and Cleland, L.G. (1994). Dietary substitution with an α-linolenic acid-rich vegetable oil increases eicosapentaenoic acid concentrations in tissues. *Am. J. Clin. Nutr.* 59, 1304–1309.

Nestel, P.J., Connor, W.E., Reardon, M.F., Connor, S., Wong, S., and Boston, R. (1984). Suppression by diets rich in fish oil of very low density lipoprotein production in man. *J. Clin. Invest.* 74, 82–89.

Oomah, B.D., and Mazza, G. (1993). Flaxseed proteins – a review. *Food Chem.* 48, 109–114.

Paubert-Braquet, M., Longchampt, M.O., Koltz, P., and Guilbaud, J. (1988). Tumor necrosis factor (TNF) primes platelet activating factor (PAF)-induced superoxide generation by human neutrophils (PMN): Consequences in promoting PMN-mediated endothelial cell (EC) damage. *Prostaglandins* 35, 803.

Phillipson, B.E., Rothrock, D.W., Connor, W.E., Harris, W.S., and Illingworth, D.R. (1985). Reduction of plasma lipids, lipoproteins and apoproteins by dietary fish oils in patients with hypertriglyceridemia. *New England J. Med.* 312, 1210–1216.

Poulton, J.E. (1989). Toxic compounds in plant foodstuffs: Cyanogens in food proteins. In Kinsella, E., and Soucie, W.G. (eds), *Toxic Compounds in Plant Foodstuffs: Cyanogens in Food Proteins*, AOCS Press, Champaign, IL, pp. 381–401.

Prasad, K. (1997a). Dietary flaxseed in prevention of hypercholesterolemic atherosclerosis. *Atherosclerosis* 132, 69–75.

Prasad, K. (1997b). Hydroxyl radical scavenging property of secoisolariciresinol diglucoside (SDG) isolated from flaxseed. *Mol. Cell. Biochem.* 168, 117–123.

Prasad, K. (1999). Reduction of serum cholesterol and hypercholesterolemic atherosclerosis by secoisolariciresinol diglucoside isolated from flaxseed. *Circulation* 99, 1355–1362.

Prasad, K. (2000). Pathophysiology of atherosclerosis. In Chang, J.B., Olsen, E.R., and Prasad, K. (eds), *Textbook of Angiology*, Springer-Verlag, New York, pp. 85–105.

Prasad, K., and Kalra, J. (1993). Oxygen free radicals and hypercholesterolemic atherosclerosis: Effect of vitamin E. *Am. Heart J.* 125, 958–973.

Prasad, K., Kalra, J., and Lee, P. (1994). Oxygen free radicals as a mechanism of hypercholesterolemic atherosclerosis: Effects of probucol. *Int. J. Angiol.* **3**, 100–112.

Prasad, K., Mantha, S.V., Kalra, J., Kapoor, R., and Kamalarajan, B.R.C. (1997a). Purpurogallin in the prevention of hypercholesterolemic atherosclerosis. *Int. J. Angiol.* **6**, 157–166.

Prasad, K., Mantha, S.V., Kalra, J., and Lee, P. (1997b). Prevention of hypercholesterolemic atherosclerosis by garlic, an antioxidant. *J. Cardiovasc. Pharmacol. Ther.* **2**, 309–320.

Prasad, K., Mantha, S.V., Muir, A.D., and Westcott, N.D. (1998). Reduction of hypercholesterolemic atherosclerosis by CDC-flaxseed with very low α-linolenic acid. *Atherosclerosis* **136**, 367–375.

Ranhotra, G.S., Gelroth, J.A., Glaser, B.K., and Potnis, P.S. (1993). Lipidemic responses in rats fed flaxseed oil and meal. *Cereal Chem.* **70**, 364–366.

Ratnayake, W.M.N., Behrens, W.A., Fischer, P.W.F., L'Abbé, M.R., Mongeau, R., and Beare-Rogers, J.L. (1992). Chemical and nutritional studies of flaxseed (variety Linott) in rats. *J. Nutr. Biochem.* **3**, 232–240.

Ross, R. (1986). The pathogenesis of atherosclerosis: An update. *New England J. Med.* **314**, 488–500.

Sanders, T.A., Sullivan, D.R., Reeve, J., and Thompson, G.R. (1985). Triglyceride-lowering effect of marine polyunsaturates in patients with hypertriglyceridemia. *Arteriosclerosis* **5**, 459–465.

Sanders, T.A.B., and Roshanai, F. (1983). The influence of different types of ω3 polyunsaturated fatty acids on blood lipids and platelet function in healthy volunteers. *Clin. Sci.* **64**, 91–99.

Shaw, J.O., Pinckard, R.N., Ferrigni, K.S., McManus, L.M., and Hanahan, D.J. (1981). Activation of human neutrophils with 1-O-hexadecyl/octadecyl-2-acetyl- *sn*-glyceryl-3-phosphorylcholine (platelet activating factor). *J. Immunol.* **127**, 1250–1255.

Steinberg, D. (1992). Antioxidants in the prevention of human atherosclerosis. *Circulation* **85**, 2338–2344.

Westcott, N.D., and Muir, A.D. (1998). Process for extracting lignans from flaxseed. Patent # 5705618.USA.

Zhu, B.Q., Smith, D.L., Sievers, R.E., Isenberg, W.M., and Parmley, W.W. (1988). Inhibition of atherosclerosis by fish oil in cholesterol-fed rabbits. *J. Am. Coll. Cardiol.* **12**, 1073–1078.

10 Flaxseed and flaxseed lignans

Effects on the progression and severity of renal failure

William F. Clark and Malcolm Ogborn

Introduction

We have been interested in nutritional modulation of renal injury for many years (Clark *et al.*, 1989; Clark *et al.*, 1990; Clark *et al.*, 1991; Holub *et al.*, 1991; Clark *et al.*, 1993a; Clark *et al.*, 1993b; Clark and Parbtani, 1994; Tomobe *et al.*, 1994; Ogborn and Sareen, 1995; Ogborn *et al.*, 1998a). This interest has led us to study dietary flaxseed and flax oil in the remnant nephron experimental model of renal injury (Ingram *et al.*, 1995), as well as flaxseed and its derived lignan in lupus nephritis (Hall *et al.*, 1993; Clark *et al.*, 1995; Clark *et al.*, 2000; Clark *et al.*, 2001) and flaxseed and flax oil in experimental polycystic kidney disease (Ogborn *et al.*, 1998b; Ogborn *et al.*, 1999). We have had the opportunity of carrying out a series of experiments in the three different domains; i.e., primary progressive renal disease, secondary vasculitis [lupus] and classic polycystic kidney disease.

Animal models

Remnant nephron model

The remnant nephron model, or the rat renal ablation model, represents kidney diseases associated with progressive glomerular scarring and decline in function (Chanutin and Ferris, 1932; Hayslett, 1979). In this model one kidney is excised and two-thirds of the remaining kidney is infarcted, resulting in functional nephrons below a critical number with an increase in single nephron filtration and associated intraglomerular hydraulic pressure (Hostetter *et al.*, 1981; Brenner, 1985). Systemic and intraglomerular hypertension, proteinuria and glomerulosclerosis follow, leading inexorably to end-stage renal failure. In this study both hemodynamic and inflammatory events are thought to interact in the development of progressive renal disease. We carried out a study to compare the effects of flax derived from α-linolenic acid supplemented as oil (no lignans) or as a crushed seed (α-linolenic acid and lignans) and its impact on the rat remnant nephron or renal ablation model (Ingram *et al.*, 1995). Rats who underwent the 5/6 nephrectomy were divided into three groups to receive either regular laboratory rat chow (RLD), 15 percent dietary flaxseed supplementation, or 15 percent flax oil supplementation. The effect of surgery on blood pressure is shown in Table 10.1 at one week following surgery. On regular laboratory chow (RLD) all animals showed a significant rise in blood pressure, which was maintained in the animals on regular laboratory rat chow throughout the experiments, and decreased

Table 10.1 Body weight, blood pressure, and plasma lipids

	Presurgery	*Week 1*	*Week 10*	*Week 20*
Body weight (g)				
RLD	275 ± 16	295 ± 23	426 ± 39*†	560 ± 66*†
Flaxseed			487 ± 39*†	544 ± 29*†
Flax oil			476 ± 44*†	534 ± 41*†
Blood pressure (mm Hg)				
RLD	143 ± 14	171 ± 19*	165 ± 22*	176 ± 22*
Flaxseed			149 ± 04	158 ± 21
Flax oil			127 ± 21†	154 ± 20
Cholesterol (mg/dL)				
RLD	42.5 ± 11.6	54.1 ± 15.5		197.2 ± 46.4*† ‡ §
Flaxseed				85.6 ± 11.6
Flax oil				65.7 ± 11.6
Triglycerides (mg/dL)				
RLD	44.3 ± 8.9	44.3 ± 17.8		150.6 ± 97.4*†
Flaxseed				88.6 ± 35.4
Flax oil				70.9 ± 26.6

Note: All rats were fed RLD for up to one week post-nephrectomy. Data are given as mean values ± SD.
* Significantly different compared with presurgery.
† Significantly different compared with week 1.
‡ Significantly different compared with flaxseed.
§ Significantly different compared with flax oil.

Table 10.2 Renal function

	Presurgery	*Week 1*	*Week 20*
GFR mL/min/kg			
RLD	9.8 ± 1.7	5.7 ± 1.6*	1.7 ± 0.8*†
Flaxseed			3.1 ± 1.4*†
Flax oil			2.7 ± 1.2*†
Proteinuria mg/24 hr)			
RLD	4 (2, 6)	32* (11, 134)	170*†‡§ (82, 352)
Flaxseed			30* (11, 89)
Flax oil			33* (6, 100)

Note: All rats were fed RLD for up to one week post-nephrectomy. Data for GFR are given as mean values ± SD. Proteinuria was abnormally distributed; therefore, geometric means and ranges are given in parentheses.
* Significantly different compared with presurgery.
† Significantly different compared with week 1.
‡ Significantly different compared with flaxseed.
§ Significantly different compared with flax oil.

to pre-levels at weeks 10 and 20 in animals receiving flaxseed and flax oil therapy (Table 10.1). The remnant nephron surgery also results in significant proteinuria which rises significantly by one week and even further by 20 weeks in those animals who are on regular laboratory rat chow, and that rise is significantly abrogated in animals at 20 weeks who are receiving either flaxseed or flax oil (Table 10.2). Inulin clearances were also measured in the animals and declined at one week post-nephrectomy and infarction and continued to decline in all groups throughout the study. By 20 weeks

Table 10.3 Glomerular histology

	Percentage of glomeruli with no expansion (Grade 0)	Percentage of glomeruli with moderate expansion (Grade 1)	Percentage of glomeruli with global expansion (Grade 2)
Mesangial expansion score			
RLD 1 wk post-nephrectomy	50 ± 8	44 ± 6	6 ± 4
RLD 20 wk post-nephrectomy	$23 \pm 13*$	$30 \pm 4*†‡$	$47 \pm 13*†‡$
Flaxseed 20 wk post-nephrectomy	32 ± 5	$53 \pm 4*$	15 ± 3
Flax oil 20 wk post-nephrectomy	$27 \pm 7*$	51 ± 7	$22 \pm 7*$

	Percentage of glomeruli with no sclerosis (Grade 0)	Percentage of glomeruli with segmental sclerosis (Grade 1)	Percentage of glomeruli with global sclerosis (Grade 2)
Glomerular sclerosis score			
RLD 1 wk post-nephrectomy	57 ± 11	38 ± 8	5 ± 7
RLD 20 wk post-nephrectomy	$8 \pm 13*$	$18 \pm 23‡§$	$74 \pm 37*‡§$
Flaxseed 20 wk post-nephrectomy	$29 \pm 9*†$	62 ± 8	8 ± 6
Flax oil 20 wk post-nephrectomy	$17 \pm 7*$	56 ± 8	24 ± 13

Note: Data are given as mean values $\pm$ SD.
* Significantly different compared with presurgery.
† Significantly different compared with week 1.
‡ Significantly different compared with flaxseed.
§ Significantly different compared with flax oil.

post-nephrectomy the inulin clearances were significantly higher in the flax diet groups compared with the regular laboratory chow group (Table 10.2).

This study also involved a detailed assessment of histologic kidney damage by looking at the amount of mesangial expansion and glomerulosclerosis. Both flaxseed and flax oil had a significantly lower percentage of glomeruli showing grade 2 mesangial expansion at the 20 week time point and, as well, both prevented the increase in glomerulosclerosis that occurred in the RLD group at week 20 (Table 10.3). Renal phospholipid content was assessed in the remnant renal tissues obtained at 20 weeks post-nephrectomy from the RLD and the flax diet groups. The total omega-3 fatty acids showed a two- to three-fold increase in the flax oil versus the flaxseed and regular laboratory chow diet groups. The total saturated fatty acids were lower and total polyunsaturated fatty acids were increased in both flax-fed groups.

Urinary thromboxane B_2 and 6-keto $PGF_{1\alpha}$ levels were measured to observe the potential effects of the α-linolenic acid and EPA from the flax oil and the flaxseed on the remnant nephron model. A progressive rise was noted in the urinary thromboxane in the regular laboratory diet group but this was not seen in the flaxseed group and was significantly decreased in the flax oil group at 20 weeks. The urinary 6-keto $PGF_{1\alpha}$ levels declined in all groups. The response of the animals in terms of plasma cholesterol and triglycerides was assessed at 20 weeks in the three dietary groups (Table 10.1). The rise in triglycerides and cholesterol noted in the RLD group was largely attenuated by the two flax diets (Table 10.1).

These experiments in the rat remnant nephron model demonstrated that dietary flaxseed and flax oil attenuate a decline in renal function and reduced glomerular injury with favorable effects on blood pressure, plasma lipids and urinary prostaglandins.

Although our data did not clearly demonstrate a specific synergistic effect of the constituents of the flaxseed diet, there was a trend in all the renal functional parameters and histology to suggest a greater benefit for flaxseed versus flax oil therapy. Thus, the benefits of the flaxseed and flax oil may have importance for all forms of renal disease since the remnant nephron model is thought to represent progression in both immune and non-immune forms of kidney disease.

Flaxseed in lupus nephritis

Systemic lupus erythematosus (SLE) is an autoimmune disease in which patients often demonstrate a characteristic clinical pattern (Dubois *et al.*, 1974; Urowitz *et al.*, 1976; Albert *et al.*, 1979; Karsh *et al.*, 1979; Ginzler *et al.*, 1982; Hosenpud *et al.*, 1984; Correia *et al.*, 1985; Balow *et al.*, 1987; Ponticelli *et al.*, 1987). Early in the disease, inflammatory reactions can take place in blood vessels throughout all parts of the body with a wide range of signs and symptoms. Late in the disease, significant morbidity and mortality occur due to kidney failure and accelerated vascular disease with heart attacks, strokes and other atherogenic complications. Throughout the entire disease process, patients suffer an increased risk of sepsis which is thought to be related to the immune dysfunction of the disorder as well as secondary to the immuno-suppressive agents used to treat it. This pattern of injury in systemic lupus erythematosus has attracted our interest in nutritional agents such as flaxseed with its potential to exert both anti-inflammatory and lipid-lowering properties potentially able to reduce the accelerated renal and vascular disease, without increasing the risk of sepsis. Flaxseed is a rich natural source of α-linolenic acid which has antihypertensive, lipid-lowering, immunosuppressive and anti-inflammatory properties. As well, it is the richest natural source of plant lignans with platelet activating factor (PAF) receptor antagonist activity. PAF has been shown to play a key role in the mediation of inflammation, immune complex deposition and alteration of glomerular permselectivity in lupus nephritis (Baldi *et al.*, 1990; Tetta *et al.*, 1990; Macconi *et al.*, 1991; Morigi *et al.*, 1991). Thus, flaxseed has the potential to inhibit various mechanisms associated with the progression of renal disease in lupus nephritis.

Flaxseed in the experimental mouse model of lupus (MRL/lpr)

We have studied the effects of dietary flaxseed supplementation in the MRL/lpr mouse model (Hall *et al.*, 1993). These mice spontaneously develop a lymphoproliferative syndrome associated with systemic autoimmune disease involving skin, heart, joint and renal injury similar to human patients with systemic lupus erythematosus (Theofilopoulos and Dixon, 1985). Human lupus nephritis, as noted, is manifested by inflammatory and atherosclerotic events with morbid and mortal sequelae. After baseline studies at nine weeks of age, two groups of mice (n = 25) received either a controlled diet or 15 percent flaxseed diet. The study end was defined as the time at which the mortality rate in one of the dietary groups approximated 50 percent. Mortality was persistently greater in the control group throughout the study than the flaxseed supplemented group ($p < 0.05$) (Figure 10.1). Proteinuria was observed at 11 weeks of age in the control diet group and at 15 weeks of age in the flaxseed group, representing a four week delay in the onset of proteinuria in the flaxseed group. The percentage of mice with proteinuria was significantly lower in the flaxseed group compared with controls at 19 weeks of age ($p < 0.0001$) and at 21 weeks of age ($p < 0.01$) (Figure 10.2).

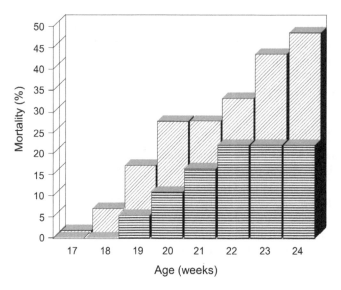

Figure 10.1 Percentage mortality in the flaxseed diet (horizontally striped bars) and the control diet (diagonally striped bars) groups. The mortality in the flaxseed-diet group was significantly lower compared with the control-diet group on Mantel-Haenszel (log-rank) survival analysis (limit of significance, P < 0.05).

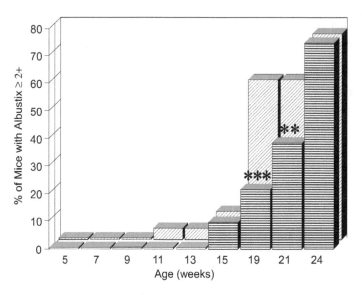

Figure 10.2 Percentage of mice with proteinuria of 1 g/L (2+) or greater on Albustix. The proteinuria was delayed by four weeks in the flaxseed group (horizontally striped bars) which had significantly lower proteinuria at 19 weeks of age (P < 0.001; ***) and 21 weeks of age (P < 0.01; **) compared with the control group (diagonally striped bars).

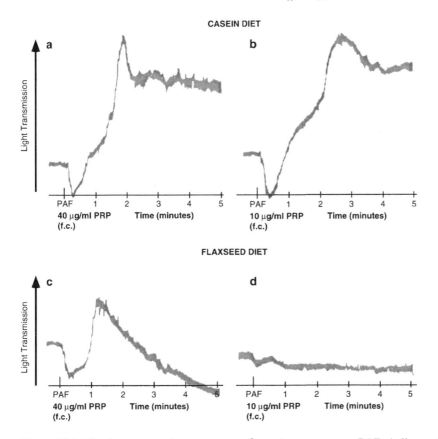

Figure 10.3 Platelet aggregation responses of rats in response to PAF challenge (10 μg and 40 μg/ml of platelet rich plasma). Rats a and b were on a control diet and rats c and d were on a flaxseed diet.

This difference was no longer apparent by 24 weeks of age. The ^{14}C-inulin clearance studies revealed that although both groups experienced a decline in GFR at 16 and 24 weeks, there was a significantly greater reduction in renal clearance for the control group at 16 weeks ($p < 0.01$) and at 24 weeks ($p < 0.02$). As well, three CD1 mice were fed the controlled diet and three a flaxseed diet for one week and then were administered an intravenous lethal dose of synthetic platelet activating factor (25 μg/kg body weight). The flaxseed diet group had a significant prolongation in their survival in terms of minutes following the PAF injection. The flaxseed diet also inhibited platelet aggregation in response to PAF, as shown in Figure 10.3. At higher concentrations of PAF (10 and 40 μg/ml of platelet rich plasma) the controlled diet fed mice showed irreversible platelet aggregation responses, whereas the flaxseed fed mice showed weaker, reversible platelet aggregation responses. At a lower dose of PAF (10 μg/ml platelet rich plasma) the flaxseed diet abrogated the platelet aggregation response (Figure 10.3). All mice gained weight throughout the study. The body weight did not differ significantly between the two groups either prior to or after the experimental diet allocation, or at the conclusion of the study.

Flaxseed dosing study in human lupus nephritis

In our previous studies we had demonstrated that flaxseed significantly preserved GFR, reduced proteinuria, lymphoproliferation and mortality in the MRL/lpr mouse model, and was much more potent than previous studies with fish oil (Felley *et al.*, 1985; Hall *et al.*, 1993). We also had studied the effects of flaxseed versus its constituents (flax oil and defatted flax) in the MRL/lpr mouse and observed that the whole flaxseed conferred superior benefit compared with its oil and defatted constituents (Parbtani and Clark, 1995). These encouraging results led us to this short-term study of the effects of three different doses of flaxseed on immune, inflammatory, rheologic, platelet and lipid parameters in lupus nephritis. The plan was to determine a dose of flaxseed which was well tolerated and would have significant effects on pathogenic mechanisms known to be involved in this progressive scarring disorder. Nine patients who met the minimum criteria of the American Rheumatism Association for the diagnosis of SLE, and who had a documented history of a positive ANA and at least one episode of elevated anti-DNA binding with proteinuria > 1 g/24 hours, red cells and red cell casts on urinalysis were enrolled, eight of whom completed the study (Clark *et al.*, 1995). After the baseline studies, patients were given 15, 30 and 45 grams of flaxseed per day sequentially at four-week intervals, followed by a five-week washout period. One patient dropped out of the study after the first four weeks due to scheduling difficulties at his workplace. Three subjects experienced difficulty ingesting the 45 grams of flaxseed daily due to increased laxation. All patients found the 15 and 30 gram flaxseed dosage well tolerated, either mixing it with orange or tomato juice or as a cereal topping. No patients had required a change in their medication during the 17 weeks of the study and three-day dietary intake records show no change in the intake of total calories, protein, fat, cholesterol, saturated, monounsaturated or polyunsaturated fatty acids. Compliance was excellent for the flaxseed dosage of 15 grams (15 ± 1 g) per day and 30 grams (30 ± 3 g) per day. However, at the 45 gram per day dosage the measurement was 43 ± 10 grams, indicating some variability as evidenced by three of the patients reporting difficulty in complying with the 45 gram per day dose due to increased laxation. This pattern of compliance was confirmed by the increase of α-linolenic acid levels and the serum phospholipids from baseline to four weeks post 15 gram flaxseed followed by a statistically significant increase with the 30 gram/day dosage and a moderate decline at the 45 gram/day dosage, with a return to baseline value after the five-week washout period.

Eicosapentaenoic acid (EPA), as well as the ratio of ω-3/ω-6 fatty acids, also increased with all flaxseed doses; however, the linoleic acid and the arachidonic acid levels were unchanged, which is compatible with the subjects not altering their diets while taking the flaxseed. The flaxseed exerted a significant cholesterol-lowering effect at 30 grams per day, reducing the total cholesterol by 11 percent and the LDL cholesterol by 12 percent. This effect was sustained over the four weeks of 45 grams per day dose (9 percent for total and 10 percent for LDL cholesterol) and the five-week washout period (7 percent for total and 11 percent for LDL cholesterol), indicating a prolonged lipid-lowering effect.

The threshold aggregation concentration of PAF was increased with 15 grams (35.9 ± 5.5 µg), 30 grams (27.1 ± 7.4 µg), and 45 grams (25.4 ± 9.3 µg) of flaxseed compared to baseline values of 16.1 ± 6.4 µg of PAF. At the end of the five-week washout period the threshold aggregation concentration for PAF approximated the baseline value (17.9 ± 5.9 μg). All of the renal function tests reflected a change with the dose of flaxseed dietary supplementation. Serum creatinine was significantly reduced with the 30 and 45 grams per day flaxseed dose, with a return toward pre-treatment values following the five-week washout period. The creatinine clearance reflected this improvement, rising with increasing flaxseed doses, while the serum urea was unchanged. Urinary protein declined with 15 and 30 grams and rose slightly with 45 grams per day, but returned to baseline level post-washout period. Three patients with a baseline proteinuria > 1 gram/24 hours showed marked reduction, particularly by the end of the 30 gram per day dose. The complements and anti-DNA antibody titres were unchanged throughout the study period. Flow cytometric studies showed there was a reduction in CD11B expression on neutrophils based on mean channel fluorescence during the 30 gram per day flaxseed dosing compared to baseline and post-five-week washout period.

This dosing study indicated that the 15 and 30 gram dose was well tolerated and produced significant changes in renal and neutrophil function and plasma lipids. These preliminary findings encouraged us to construct a larger randomized controlled prospective crossover study of flaxseed at 30 gram dose in patients with lupus nephritis.

Flaxseed in lupus nephritis: a two-year crossover study

We contacted forty patients with systemic lupus erythematosus and confirmed nephritis and asked them to participate in a randomized controlled crossover trial of flaxseed. Twenty-three agreed to be randomized to receive 30 grams of ground flaxseed daily or no flaxseed (no placebo) for one year followed by a twelve-week washout period and the reverse treatment for one year (Clark *et al.*, 2000). Eight patients dropped out of the study and of the 15 remaining subjects flaxseed sacket count and serum phospholipid levels indicated only 9 of the 15 were adherent to the flaxseed diet. Plasma lipids were unaltered by the flaxseed supplementation whereas serum creatinine in the compliant flaxseed patients declined from a mean of 0.97 ± 0.31 mg/dl to a mean of 0.94 ± 0.30 mg/dl and rose on the no flaxseed to a mean of 1.03 ± 0.28 mg/dl (p < 0.081). Of the 15 patients who completed the study, similar changes were noted (p < 0.1). The 9 compliant patients had lower serum creatinines at the end of the study than the 17 patients who refused to participate, of the 40 patients who had been asked to enrol in the study (p < 0.05). The microalbumin at baseline declined in both no flaxseed and flaxseed time periods but there was a trend for a greater drop in the flaxseed group (p < 0.2).

This study indicated that flaxseed exerted a renoprotective effect but our interpretation was affected by underpowering due to poor adherence and potential Hawthorne effects. The mention of the Hawthorne effect is that we were seeing a statistically significant difference between those who complied well with the study and those who did not even wish to participate and thus we were predisposing ourselves to a positive result in which compliant subjects are more likely to have a better outcome than those who do not wish to comply. Needless to say, our interpretation of this particular study was that the flaxseed, at a 30 gram dose, does not seem to be well tolerated if patients have to take it for a prolonged period of time (38 percent adherence rate). This led to research of substances derived from the flaxseed which might be more palatable. The isolation of the lignan precursor derived from flax by Drs. Westcott and Muir allowed us to construct a study of the lignan precursor in the MRL/lpr lupus mouse (Westcott and Muir, 1998; Clark *et al.*, 2001).

Lignan precursor derived from flax in the MRL/lpr lupus mouse

Flaxseed has demonstrated renoprotective effects in the MRL/lpr lupus mouse as well as in human lupus nephritis. The results of our two-year crossover study demonstrated a modest benefit; however, the 38 percent adherence rate clearly demonstrated that the flaxseed was not highly palatable for long-term treatment for patients with lupus nephritis. This encouraged us to construct a study which looked at the lignan derived from flaxseed to see if it would exert similar renoprotection in the aggressive MRL/lpr lupus mouse model to that of whole flax (Clark *et al.*, 2001). The lignan precursor avoids the undesirable side-effects of flax (i.e., increased laxation, cyanogenic glycosides, and caloric loads). Experiments were conducted in 141 MRL/lpr mice who were randomly assigned to receive saline gavage, 600, 1200 or 4800 µg of lignan. Absorption studies indicated that the lignan precursor was converted in the gut as expected, although the majority of it circulated in the form of the converted metabolite secoisolariciresinol (SDG) with a much smaller percentage in the enterodiol and enterolactone form. The SDG did demonstrate a protective effect against the PAF lethal challenge. PAF has been implicated in inflammatory and mitogenic events and, as well, is a known inducer of proteinuria by changing glomerular permeability due to electrostatic forces. Also, PAF has been shown to directly correlate with disease activity in both experimental and human lupus nephritis and also synthetic anti-PAF agents have been shown to provide protection from progressive renal impairment in this model of nephritis (Baldi *et al.*, 1990; Tetta *et al.*, 1990; Macconi *et al.*, 1991; Morigi *et al.*, 1991). The renoprotection afforded in terms of preventing proteinuria and the preservation of isotope GFR (Figures 10.4 and 10.5) was gradually reduced throughout the study time. This renoprotection was also reflected in terms of kidney size. A larger kidney size would be expected in animals who underwent hyperfiltration secondary to renal injury. In fact, the kidney weights do tend to reflect this renoprotection at 14 weeks whereas there is no significant change in renal size from baseline for both the low and high

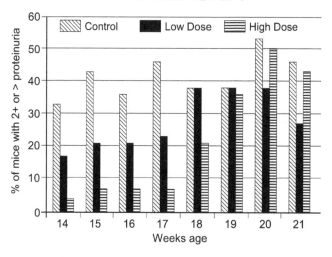

Figure 10.4 The effect of two dose levels of SDG on proteinuria in mice.

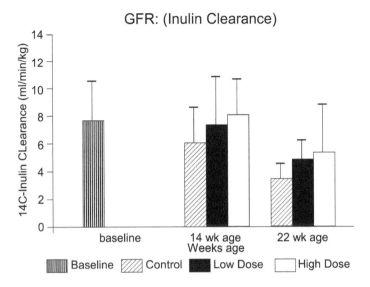

Figure 10.5 Inulin clearance rates as an indicator of GFR in mice fed low or high doses of SDG.

dose SDG mice. However, by 22 weeks all animals demonstrated a similar degree of increase in renal size. The renoprotection noted for the SDG was similar to that previously noted for the 15 percent dietary supplementation with flaxseed in the MRL/lpr mouse model. The SDG can be administered orally or intravenously. This use avoids the undesirable side-effects of flaxseed, i.e., increased laxation, cyanogenic glycosides and caloric loads. This study shows that SDG, a lignan precursor derived from flaxseed, is well tolerated and provides renoprotection similar to whole flaxseed in the aggressive MRL/lpr lupus mouse model.

Polycystic kidney disease – humans and experimental animals

Polycystic kidney disease (PKD) is the fourth most common cause of end-stage renal disease, with treatment costs exceeding a billion dollars per year in the USA with proportionately similar costs in other developed countries (Ogborn, 1994). Tubular dilatation is associated with increased rates of both epithelial proliferation (Nadasdy *et al.*, 1995) and apoptosis (Woo, 1995). These components may be seen in many forms of renal injury and PKD is now generally considered to be a disorder caused by dysregulation of renal repair and regeneration processes (Rankin *et al.*, 1992). Although disruption of architecture by cystic change in a minority of nephrons may contribute to the development of chronic renal failure in PKD, human cystic diseases demonstrate interstitial inflammation, fibrosis and nephron loss similar to that seen in many forms of chronic renal disease and which are the most important correlate with progression to chronic renal failure (Remuzzi *et al.*, 1997).

The Han:SPRD-*cy* rat is a model of PKD that shares autosomal dominant inheritance, progression through early adult life and sexual dimorphism with human disease (Cowley *et al.*, 1993). The disease is characterized by progressive dilatation of nephrons in young animals, associated with marked interstitial inflammation and fibrosis with associated nephron loss in older animals (Gretz *et al.*, 1995). Unlike human PKD, Han:SPRD-*cy* rat PKD has proved amenable to treatment. Modification of this model of PKD has been achieved with a variety of environmental manipulations including dietary protein restriction (Ogborn and Sareen, 1995), angiotensin converting enzyme inhibition or angiotensin receptor blockade, salt loading (Keith *et al.*, 1994), methyl-prednisolone therapy (Gattone *et al.*, 1995), or hypocholesterolemic therapy (Gile *et al.*, 1995).

Substitution of soy protein into the diet of Han:SPRD-*cy* rats results in slower disease progression and dramatic reduction in interstitial inflammation and fibrosis and is associated with decreased production of long chain ω-6 polyunsaturated fatty acid species (Ogborn *et al.*, 1998a; Ogborn *et al.*, 2000). As flaxseed has shown beneficial results in the renal ablation model of chronic renal failure (Ingram *et al.*, 1995), and has been claimed to produce benefits to general health comparable to soy, we undertook a series of studies to test the hypothesis that dietary flaxseed and derivatives would modify the clinical course, renal pathology and renal biochemistry of male Han:SPRD-*cy* rats.

Flaxseed dietary supplement in the Han:SPRD-cy rat

In our first study, experimental animals received a diet consisting of 80 percent rat chow from the same batch as the control animals to which was added 10 percent by weight corn starch and 10 percent by weight ground whole flaxseed, producing a diet that was 22.5 percent protein, 55 percent digestible carbohydrate and 5 percent lipid by weight (Ogborn *et al.*, 1999). Affected Han:SPRD-*cy* animals fed both flaxseed supplemented and control diets for eight weeks thrived with mean weights at sacrifice of 381 ± 17 GM and 369 ± 18 GM, respectively (p = ns), demonstrating that any benefit derived from flaxseed was not due to protein or energy deprivation. Flaxseed had no significant effect on total renal volume in Han:SPRD-*cy* rats (8.22 ml/kg versus 8.54 ml/kg). Littermates with normal kidneys also did not show any difference in renal volume in response to diet (4.68 ml/kg versus 4.58 ml/kg). Histologic studies revealed that flaxseed feeding was associated with a modest reduction in cystic change (1.78 ml/kg versus 2.03 ml/kg; p = 0.02). Markers of tubular remodeling, including frequency of both apoptotic nuclei and nuclei expressing PCNA, did not differ between the two treatment groups. Renal fibrous volume, measured by methyl blue staining that corresponded to interstitial type III collagen (Ogborn *et al.*, 1999), however, was reduced in flaxseed supplemented animals (0.60 ml/kg versus 0.93 ml/kg; p = 0.0009). A parallel decrease was seen in the number of macrophages infiltrating the kidney (13.8 cells/hpf versus 16.7 cells/hpf; p = 0.026). Serum creatinine was significantly lower in flaxseed-fed Han:SPRD-*cy* animals (69 μmol/l versus 81 μmol/l; p = 0.02). Serum creatinine in normal animals was uninfluenced by diet (flaxseed 67 ± 14 μmol/l versus control 69 ± 19 μmol/l). The diet did not, however, have any hypocholesterolemic effect on Han:SPRD-*cy* animals. Analysis for fatty acid content from affected kidneys revealed that flaxseed supplementation was associated with a small reduction in renal palmitate content, a modest but significant reduction in arachidonic acid, and significant increases in ω3 unsaturated precursors to prostanoid synthesis compared to controls (Figure 10.6). The ratio of ω3:ω6-unsaturated fatty acids was significantly increased in renal tissue from animals fed the flaxseed diet (0.30 versus 0.11; p < 0.0001).

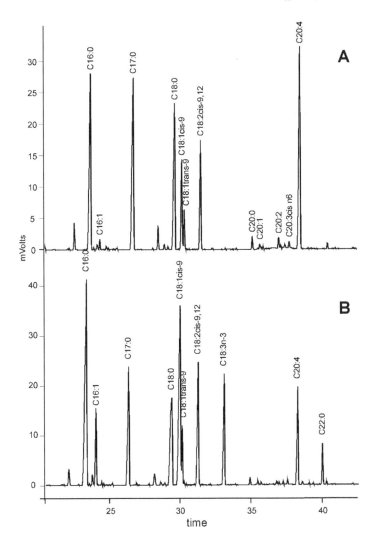

Figure 10.6 GLC of free fatty acid content of kidney tissue from rats fed a flax oil supplemented diet (B) or a control diet (A). The free fatty acids were extracted
from tissue samples using a modified Folch extraction and chromatographed
on a 30m Zebron ZB-Wax column with FID detection.

The ω3-polyunsaturated fatty acid inhibitor of macrophage activity, docosahexaenoic
acid (DHA) was reduced in animals fed a flaxseed diet.

We then applied nuclear magnetic resonance spectroscopic techniques to study
biochemical changes that occurred in renal tissue in association with flaxseed feeding
to determine if there were any factors independent of lipid effects that might be associated with a reduction in renal injury (Ogborn *et al.*, 1998b). We had previously shown
that expression of disease in rats on a standard lab chow diet was associated with increased
excretion and tissue loss of anions involved in the citric acid cycle (Ogborn *et al.*,
1997). There was also significant reduction in tissue content of osmolytes, particularly

betaine, critical to cell volume regulation in the hypertonic regions of the renal medulla. Amelioration of the disease by soy protein feeding was associated with an increase in renal betaine and succinate content above the levels normally seen in healthy animals (Ogborn *et al.*, 1999). Affected animals on either diet demonstrated increased excretion of citric acid cycle metabolites and reduced excretion of creatinine consistent with declining renal function which we have described in previous studies. Animals on flaxseed diet, however, despite somewhat milder disease, demonstrated significantly greater citrate excretion (p = 0.0004). This result was not associated with a significant change in urinary ammonium excretion, suggesting that this was not a response to relative alkalosis induced by the dietary modification.

^{1}H-NMR analysis of tissue from affected animals revealed that flaxseed feeding was associated with a significantly higher content of glutamate, succinate and betaine, reminiscent of the effects seen with a soy protein-based diet.

Flax oil in the Han:SPRD-cy rat

We have recently undertaken studies to test individual components within flaxseed for benefits in the Han:SPRD-*cy* rat. A recently completed study involved the use of lignan-poor flax oil as the lipid source of a synthetic diet based upon 20 percent casein as the protein source. This synthetic diet is associated with accelerated disease progression in male Han:SPRD-*cy* rats. Animals receiving the flax oil-based diet had less cystic change (2.02 ml/kg versus 2.97, p < 0.0001), less fibrosis (0.54 ml/kg versus 0.74, p = 0.0004) and lower numbers of renal macrophages (19.5 per high power video field versus 26.5, p = 0.0006). The flax oil diet resulted in hepatic enrichment of α-linolenic acid, eicosapentaenoic acid and docohosahexaenoic acid (DHA), with associated depletion of linoleic, γ-linolenic and arachidonic acids. The lipid profile of the kidneys was similar to that of liver except that there was no significant difference in arachidonic acid content. Serum creatinine tended to be lower in flax oil-fed animals (189 μmol/l versus 205) but this did not reach statistical significance. Cholesterol and triglycerides were not different between the two groups.

These studies have confirmed that intact flaxseed may play a significant role in renal remodeling in the Han:SPRD-*cy* model of PKD. Future studies in this animal model will explore the contribution of lignans, interactions between flax components and establish dose response relationships for this interesting group of nutritional interventions.

References

Albert, D.A., Hadler, N.M., and Ropes, M.W. (1979). Does corticosteroid therapy affect the survival of patients with systemic lupus erythematosus? *Arthr. Rheum.* 22, 945–953.

Baldi, E., Emancipator, S.N., Hassan, M.O., and Dunn, M.J. (1990). Platelet activating factor receptor blockade ameliorates murine systemic lupus erythematosus. *Kidney Int.* 38, 1030–1038.

Balow, J., Austin III, H.A., Tsokos, G.C., Antonobych, T.T., Steinberg, A.D., and Klippel, J.H. (1987). Lupus nephritis. *Ann. Int. Med.* 106, 79–94.

Brenner, B.M. (1985). Nephron adaptation to renal injury or ablation. *Am. J. Physiol.* 249, F324–F327.

Chanutin, A., and Ferris, E.B. (1932). Experimental renal insufficiency produced by partial nephrectomy. I. Control diet. *Arch. Intern. Med.* 49, 767–787.

Clark, W.F., Kortas, C., Heidenheim, P., Garland, J., Spanner, E., and Parbtani, A. (2001). Flaxseed in lupus nephritis: A two year crossover study. *J. Am. Coll. Nutr.* 20, 143–148.

Clark, W.F., Muir, A.D., Westcott, N.D., and Parbtani, A. (2000). A novel treatment for lupus nephritis: Lignan precursor derived from flax. *Lupus* 9, 429–436.

Clark, W.F., and Parbtani, A. (1994). Omega-3 fatty acid supplementation in clinical and experimental lupus nephritis. Symposium on essential fatty acid deficiencies and ω-3 unsaturated fat dietary supplementation in glomerulonephritis: Basis and practice. *Am. J. Kidney Dis.* 23, 644–647.

Clark, W.F., Parbtani, A., Huff, M.W., Reid, B., Holub, B.J., and Falardeau, P. (1989). Omega-3 fatty acid dietary supplementation in systemic lupus erythematosus. *Kidney Int.* 36, 653–660.

Clark, W.F., Parbtani, A., Huff, M.W., Spanner, E., De Salis, H., Chin-Yee, I., Philbrick, D.J., and Holub, B.J. (1995). Flaxseed: A potential treatment for lupus nephritis. *Kidney Int.* 48, 475–480.

Clark, W.F., Parbtani, A., Naylor, C.D., Levinton, C.M., Muirhead, N., Spanner, E., Huff, M.W., Philbrick, D.J., and Holub, B.J. (1993a). Fish oil in lupus nephritis: Clinical findings and methodological implications. *Kidney Int.* 44, 75–86.

Clark, W.F., Parbtani, A., Philbrick, D.J., Holub, B.J., and Huff, M.W. (1991). Chronic effects of ω-3 fatty acids (fish oil) in a rat 5/6 renal ablation model. *J. Am. Soc. Nephrol.* 1, 1343–1353.

Clark, W.F., Parbtani, A., Philbrick, D., McDonald, J.W.D., Smallbone, B., Reid, B., Holub, B.J., and Kreeft, J. (1990). Comparative efficacy of dietary treatments on renal function in rats with sub-total nephrectomy: Renal polyunsaturated fatty acid incorporation and prostaglandin excretion. *Clin. Nephrol.* 33, 25–34.

Clark, W.F., Parbtani, A., Philbrick, D.J., Spanner, E., Huff, M.W., and Holub, B.J. (1993b). Dietary protein restriction versus fish oil supplementation in the chronic remnant nephron model. *Clin. Nephrol.* 39, 293–304.

Correia, P., Cameron, J.S., Lian, J.D., Hicks, J., Ogg, C.S., Williams, D.G., Chantler, C., and Haycock, D.G. (1985). Why do patients with lupus nephritis die? *Brit. Med. J.* 190, 126–131.

Cowley, B.D., Gudapaty, S., Kraybill, A.L., Barash, B.D., Harding, M.A., Calvet, J.P., and Gattone, V.H. (1993). Autosomal-dominant polycystic kidney disease in the rat. *Kidney Int.* 43, 522–534.

Dubois, E.L., Wierzchowiecki, M., Cox, M.B., and Weiner, J.M. (1974). Duration and death in systemic lupus erythematosus. An analysis of 249 cases. *JAMA* 227, 1399–1402.

Felley, V.E., Ferretti, A., Izui, S., and Strom, T.B. (1985). A fish oil diet rich in eicosapantaenoic acid reduces cycloosygenase metabolites and suppresses lupus in MRL.lpr mice. *J. Immunol.* 134, 1914–1919.

Gattone, V.H., Cowley, B.D., Barash, B.D., Nagao, S., Takahashi, H., Yamaguchi, T., and Grantham, J.J. (1995). Methylprednisolone retards the progression of inherited polycystic kidney disease in rodents. *Am. J. Kidney Dis.* 25, 302–313.

Gile, R.D., Cowley, B.D., Gattone, V.H., O'Donnell, M.P., Swan, S.K., and Grantham, J.J. (1995). Effect of lovastatin on the development of polycystic kidney disease in the Han:SPRD rat. *Am. J. Kidney Dis.* 26, 501–507.

Ginzler, E.M., Diamond, H.S., Weiner, M., Schlesinger, M., Fries, J.F., Wasner, C., Medsger, T.S.J., Ziegler, G., Klippel, J.H., Hadler, N.M., Albert, D.A., Hess, E.V., Spencer-Green, G., Grayzel, A., Worth, D., Hahn, B.H., and Barnett, E.V. (1982). A multicenter study of outcome in systemic lupus erythematosus. 1. Entry variables as predicters of prognosis. *Arthr. Rheum.* 25, 601–611.

Gretz, N., Ceccherini, I., Kranzlin, B., Kloting, I., Devoto, M., Rohmeiss, P., Hocher, B., Waldherr, R., and Romeo, G. (1995). Gender-dependent disease severity in autosomal polycystic kidney disease of rats. *Kidney Int.* 48, 496–500.

Hall, A.V., Parbtani, A., Clark, W.F., Spanner, E., Keeney, M., Chin-Yee, I., Philbrick, D.J., and Holub, B.J. (1993). Abrogation of MRL/lpr lupus nephritis by dietary flaxseed. *Am. J. Kidney Dis.* 22, 326–332.

Hayslett, J.P. (1979). Functional adaptation to reduction in renal mass. *Physiol. Rev.* 509, 137–164.

Holub, B.J., Philbrick, D.J., Parbtani, A., and Clark, W.F. (1991). Dietary lipid modification of renal disorders and ether phospholipid metabolism. *Biochem. Cell Biol.* 69, 485–489.

Hosenpud, J.D., Montanaro, A., Hart, M.V., Haines, J.E., Specht, H.D., Bennett, R.M., and Kloster, F.E. (1984). Myocardial perfusion abnormalities in asymptomatic patients with systemic lupus erythematosus. *Am. J. Med.* 77, 286–292.

Hostetter, T.H., Olson, J.L., Rennke, H.G., Venkatachalam, M.A., and Brenner, B.M. (1981). Hyperfiltration in remnant nephrons. A potentially adverse response in renal ablation. *Am. J. Physiol.* 241, F85–F93.

Ingram, A.J., Parbtani, A., Clark, W.F., Spanner, E., Huff, M.W., Philbrick, D.J., and Holub, B.J. (1995). Effects of flaxseed and flax oil diets in a Rat-5/6 Renal Ablation Model. *Am. J. Kidney Dis.* 25, 320–329.

Karsh, J., Klippel, J.H., Balow, J.E., and Decker, J.L. (1979). Mortality in lupus nephritis. *Arthr. Rheum.* 22, 764–769.

Keith, D.S., Torres, V.E., Johnson, C.M., and Holley, K.E. (1994). Effect of sodium chloride, enalapril, and losartan on the development of polycystic kidney disease in Han:SPRD rats. *Am. J. Kidney Dis.* 24, 491–498.

Macconi, D., Noris, M., Benfenati, E., Quaglia, R., Pagliarino, G., and Remuzzi, G. (1991). Increased urinary excretion of platelet activating factor in mice with lupus nephritis. *Life Sci.* 48, 1429–1437.

Morigi, M., Macconi, D., Riccardi, E., Boccardo, P., Zilio, P., Bertani, T., and Remuzzi, G. (1991). Platelet-activating factor receptor blocking reduces proteinuria and improves survival in lupus autoimmune mice. *J. Pharmacol. Exp. Ther.* 258, 601–606.

Nadasdy, T., Laszik, Z., Lajoie, G., Blick, K.E., Wheeler, D.E., and Silva, F.G. (1995). Proliferative activity of cyst epithelium in human renal cystic diseases. *J. Am. Soc. Nephrol.* 5, 1462–1468.

Ogborn, M.R. (1994). Polycystic kidney disease – a truly pediatric problem. *Pediatr. Nephrol.* 8, 762–767.

Ogborn, M.R., Bankovic Calic, N., Shoesmith, C., Buist, R., and Peeling, J. (1998a). Soy protein modification of rat polycystic kidney disease. *Am. J. Physiol.* 274, F541–F549.

Ogborn, M.R., Nitschmann, E., Bankovic Calic, N., Buist, R., and Peeling, J. (1998b). The effect of dietary flaxseed supplementation on organic anion and osmolyte content and excretion in rat polycystic kidney disease. *Biochem. Cell Biol.* 76, 553–559.

Ogborn, M.R., Nitschmann, E., Weiler, H.A., and Bankovic Calic, N. (2000). Modification of polycystic kidney disease and fatty acid status by soy protein diet. *Kidney Int.* 57, 159–166.

Ogborn, M.R., Nitschmann, E., Weiler, H., Leswick, D., and Bankovic Calic, N. (1999). Flaxseed ameliorates interstitial nephritis in rat polycystic kidney disease. *Kidney Int.* 55, 417–423.

Ogborn, M.R., and Sareen, S. (1995). Amelioration of polycystic kidney disease by modification of dietary protein intake in the rat. *J. Am. Soc. Nephrol.* 6, 1649–1654.

Ogborn, M.R., Sareen, S., Prychitko, J., Buist, R., and Peeling, J. (1997). Altered organic anion and osmolyte content and excretion in rat polycystic kidney disease: An NMR study. *Am. J. Physiol.* 272, F63–F69.

Parbtani, A., and Clark, W.F. (1995). Flaxseed and its components in renal disease. In Cunnane, S.C., and Thompson, L.U. (eds), *Flaxseed in Human Nutrition*, Vol., AOCS Press, Champaign, Il, pp. 244–260.

Ponticelli, C., Zucchelli, P., Moroni, G., Cagnoli, L., Banfi, G., and Pasquali, S. (1987). Long-term prognosis of diffuse lupus nephritis. *Clin. Nephrol.* 28, 263–271.

Rankin, C.A., Grantham, J.J., and Calvet, J.P. (1992). C-fos expression is hypersensitive to serum-stimulation in cultured cystic kidney cells from the C57BL/6J-cpk mouse. *J. Cell Physiol.* 152, 578–586.

Remuzzi, G., Ruggenenti, P., and Benigni, A. (1997). Understanding the nature of renal disease progression. *Kidney Int.* 51, 2–15.

Tetta, C., Bussolino, F., Modena, V., Montrucchio, G., Segoloni, G., Pescarmona, G., and Camussi, G. (1990). Release of platelet-activating factor in systemic lupus erythematosus. *Int. Arch. Allergy Appl. Immunol.* 91, 244–256.

Theofilopoulos, A.N., and Dixon, F.J. (1985). Murine models of systemic lupus erythematosus. *Adv. Immunol.* 37, 269–390.

Tomobe, K., Philbrick, D., Aukema, H.M., Clark, W.F., Ogborn, M.R., Parbtani, A., Takahashi, H., and Holub, B.J. (1994). Early dietary protein restriction slows disease progression and lengthens survival in mice with polycystic kidney disease. *J. Am. Soc. Nephrol.* 5, 1355–1360.

Urowitz, M.B., Bookman, A.A., Koehler, B.E., Gordon, D.A., Smythe, H.A., and Ogryzlo, M.A. (1976). The bimodal mortality pattern of systemic lupus erythematosus. *Am. J. Med.* 60, 221–225.

Westcott, N.D., and Muir, A.D. (1998). Process for extracting lignans from flaxseed. Patent # 5705618.USA.

Woo, D. (1995). Apoptosis and loss of renal tissue in polycystic kidney diseases [see comments]. *New England J. Med.* 333, 18–25.

11 Mammalian metabolism of flax lignans

Alister D. Muir and Neil D. Westcott

Introduction

The biological activity of flax lignans was discovered in a manner that could be considered to be the reverse of the normal discovery process. Although the chemical identity of the flax lignans was established in 1956 (Bakke and Klosterman, 1956), it was not until the late 1970s that investigators studying steroidal hormone metabolism observed two unknown compounds that exhibited a cyclical excretion pattern during the menstrual cycle of the vervet monkey (Setchell *et al.*, 1980a) and that the levels of these compounds were often higher in female urine during early pregnancy (Setchell and Adlercreutz, 1979; Setchell *et al.*, 1980b; Stitch *et al.*, 1980; Setchell *et al.*, 1981a). Later the connection was made between the appearance of these phenolic metabolites in mammalian urine and lignans in the diet (Setchell *et al.*, 1981b; Axelson *et al.*, 1982). These metabolites (with a hydroxyl group at the *meta* position) were subsequently termed mammalian lignans to distinguish them from plant lignans where this first hydroxyl group was in the *para* position (Axelson and Setchell, 1981). So it could be said that the biologically active molecules were discovered before their origins were determined.

Although the conversion of plant lignans, including the flax lignans, into mammalian lignans has been known for more than 20 years, there are still many aspects of this metabolic pathway that are poorly understood and many questions still remain to be answered. In this chapter we will summarize what is known about this metabolic pathway and identify areas for future research.

Materials and methods

In order to understand the current state of knowledge of lignan metabolism, one must first understand the nature and limitations of the analytical methods used by researchers in their studies of lignan metabolism. The native plant form (ester linked complex) of the flax lignans is also an important issue that is often overlooked when studies to investigate the metabolism of these compounds are being designed. In flaxseed the proportion of secoisolariciresinol 1 that exists in a free or non-esterified form is less than 1 percent of the total lignan content and the stored form of the trace levels of the minor lignans that have been reported is largely unknown (see Chapter 3).

Analysis of lignans

The methods for isolation, detection and quantification of the mammalian lignans, enterodiol 2 and enterolactone 3, have evolved rapidly since the discovery of these compounds

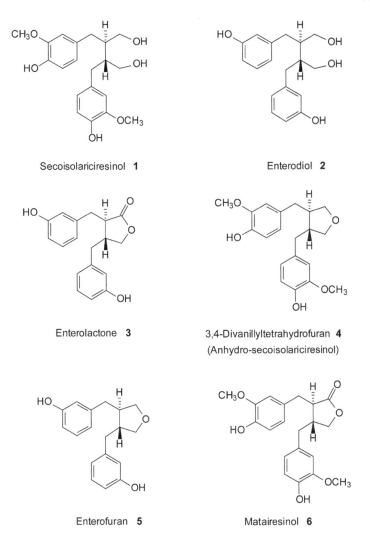

Figure 11.1 Secoisolariciresinol, mammalian lignans and related compounds.

in the late 1970s. The methods have generally employed some form of mass spectro-scopic analysis (Setchell *et al.*, 1981a; Adlercreutz *et al.*, 1993), driven for the most part by the low levels of these compounds present in many of the tissues under inves-tigation. Mammalian lignan levels in urine are often present at much higher concen-trations than in plasma, and given that it is usually possible to get a much larger urine sample than a plasma sample, a greater range of detection systems is possible. The mammalian lignans were first encountered during the GC analysis of steroidal hor-mones and the principal approaches to the analysis of these compounds have evolved from the classical techniques for the analysis of hormones. However, none of the compounds, either plant or mammalian lignans or any of their naturally occurring con-jugates, are sufficiently volatile for GC analysis and consequently conversion to aglycones and derivatization is necessary for GC and GC-MS analysis.

Initially chromatography was on packed columns, and later on capillary GC columns were used (Fotsis *et al.*, 1982) and detection quickly shifted from classical GC detectors to mass spectroscopic detectors that generally allow much greater sensitivity and selectivity. Quantitation has always been an issue and it was not till the advent of isotope dilution methods (Adlercreutz *et al.*, 1994; Adlercreutz *et al.*, 1995a; Mazur *et al.*, 1996) that reliable quantitative methods became routine. The difficulty with these methods is that only the compounds for which isotopically labeled standards exist are detected. Coulometric array detection methods have recently been developed to increase the sensitivity of HPLC analysis of lignans and other phytoestrogens (Gamache and Acworth, 1998). However, this technique suffers from the same limitations as the GC isotope dilution methods in that it only detects compounds for which authentic standards are available. Recently, soft ionization LC-MS and LC-MS-MS methods have become available (Muir *et al.*, 2000a; Muir *et al.*, 2001a) which greatly simplify the sample preparation and allow detection of the intact glycosides and conjugates.

An immunoassay for the mammalian lignan enterolactone has also been developed to facilitate determination of this compound in plasma (Adlercreutz *et al.*, 1999; Uehara *et al.*, 2000). Although this technique is now being used to measure plasma enterolactone levels in diet–disease correlation studies in humans and in specific diet intervention studies, immunoassays are not yet available for the other lignans found in plasma and urine.

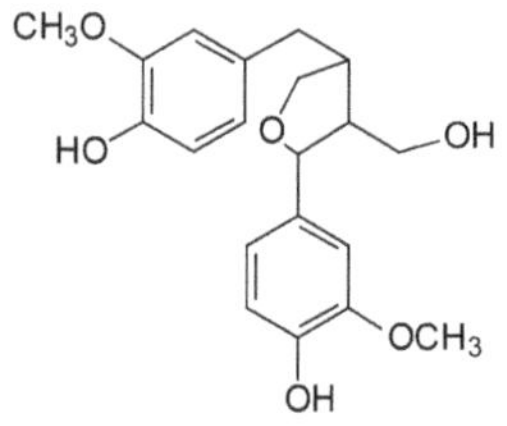

Lariciresinol 7

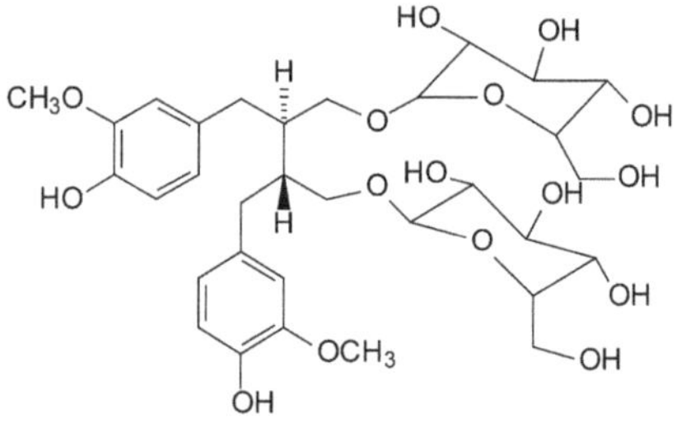

Secoisolariciresinol diglucoside (SDG) 8

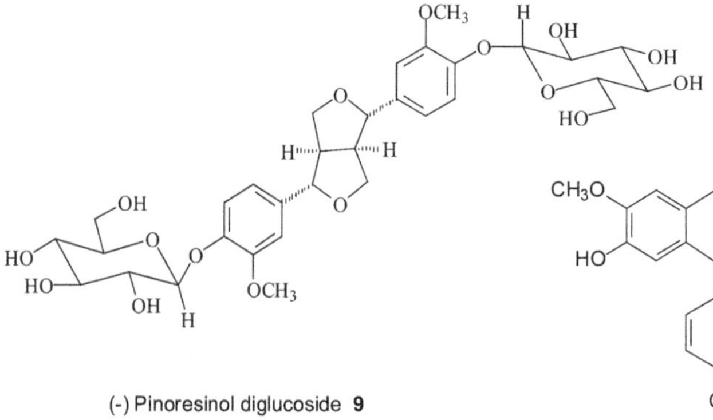

(-) Pinoresinol diglucoside 9

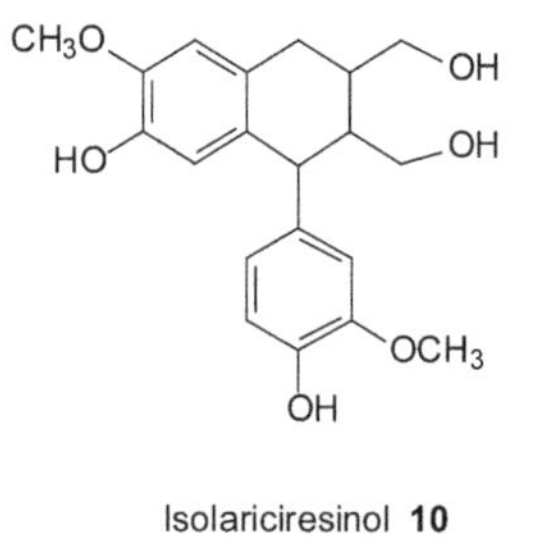

Isolariciresinol 10

Figure 11.2 Lignans isolated from flaxseed.

Analytical artifacts

In any analytical procedure there is always the possibility that the compound of interest will be degraded, isomerized, or irreversibly absorbed. In the analysis of flax lignans there are several possibilities. Secoisolariciresinol can be readily dehydrated to give 3,4-divanillyltetrahydrofuran (anhydrosecoisolariciresinol) 4 under acid conditions (Lapteva *et al.*, 1971), typically used to cleave the sugars from glycosides and conjugates, particularly when analyzing plant lignans. Enterofuran 5, a similar dehydration of enterodiol, has recently been identified as another degradation product generated during the preparation of samples for GC-MS analysis (Liggins *et al.*, 2000; Heinonen *et al.*, 2001). Treatment of plant extracts with base will cause the lactone ring of matairesinol 6 to open and prolonged exposure of any lignan to strong acidic or basic conditions will cause degradation of the target compound. Lariciresinol 7 can readily isomerize to isolariciresinol in the presence of HCl (pH 2.0) or formic acid (pH 1.0) at room temperature (Lapteva *et al.*, 1971).

From plant to the small intestine

The lignans found in flaxseed occur as an ester linked complex with hydroxy-methyl-glutarate (HMGA), a series of cinnamic acids and cinnamic acid glycosides where the secoisolariciresinol diglucoside (SDG) 8 is linked to this complex through an ester bond between the glucose of SDG and the HMGA (see Chapter 3 for details) (Bakke and Klosterman, 1956; Westcott and Muir, 1999; Muir and Westcott, 2000; Westcott and Muir, 2000; Kamal-Eldin *et al.*, 2001). The principal lignan found in flaxseed is secoisolariciresinol diglucoside. However, trace amounts of matairesinol, 0.4–1.2 percent of the secoisolariciresinol content (Mazur *et al.*, 1996; Liggins *et al.*, 2000), pinoresinol (Meagher *et al.*, 1999), pinoresinol diglucoside 9 (Qiu *et al.*, 1999), and isolariciresinol 10 (Meagher *et al.*, 1999) have also been reported. Recent studies have suggested that all of these compounds except isolariciresinol could be precursors for mammalian lignans (Heinonen *et al.*, 2001).

Conversion of plant lignans to mammalian lignans

The key first step in the metabolism of flax lignans is the release of the lignan aglycone. Although the major lignan in flaxseed is a diglucoside, SDG 10 is not readily hydrolyzed by almond β-glucosidase; the sugars can be readily cleaved by β-glucuronidase isolated from *Helix pomatia* (Obermeyer *et al.*, 1995; Muir *et al.*, 2000b). Although the β-glucuronidase isolated from *E. coli* is capable of hydrolyzing mammalian lignan conjugates (Adlercreutz *et al.*, 1994), we found that it was unable to release any significant amount of secoisolariciresinol from the flax lignan complex or from SDG (Muir *et al.*, 2001b). There are, however other sources of β-glucosidase and β-glucuronidase activity in the gut (Hawksworth *et al.*, 1971) and it remains to be determined which of these sources of enzyme activity is responsible for releasing secoisolariciresinol from flaxseed. Recent studies have shown that rats fed flaxseed or SDG had significantly higher levels of cecal β-glucuronidase activity compared to control animals and this was correlated with a colon cancer protective effect (Jenab and Thompson, 1996; Jenab *et al.*, 1999). Although β-glucuronidase activity is present throughout the GI tract (Marsh *et al.*, 1953), the level of activity in the small intestine of humans and rabbits is very low compared to that of rats and mice, which have much higher microbial populations in the small intestine (Hawksworth *et al.*, 1971). The significance of this in relation to flax lignan metabolism is not known. However, it does suggest that extrapolation of metabolic patterns observed in rats and mice to humans should be verified by carefully controlled studies in animal models that more closely model the human digestive tract.

One of the difficulties encountered in interpreting the results from many of the animal feeding experiments is the lack of controlled diets. Many of the animal chow diets used also contain other sources of plant lignans and isoflavones. This can be illustrated in a recent study of mammalian lignan metabolism where large amounts of daidzein and equol were found in the bile of enterolactone dosed rats (Niemeyer *et al.*, 2000). In this same study, a major peak corresponding to 5-hydroxy-enterolactone was detected. Since this was not an expected metabolite and large amounts of daidzein were also found in this sample, it was not possible to determine its metabolic origin.

Colonic metabolism

Metabolism of flax lignans by fecal flora

The first evidence that gastrointestinal flora were involved in the conversion of plant lignans into mammalian lignans was provided by Axelson and Setchell (1981). The normal levels of the mammalian lignans in rat urine were found to have decreased to undetectable levels in "germ-free" rats. The administration of antibiotics and the subsequent drop in urinary levels also indicated a bacterial role in their production (Axelson and Setchell, 1981; Setchell *et al.*, 1981b), and subsequent studies suggested *Clostridia* bacteria might be the main type of bacteria involved in mammalian lignan synthesis (Setchell *et al.*, 1981b; Borriello *et al.*, 1985).

In 1985, Borriello *et al.* were able to demonstrate that incubation of fecal cultures with flaxseed resulted in the production of enterodiol and enterolactone under both aerobic and anaerobic conditions. Their experiments indicated that the cultures were converting the plant lignans into enterodiol and then subsequently to enterolactone, and not the reverse, as had been postulated (Setchell *et al.*, 1980b). This was confirmed when cultures were able to convert enterodiol to enterolactone but sterile fecal cultures could not, and no conversion was observed when enterolactone was supplied. The role of the gut flora was further defined by the observation that a fecal flora sample from a human subject receiving the oral antibiotic metronidazole was not able to convert secoisolariciresinol to either mammalian lignan but was able to convert enterodiol to enterolactone (Borriello *et al.*, 1985), suggesting that more than one species of bacteria was involved in the conversion process.

In 2000, Wang *et al.* reported extensive studies on the metabolism of secoisolariciresinol diglucoside. However, the enzymes responsible for the initial conversion to secoisolariciresinol are still unclear. They demonstrated that the removal of the sugars could take place under aerobic conditions, whereas all subsequent metabolism only occurred under anaerobic conditions. Two strictly anaerobic bacterial strains were identified in this study. A gram positive elliptical coccus (*Peptostreptococcus* sp.) was shown to be responsible for the sequential demethylation of secoisolariciresinol 1 to 3-demethyl-(-)-secoisolariciresinol 11 and didemethylsecoisolariciresinol 13. This bacterium could also demethylate 2-(3-hydroxybenzyl)-3-(4-hydroxy-3-methoxybenzyl)

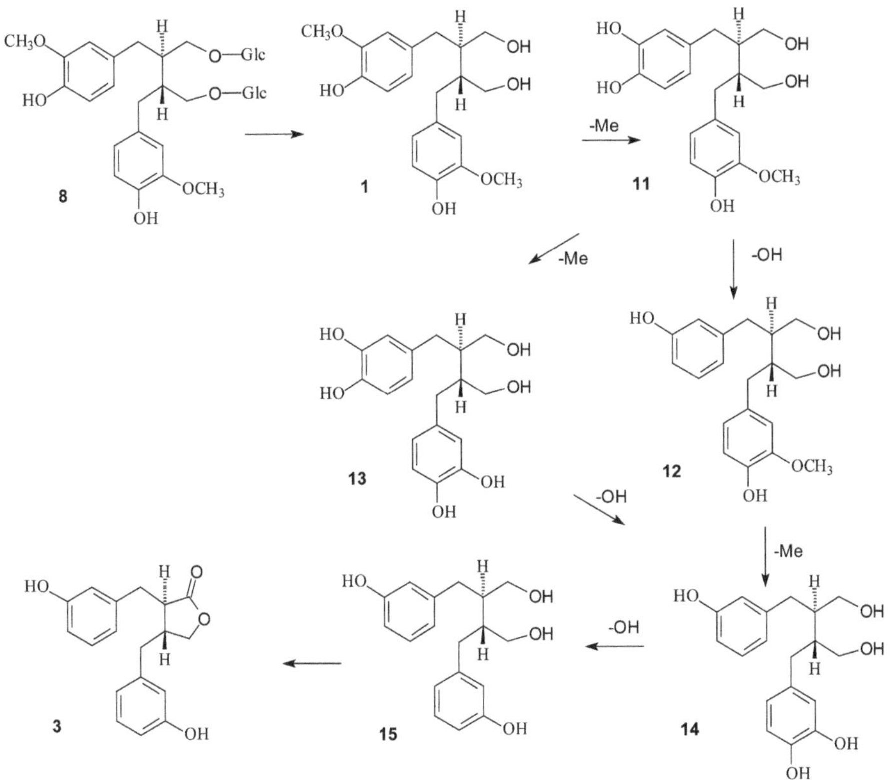

Figure 11.3 Metabolism of secoisolariciresinol diglucoside.

butane-1,4-diol 12 which was shown to result from the dehydroxylation of compound 11. The second bacterium was identified as a gram positive non-sporeforming rod (*Eubacterium* sp.) species that was responsible for the dehydroxylation of 3-demethyl-(-)-secoisolariciresinol 11, and compounds 13 and 14 (Wang *et al.*, 2000). The time course experiments suggested that the demethylation products appeared prior (20–30h) to the dehydroxylation products (48h). This is the opposite of the earlier observations of Setchell and co-workers (Borriello *et al.*, 1985).

One of the difficulties in assessing the role of different microorganisms in the metabolism of flaxseed lignans is the very limited number of experimental systems. Emerging evidence suggests that other factors such as the fat and fiber content of the diet of subjects (human or animal) providing the fecal or plasma samples influence the metabolic patterns (Setchell and Adlercreutz, 1988; Adlercreutz, 1999).

Interconversion of enterolactone and enterodiol

When the mammalian lignans were first identified, the relative concentrations of enterolactone and enterodiol were observed to vary considerably, relative to one another. Early feeding experiments suggested that enterolactone was converted to enterodiol (Setchell *et al.*, 1980b). Subsequently Borriello *et al.* (1985) demonstrated the conversion of plant lignans into enterodiol and then subsequently to enterolactone, and not the reverse, as had been postulated (Setchell *et al.*, 1980b). Similar, more detailed observations were recently published by Wang *et al.* (2000). They observed that fecal cultures were able to convert enterodiol to enterolactone but sterile fecal cultures were not, and no conversion was observed when enterolactone was supplied. Relatively few feeding trials with enterolactone or enterodiol have been undertaken since then. In a recent study, Niemeyer *et al.* (2000) observed low levels of enterolactone and enterolactone metabolites in the urine of rats receiving oral doses of enterodiol. However, the presence of trace levels of both enterolactone and enterodiol in the urine of control animals makes it difficult to draw conclusions from these experiments.

Liver metabolism and enterohepatic circulation

Although there are relatively few studies of the conversion of plant lignans into mammalian lignans by the gut microflora, this metabolic process is widely cited in the literature. The subsequent metabolism of mammalian lignans and plant lignans that are absorbed by the body and pass into the blood is less well known and understood. The conjugation of the mammalian lignans to produce glucuronides and sulfates was known from the earliest days of the studies of these compounds, but the enzymes involved and their substrate specificity are not known. It is presumed that these conjugation reactions occur in the liver but this has not yet been established experimentally.

Enterohepatic circulation

Mammalian lignans were discovered during studies of steroid metabolism and it was immediately suspected that these compounds would undergo enterohepatic circulation in the same way that the steroidal hormones do (Figure 11.4). The first experimental evidence that mammalian lignans underwent enterohepatic circulation was provided by Axelson and Setchell (1981), who demonstrated that when bile fistula were inserted in female rats, the urinary excretion of mammalian lignans decreased significantly.

In a recent study where rye (another source of plant lignans) was fed to pigs, more than 95 percent of the conjugated lignans found in the ileum were plant lignans, while the mammalian lignans represented between 20 and 40 percent of the unconjugated lignans (Glitsø *et al.*, 2000) as determined by isotope dilution GC-MS. The presence of the plant lignan conjugates in the ileum was presumed to be the result of enterohepatic circulation of the plant lignans, indicating that the plant lignans themselves may also be biologically active. An interesting and confusing observation in this study was the finding that the level of fecal excretion of lignans was significantly higher than expected from the dietary analysis. This finding suggests that combined enzyme and acid hydrolysis based methods used in this study are significantly underestimating the lignan content of rye and presumably also of flaxseed. This is consistent with the results we obtained in our own work where these assays also significantly underestimated the lignan content of flaxseed.

Oxidative metabolism of mammalian lignans

Jacobs and Metzler (1999) studied the metabolism of enterolactone and enterodiol by aroclor-induced rat liver microsomes and found a series of mono-hydroxylated

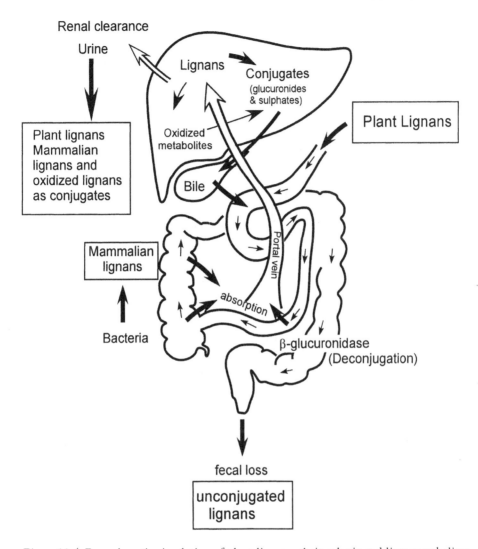

Figure 11.4 Enterohepatic circulation of plant lignans, their colonic and liver metabolites. The sites of metabolism, absorption and secretion are shown schematically as the precise sites are not known. (Adapted from: Setchell *et al.*, 1982).

derivatives. A total of six aromatic mono-hydroxylated derivatives of enterolactone and the corresponding derivatives of enterodiol were detected with the hydroxylation occurring either ortho or para to the aromatic hydroxyl group (Figures 11.5 and 11.6). In addition, a series of aliphatic mono-hydroxylated derivatives were also detected. However, the precise structure of these compounds was not determined (Niemeyer *et al.*, 2000). In addition to the detection of these compounds in microsomes, these metabolites have now been found in the bile and urine of rats receiving either intraduodenal or oral doses of enterolactone, enterodiol or flaxseed (Niemeyer *et al.*, 2000). Only the aromatic hydroxylated derivatives were found in human urine (Jacobs *et al.*, 1999), although both aromatic and aliphatic hydroxylated metabolites were found

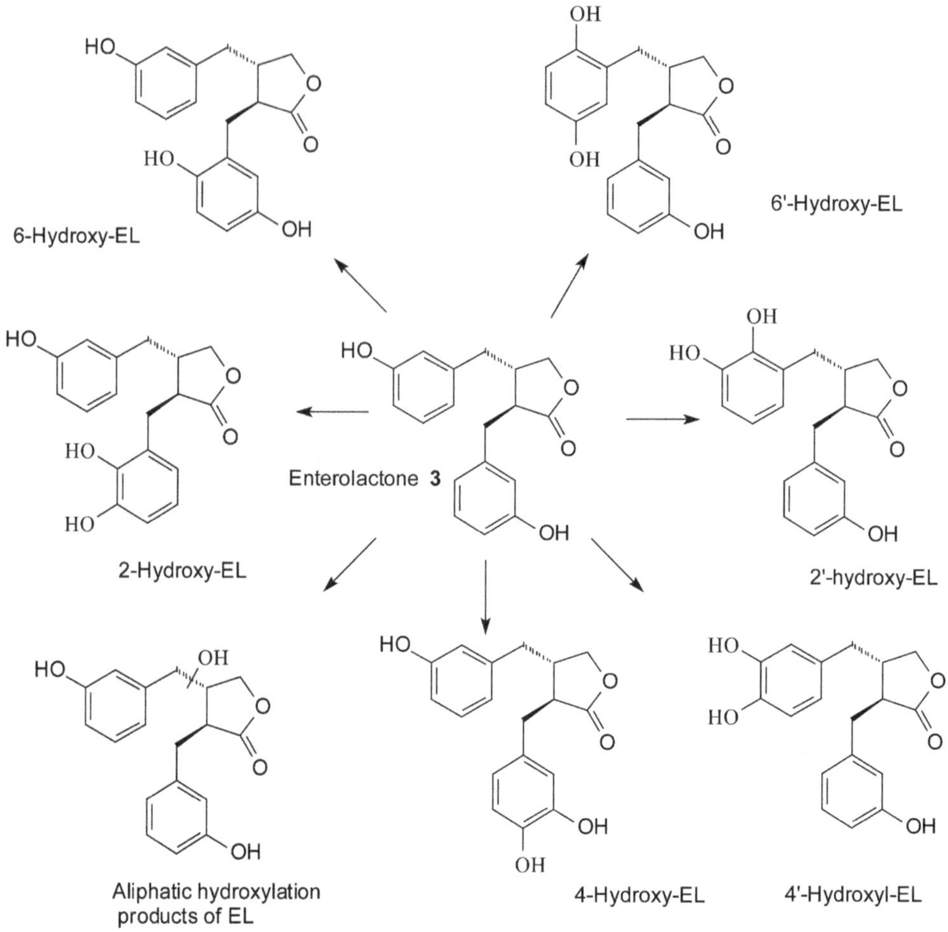

Figure 11.5 Oxidation products of enterolactone.

in rat urine (Niemeyer *et al.*, 2000). The 2-hydroxy and 4-hydroxy derivatives are catechols and appear to undergo further metabolism to the corresponding monomethylated derivatives (Niemeyer *et al.*, 2000). Some or all of these oxidative metabolites may be biologically active. However, the levels of these metabolites are relatively low, accounting for approximately 3 percent of the administered dose.

Conjugation (excretion) reactions

Although it was known from the time of their discovery that enterolactone and enterodiol were excreted as conjugates (Axelson and Setchell, 1980), the precise identity of all the mammalian lignan conjugates has yet to be definitively established. Axelson and Setchell (1980) established the presence of a mono-glucuronide of enterolactone based on GC-MS analysis of the methyl ester-TMS ether derivative and observed the

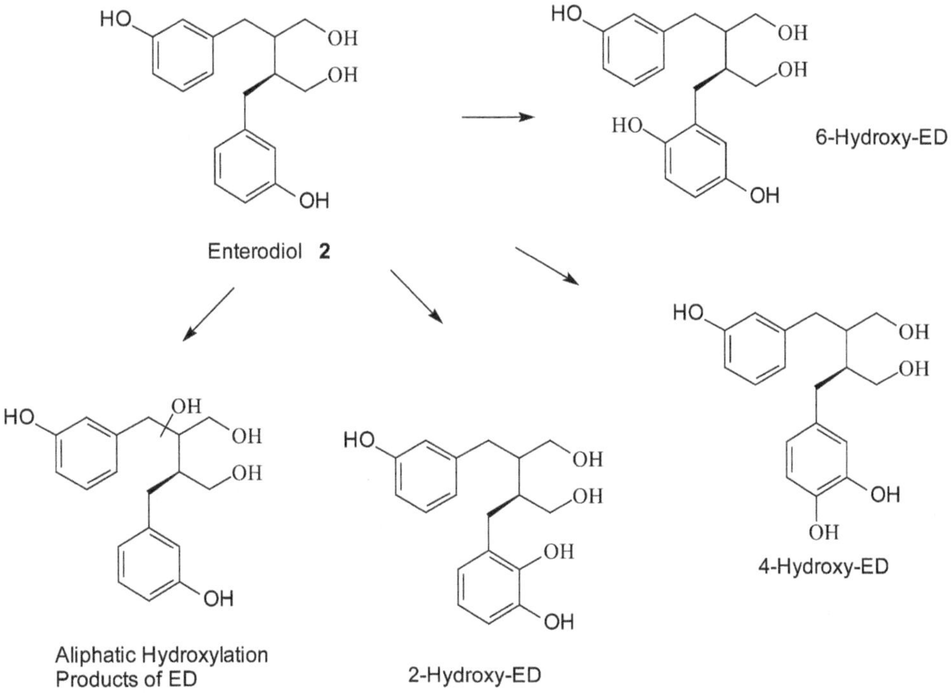

Figure 11.6 Oxidation products of enterodiol.

corresponding increase in molecular weight expected when deuterated TMS derivatives were analyzed. Mono-glucuronides of enterodiol were detected but definitive identification was not achieved. The location of the glucuronide moiety was not determined in either case. The presence of mono- and disulphates of enterolactone and enterodiol was deduced based on their anion exchange chromatographic mobilities, conversion to aglycones by a mild solvolytic procedure (Murray *et al.*, 1977) and by the formation of TMS ether derivatives of the aglycones. The latter reaction is a characteristic reaction of an aromatic sulphate.

In a later study (Adlercreutz *et al.*, 1995b), DEAE-Ac⁻ chromatography was used to separate the different conjugate groups which were then hydrolyzed to their respective aglycones. In human urine, enterolactone, enterodiol and matairesinol were mainly excreted as mono-glucuronides (73–94%), with lesser amounts of monosulfates, sulfoglucuronides, diglucuronides and disulfates. Only trace levels of non-conjugated lignans were detected (Axelson and Setchell, 1980; Adlercreutz *et al.*, 1995b).

Brunner *et al.* (1986) used a solubilized rabbit liver microsomal UDP-glucuronyltransferase to produce glucuronides of synthetic enterolactone, enterodiol, and secoisolariciresinol. In this system they were able to detect the presence of both phenolic hydroxy and primary hydroxyl glucuronides of enterodiol, while incubation with enterolactone yielded two glucuronides which were presumed to be a consequence of the racemic nature of the synthetic enterolactone. Further characterization was not possible at that time due to lack of material.

Genotoxicity of mammalian and plant lignans

While much of the focus is on the estrogenic activity of mammalian lignans, these compounds may have other biological effects including genotoxicity, which may not be beneficial. Kulling *et al.* (1998) have shown that while the isoflavone genistein exhibits mutagenic activity in cultured cells, the mammalian lignans enterolactone and enterodiol, along with their presumed plant precursors secoisolariciresinol and matairesinol, did not exhibit any mutagenic activity at concentrations of 100μM.

References

Adlercreutz, H. (1999). Phytoestrogens. State of the art. *Environmental Toxicology and Pharmacology* 7, 201–207.

Adlercreutz, H., Fotsis, T., Kurzer, M.S., Wähälä, K., Mäkelä, T., and Hase, T. (1995a). Isotope dilution gas chromatographic-mass spectrometric method for the determination of unconjugated lignans and isoflavanoids in human feces, with preliminary results in omnivorous and vegetarian women. *Anal. Biochem.* 225, 101–108.

Adlercreutz, H., Fotsis, T., Lampe, J., Wähälä, K., Mäkelä, T., Brunow, G., and Hase, T. (1993). Quantitative determination of lignans and isoflavonoids in plasma of omnivorous and vegetarian women by isotope dilution gas chromatography-mass spectrometry. *Scand. J. Clin. Lab. Invest.* 53, 5–18.

Adlercreutz, H., Fotsis, T., Watanabe, S., Lampe, J., Wähälä, K., Mäkeläa, T., and Hase, T. (1994). Determination of lignans and isoflavonoids in plasma by isotope dilution gas-chromatography-mass spectrometry. *Cancer Detect. Prev.* 18, 259–271.

Adlercreutz, H., Lapcík, O., Hampl, R., Wähälä, K., Al-Maharik, N., Wang, C.-J., and Mikola, H. (1999). Immunoassay of phytoestrogens in human plasma. *J. Med. Food* 2, 131–133.

Adlercreutz, H., van der Widlt, J., Kinzel, J., Attalla, H., Wähälä, K., Mäkelä, T., Hase, T., and Fotsis, T. (1995b). Lignan and isoflavonoid conjugates in human urine. *J. Steroid Biochem. Mol. Biol.* 52, 97–103.

Axelson, M., and Setchell, K.D.R. (1980). Conjugation of lignans in human urine. *FEBS Lett.* 122, 49–53.

Axelson, M., and Setchell, K.D.R. (1981). The excretion of lignans in rats – evidence for an intestinal bacterial source for this new group of compounds. *FEBS Lett.* 123, 337–342.

Axelson, M., Sjövall, J., Gustaisson, B.E., and Setchell, K.D.R. (1982). Origin of lignans in mammals and identification of a precursor from plants. *Nature* 298, 659–670.

Bakke, J.E., and Klosterman, H.J. (1956). A new diglucoside from flaxseed. *Proc. N. Dakota Acad. Sci.* 10, 18–22.

Borriello, S.P., Setchell, K.D.R., Axelson, M., and Lawson, A.M. (1985). Production and metabolism of lignans by the human fecal flora. *J. Appl. Bacteriol.* 58, 37–43.

Brunner, G., Tegtmeier, F., Kirk, D.N., Wynn, S., and Setchell, K.D.R. (1986). Enzymatic synthesis and chromatographic purification of lignan glucuronides. *Biomed. Chromatog.* 2, 89–92.

Fotsis, T., Heikkinen, R., Adlercreutz, H., Axelson, M., and Setchell, K.D.R. (1982). Capillary gas chromatographic method for the analysis of lignans in human urine. *Clin. Chim. Acta.* 121, 361–371.

Gamache, P.H., and Acworth, I.N. (1998). Analysis of phytoestrogens and polyphenols in plasma, tissue and urine using HPLC with coulometric array detection. *Proc. Soc. Expt. Biol. Med.* 217, 274–280.

Glitsø, L.V., Mazur, W.M., Adlercreutz, H., Wähälä, K., Mäkelä, T., Sandström, B., and Bach Knudsen, K.E. (2000). Intestinal metabolism of rye lignans in pigs. *Br. J. Nutr.* 84, 429–437.

Hawksworth, G., Drasar, B.S., and Hill, M.J. (1971). Intestinal bacteria and the hydrolysis of glycosidic bonds. *J. Med Microbiol.* 4, 451–459.

Heinonen, S., Nurmi, T., Liukkonen, K., Poutanen, K., Wähälä, K., Deyama, T., Nishibe, S., and Adlercreutz, H. (2001). *In vitro* metabolism of plant lignans: New precursors of mammalian lignans enterolactone and enterodiol. *J. Agric. Food Chem.* **49**, 3178–3186.

Jacobs, E., Kulling, S.E., and Metzler, M. (1999). Novel metabolites of the mammalian lignans enterolactone and enterodiol in human urine. *J. Steroid Biochem. Mol. Biol.* **68**, 211–218.

Jacobs, E., and Metzler, M. (1999). Oxidative metabolism of the mammalian lignans enterolactone and enterodiol by rat, pig, and human liver microsomes. *J. Agric. Food Chem.* **47**, 1071–1077.

Jenab, M., Richard, S.E., Orcheson, L.J., and Thompson, L.U. (1999). Flaxseed and lignans increase cecal β-glucuronidase activity in rats. *Nutr. Cancer* **33**, 154–158.

Jenab, M., and Thompson, L.U. (1996). The influence of flaxseed and lignans on colon carcinogenesis and ß-glucuronidase activity. *Carcinogenesis* **17**, 1343–1348.

Kamal-Eldin, A., Peerlkamp, N., Johnsson, P., Andersson, R., Andersson, R.F., Lundgren, L.N., and Aman, P. (2001). An oligomer from flaxseed composed of secoisolariciresinoldiglucoside and 3-hydroxy-3-methyl glutaric acid residues. *Phytochemistry* **58**, 587–590.

Kulling, S.E., Jacobs, E., Pfeiffer, E., and Metzler, M. (1998). Studies on the genotoxicity of the mammalian lignans enterolactone and enterodiol and their metabolic precursors at various endpoints *in vitro*. *Mutation Research* **416**, 115–124.

Lapteva, K.I., Tyukavkina, N.Y., and Ryzhova, L.I. (1971). Lignan compounds from the wood of *Larix dahurica* and *L. sibirica*. *Chem. Nat. Compd.* **7**, 802.

Liggins, J., Grimwood, R., and Bingham, S.A. (2000). Extraction and quantification of lignan phytoestrogens in food and human samples. *Anal. Biochem.* **287**, 102–109.

Marsh, C.A., Alexander, F., and Levvy, G.A. (1953). Glucuronide decomposition in the digestive tract. *Nature* **170**, 163.

Mazur, W., Fotsis, T., Wähälä, K., Ojala, S., Salakka, A., and Adlercreutz, H. (1996). Isotope dilution gas chromatographic-mass spectrometric method for the determination of isoflavonoids, coumestrol, and lignans in food samples. *Anal. Biochem.* **233**, 169–180.

Meagher, L.P., Beecher, G.R., Flanagan, V.P., and Li, B.W. (1999). Isolation and characterization of the lignans, isolariciresinol and pinoresinol, in flaxseed meal. *J. Agric. Food Chem.* **47**, 3173–3180.

Muir, A.D., Ballantyne, K., Reschny, K., and Westcott, N.D. (2000a). LC-MS analysis of lignan metabolites. In *Proceedings of the Flax Institute of the USA*, Flax Institute of the USA, Fargo, N.D., pp. 33–38.

Muir, A.D., Prasad, K., Thompson, K., and Westcott, N.D. (2001a). LC-MS/MS analysis of plant lignan metabolites in plasma and urine. *Whole Grain and Human Health*. Porvoo, Finland.

Muir, A.D., Thompson, K., Reschny, K., Prasad, K., and Westcott, N.D. (2001b). Flax lignans – enzymatic breakdown of the SDG lignan complex. *2nd International Conference and Exhibition on Nutracenuticals and Functional Foods. Portland, OR.*

Muir, A.D., and Westcott, N.D. (2000). Quantitation of the lignan secoisolariciresinol diglucoside in baked goods containing flaxseed or flaxmeal. *J. Agric. Food Chem.* **48**, 4048–4052.

Muir, A.D., Westcott, N.D., Ballantyne, K., and Northrup, S. (2000b). Flax lignans – recent developments in the analysis of lignans in plant and animal tissues. In *Proc. Flax Institute of the USA*, Vol. 58, Flax Institute of the USA, Fargo, ND, pp. 23–32.

Murray, S., Baillie, T.A., and Davies, D.S. (1977). A non-enzymic procedure for the quantitative analysis of (3-methoxy-4-sulphoxyphenyl)ethylene glycol (MHPG sulphate) in human urine using stable isotope dilution and gas chromatography-mass spectrometry. *J. Chrom.* **143**, 541–551.

Niemeyer, H.B., Honig, D., Lange-Böhmer, A., Jacobs, E., Kulling, S.E., and Metzler, M. (2000). Oxidative metabolites of the mammalian lignans enterodiol and enterolactone in rat bile and urine. *J. Agric. Food Chem.* **48**, 2910–2919.

Obermeyer, W.R., Musser, S.M., Betz, J.M., Casey, R.E., Pohland, A.E., and Page, S.W. (1995). Chemical studies of phytoestrogens and related compounds in dietary supplements: Flax and chaparral. *Proc. Soc. Expt. Biol. Med.* **208**, 6–12.

Qiu, S.-X., Lu, Z.-Z., Luyengi, L., Lee, S.-K., Pezzuto, J.M., Farnsworth, N.R., Thompson, L.U., and Fong, H.H.S. (1999). Isolation and characterization of flaxseed (*Linum usitatissimum*) constituents. *Pharma. Biol.* **37**, 1–7.

Setchell, K.D.R., and Adlercreutz, H.J. (1979). The excretion of two new phenolic compounds (180/442 and 180/410) during the human menstrual cycle and in pregnancy. *J. Steroid Biochem.* **11**, 15–16.

Setchell, K.D.R., and Adlercreutz, H. (1988). Mammalian lignans and phytoestrogens, recent studies on their formation, metabolism and biological role in health and disease. In Rowland, I.R. (ed.) *Role of the Gut Flora in Toxicity and Cancer.* Academic Press, New York, pp. 315–345.

Setchell, K.D.R., Bull, R., and Adlercreutz, H. (1980a). Steroid excretion during the reproductive cycle and pregnancy of the vervet monkey (*Ceropithecus aethiopus pygerythrus*). *J. Steroid Biochem.* **12**, 375–384.

Setchell, K.D.R., Lawson, A.M., Axelson, M., and Adlercreutz, H. (1981a). The excretion of two new phenolic compounds during the human menstral cycle and in pregnancy. In Adlercreutz, H., Bulbrook, R.D., van der Molen, H.J., Vermeulen, A., and Sciarra, F. (eds), *Endocrinological Cancer, Ovarian Function and Disease. Proc IX meeting Int. Study Group for Steroid Hormones, Rome, Dec 5–7, 1979,* Vol. 515, Excerpta Medica, Amsterdam, pp. 207–215.

Setchell, K.D.R., Lawson, A.M., Borriello, S.P., Adlercreutz, H., and Axelson, M. (1982). Formation of lignans by intestinal microflora. In Malt, R.A., and Williamson, R.C.N. (eds) *Falk Symposium 31, Colonic Carcinogenesis.* MTP Press Ltd., Lancaster, pp. 93–97.

Setchell, K.D.R., Lawson, A.M., Borriello, S.P., Harkness, R., Gordon, H., Morgan, D.M.L., Kirk, D.N., Adlercreutz, H., Anderson, L.C., and Axelson, M. (1981b). Lignan formation in man–microbial involvement and possible roles in relation to cancer. *Lancet* ii, 4–7.

Setchell, K.D.R., Lawson, A.M., Mitchell, F.L., Adlercreutz, H., Kirk, D.N., and Axelson, M. (1980b). Lignans in man and animal species. *Nature* **287**, 740–742.

Stitch, S.R., Toumba, J.K., Groen, M.B., Funke, C.W., Leemhuis, J., Vink, G.F., and Woods, G.F. (1980). Excretion, isolation and structure of a new phenolic constituent in female urine. *Nature* **287**, 738–740.

Uehara, M., Lapčík, O., Hampl, R., Al-Maharik, N., Mäkelä, T., Wähälä, K., Mikola, H., and Adlercreutz, H. (2000). Rapid analysis of phytoestrogens in human urine by time-resolved fluoroimmunoassay. *J. Steroid Biochem. Mol. Biol.* **72**, 273–282.

Wang, L.-Q., Meselhy, M.R., Li, G.-W., and Hattori, M. (2000). Human intestinal bacteria capable of transforming secoisolariciresinol diglucoside to mammalian lignans, enterodiol and enterolactone. *Chem. Pharm. Bull.* **48**, 1606–1610.

Westcott, N.D., and Muir, A.D. (1999). Process for extracting and purifying lignans and cinnamic acid derivatives from flaxseed. Patent # 702410. Australia.

Westcott, N.D., and Muir, A.D. (2000). Overview of flax lignans. *INFORM* **11**, 118–121.

12 Flaxseed constituents and human health

Alister D. Muir and Neil D. Westcott

Introduction

The impact of the two major components in flaxseed (α-linolenic acid and the lignan SDG) on human health has been addressed in the preceding chapters. This chapter will focus on the biological activity of the other components of flaxseed and related members of the genus *Linum*. Flaxseed is a rich source of dietary fiber, soluble fiber or gums which have been implicated in a number of beneficial health effects. Flax seed also contains a number of components (anti-nutrients) that have negative health implications including cyanogenic glycosides, phytic acid, a vitamin B_6 antagonist, and antigenic carbohydrates. Heavy metals such as cadmium are also known to accumulate in flax, particularly in the seed. Within the genus *Linum*, there are species which accumulate a different class of lignans from those found in *L. usitatissimum*. These lignans have potential value as pharmaceutical precursors. Flax is also known to accumulate β-hydroxy-β-methylglutaric acid 1 (HMGA) (Klosterman and Smith, 1954) in the seed. Although HMGA may function as a competitive inhibitor for HMG-Coenzyme A, its contribution to the health benefits of flaxseed has not yet been demonstrated, and indeed it may be difficult to distinguish the effects of the HMGA from other components in flax.

Anti-nutrients

Flaxseed contains significant levels of phytic acid 2 (23–33 g/kg meal) (Oomah *et al.*, 1996), with cultivar, location and year effects all having significant effects on levels. Phytic acid can bind calcium, magnesium, zinc and iron and decrease their absorption from the GI tract.

Vitamin B_6 antagonist

It had been shown that flax meal contained a vitamin B_6 antagonist that caused poor growth and vitamin B_6 deficiencies in chicks. The toxic effects could be countered by prompt administration of pyridoxine (Kratzer and Williams, 1948; Kratzer *et al.*, 1954). The toxic effect was found to be extractable with 70 percent aqueous alcohol (Evenstad *et al.*, 1965) and subsequently identified as glutamate of the non-protein amino acid 1-amino-D-proline and given the trivial name linatine 3 (Klosterman *et al.*, 1967). Although it is possible that there could be a toxic effect in humans, linatine does not appear to have any detrimental effect on mature poultry and a toxic effect

1

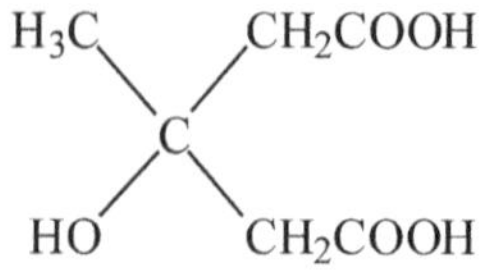

$$H_3C \quad CH_2COOH$$
$$HO \quad CH_2COOH$$

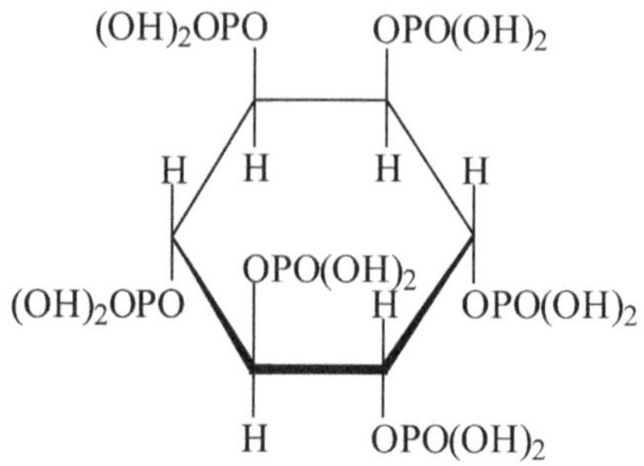

2

(OH)$_2$OPO   OPO(OH)$_2$

H   H   H   H

OPO(OH)$_2$

(OH)$_2$OPO   H   OPO(OH)$_2$

H   OPO(OH)$_2$

3

$$NH_2$$
$$O=CCH_2CH_2CHCOOH$$
$$NH$$
$$N \quad COOH$$

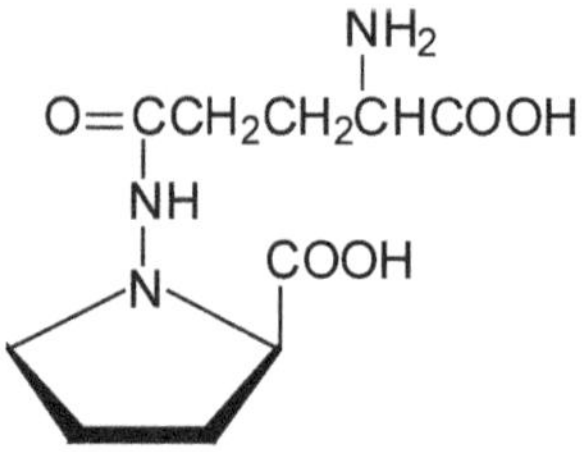

Figure 12.1 Antinutritive compounds in flaxseed.

has yet to be demonstrated in mammals (Klosterman, 1974). Originally isolated from the seed, linatine was subsequently isolated from all parts of the immature flax plant (Nugent, 1971).

Gums

Several studies have suggested that the dietary fiber content of flaxseed may lower serum total cholesterol and LDL-cholesterol. However, the design of many of these studies does not allow the lipid lowering effect to be definitively assigned to the gum component. In some studies whole flaxseed was consumed (Bierenbaum *et al.*, 1993), while others have examined the effect of isolated gums (Cunnane *et al.*, 1993). However, most of the benefit attributed to flaxseed gums is based on extrapolation from other sources of dietary fiber.

Isolated flax gums have been shown to have a hypoglycaemic effect in the blood glucose response test (Cunnane *et al.*, 1993) and to reduce postprandial blood glucose responses typical of viscous fiber (Wolever and Jenkins, 1993).

It is difficult to evaluate the biological activity of flax gums in large part because two other biologically active molecules (lignans and cyanogenic glycosides) (Cunnane *et al.*, 1993) are known to co-extract with the gum when attempts are made to produce gum fractions for biological evaluation.

Partially defatted flaxseed reduced total cholesterol (4.6 ± 1.2%; P = 0.001), LDL cholesterol (7.6 ± 1.8%; P = 0.001), apolipoprotein B (5.4 ± 1.4%; P = 0.001), and apolipoprotein A-I (5.8+ ± 1.9%; P = 0.005), but had no effect on serum lipoprotein ratios at week three compared with the control (Jenkins *et al.*, 1999).

Flaxseed is known to have a significant bulk laxative effect (Schilcher *et al.*, 1986; Tarpila and Kivinen, 1997) which is usually attributed to the gum content.

Proteins

Relatively little is known about the biological properties of flax proteins. Trypsin inhibitor activity in flax meal was first reported in 1983 (Madhusudhan and Singh, 1983). However, no amylase inhibitor or haemagglutinating activity was reported for flax meal. Bhatty (1993) demonstrated that flax meal contained low levels of trypsin inhibitor activity, typically less than 3 percent of the activity of soyabean and that commercial meals contained less than laboratory prepared meals. These levels are not thought to have any significance for human and livestock nutrition. Flax proteins may also cause allergic reactions when inhaled as dust (Bernstein and Bernstein, 1988). See section on Anaphylaxis and allergic reactions below, p. 246.

A series of immunosuppressive cyclic peptides cyclolinopeptide A-E have been isolated from flaxseed (Kaufmann and Tobschirbel, 1959; Morita *et al.*, 1997; Morita *et al.*, 1999). However, it is not yet clear if these peptides have any biological activity when consumed as part of the diet.

Lignans other than SDG

Podophyllotoxin

Podophyllotoxin 4 and more recently the semisynthetic derivatives etoposide and teniposide have important anticancer properties and are used in the treatment of certain cancers. Etoposide was approved by the FDA for treating testicular cancer in 1983 and is one of the most active agents in the treatment of testicular teratoma, Hodgkin's and non-Hodgkin's lymphomas and small-cell lung cancer (Wang and Lee, 1997). Several related cyclolignans have also been shown to have antiviral activity (Gordaliza *et al.*, 1994). The traditional source of podophyllotoxin is the rhizomes of *Podophyllum hexandrum* and *P. peltatum*. Both of these species are now endangered as a result of wildcrafting and alternative sources are being sought. Many species of *Linum* (Broomhead and Dewick, 1988; Wichers *et al.*, 1991; Konuklugil, 1996) accumulate podophyllotoxin and related compounds including 5-methoxypodophyllotoxin 5 and β-peltatin methyl ether 6. Many of these compounds are cytotoxic (Berlin *et al.*, 1986; van Uden *et al.*, 1992) and continue to be used as starting material for the semisynthesis of new anticancer derivatives (Wang and Lee, 1997).

Acute toxicology

Flaxseed is essentially free of compounds that can cause acute toxicity. However, like almost all plant material in the human diet, there is a risk of acute allergic reactions to compounds present in the seed, for a very small proportion of the population.

4

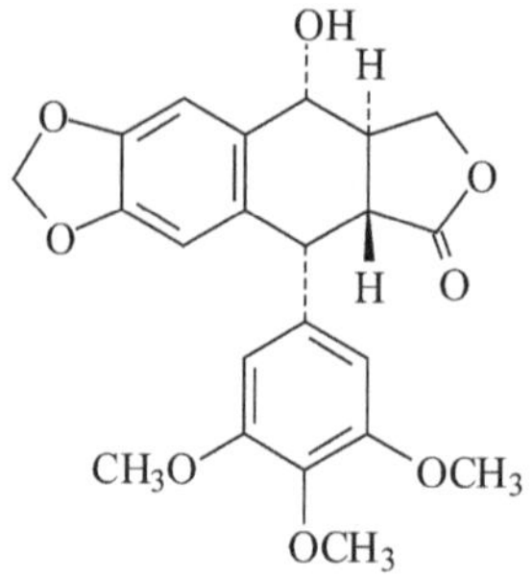

5

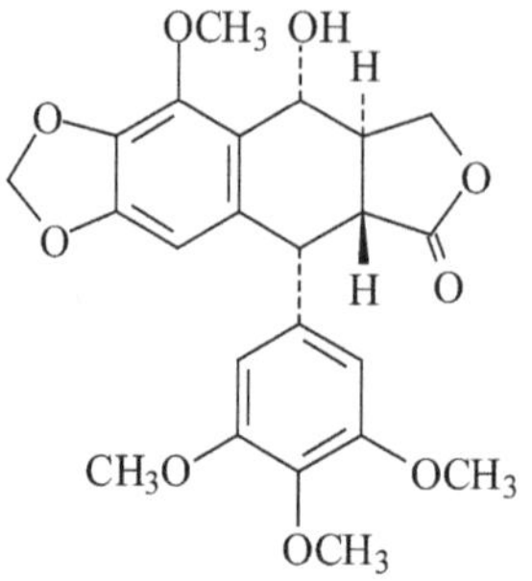

6

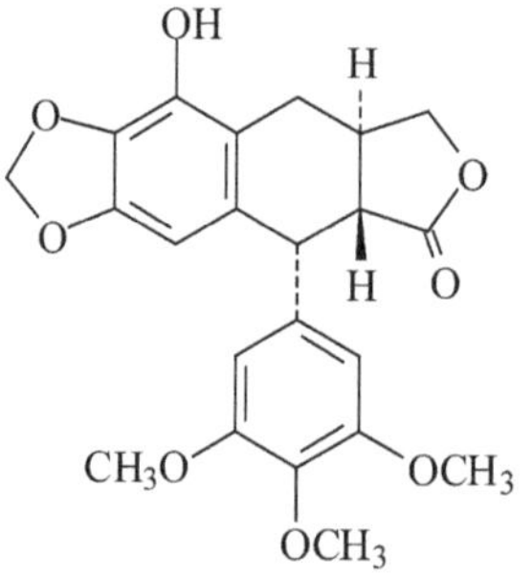

Figure 12.2 Podophyllotoxin lignans found in *Linum* spp.

Anaphylaxis and allergic reactions

Although relatively rare, there are a few reports of anaphylaxis reactions to components in flax when ingested. In 1995, Muthiah *et al.* (1995) reported a case of anaphylaxis reaction to carbohydrate components in flax and a second IgE-mediated case was reported in 1998 (Lezaun *et al.*, 1998). In 1996, Alonso *et al.* (1996) reported a case of an anaphylactic reaction after the intake of flax oil.

Exposure to dust resulting from the processing of flax fiber is known to cause non-IgE mediated byssinosis (Nowier *et al.*, 1975). IgE-mediated asthma has also been reported in workers processing flax oil (Bernstein and Bernstein, 1988).

Flaxseed is also reported to contain potent allergens but relatively little is known about the allergenic principles (William and Kenneth, 1988).

Chronic toxicology

Cyanogenic glycosides

Flax has long been known to accumulate cyanogenic glycosides in the seed. These compounds of themselves are not generally thought to be toxic, but rather to release HCN when subjected to enzymatic hydrolysis. All plants that accumulate cyanogenic glycosides also produce the enzymes necessary for hydrolysis of these glycosides (Fan and Conn, 1985). In intact plant tissue these enzymes are located in cellular structures that segregate the enzyme from the glycoside. However, when tissue damage occurs then the enzyme and the glycoside can come in contact and the release of HCN can occur. HCN inhibits cytochrome oxidase and other respiratory enzymes. For a recent review on the toxicology of cyanogenic glycosides see Majak and Benn (1994).

While cyanogenic glycosides have long been recognized for their chronic toxic effects, the cyanogenic glycoside content has also become associated with the protective effect of flaxseed meal against selenium toxicity first observed in 1941 (Moxon, 1941; Olson and Halverson, 1954). Subsequent experimentation demonstrated that the protective component could be extracted with hot 50 percent aqueous alcohol (Halverson *et al.*, 1955; Jensen and Chang, 1976). The cyanogenic glycosides linustatin 7 and neolinustatin 8 were subsequently isolated (Palmer *et al.*, 1980; Smith *et al.*, 1980) and shown to be responsible for the protective effect. There appears to be some specificity for the cyanogenic glycoside with the disaccharides linustatin and neolinustatin exhibiting greater protective effect than the monosaccharide linamarin 9 (Palmer, 1995). However, since it would appear that it is the reaction of the CN^- ion with Se to form the soluble $SeCN^-$ ion that is responsible for the detoxification effect, the response to the different cyanogenic glycosides more likely reflects the substrate specificity of the intestinal β-glucosidases.

Attempts to demonstrate a protective effect in large animals have met with mixed results, with beneficial effects observed in swine (Wahlstrom *et al.*, 1956), and no effect in range cattle (Dinkel *et al.*, 1957). Recent experiments with sheep have suggested that high levels of cyanogenic glycosides could adversely affect selenium status when selenium levels are at or below minimum levels and increase the incidence of nutritional myopathy (Gutzwiller, 1993).

There is some concern about the long-term consumption of flaxseed in relation to human health. Long-term consumption of flaxseed has been shown to increase plasma levels and urinary excretion of thiocyanate (Schulz *et al.*, 1983). Cunnane *et al.* (1993) reported a two-fold increase in urinary excretion of thiocyanate but the difference was not statistically significant. The effect of these levels on human health is not well established, but the combination of iodine deficiency and high plasma thiocyanate levels has been implicated in increased levels of goiter and cretinism (Ermans *et al.*, 1983), and elevated plasma levels of thiocyanate are common in smokers and have been associated with increased numbers of lower birth weight babies in women who smoke (Hauth *et al.*, 1984). It would also appear that when consumed as a component of food, up to 25 percent of the cyanogenic glycosides can be excreted intact (Carlsson *et al.*, 1999). Cooking of ground flaxseed in muffins appears to significantly reduce the level of detectable cyanogenic glycosides (Cunnane *et al.*, 1993).

7

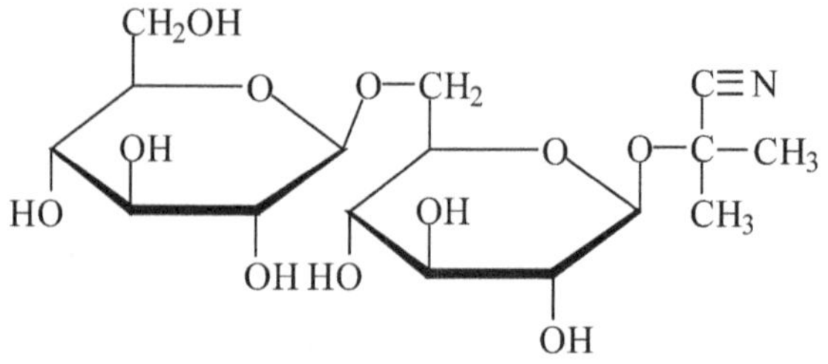

8

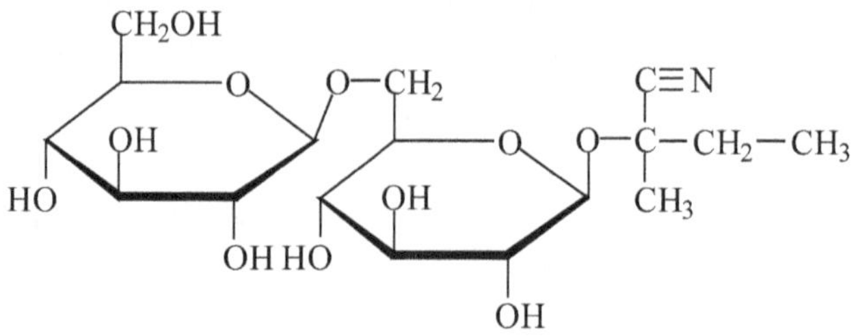

9

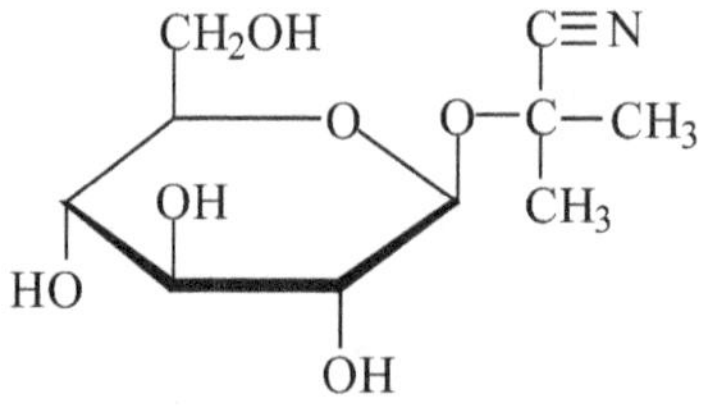

Figure 12.3 Cyanogenic glycosides found in flax.

Kolodziejczyk and Fedec (1995) suggested that cold pressed flax meal may not be subjected to sufficient heat to inactivate endogenous glucosidases that can hydrolyze the cyanogenic glycosides when the meal is hydrated. However, there does not appear to be any experimental evidence to support this concern.

Cadmium accumulation

Cadmium can accumulate in the kidney cortex and cause renal tubular dysfunction (Elinder and Järup, 1996; Jin *et al.*, 1998) and chronic exposure to high levels of cadmium causes a condition known as Itai itai disease, named for an area in Japan where cadmium contamination affected a signficant number of individuals. For a recent review of the health effects of cadmium exposure see Järup *et al.* (1998). Flax, along with sunflower and durum wheat (McLaughlin *et al.*, 1999), is a crop species that can absorb and translocate cadmium to edible crop parts even when grown on uncontaminated alkaline soils (Li *et al.*, 1997). Screening studies have shown that there is considerable genetic variability present for cadmium accumulation (Li *et al.*, 1997) and breeding programs are currently underway to develop low cadmium accumulating varieties.

References

Alonso, L., Marcos, M.L., Blanco, J.G., Navarro, J.A., Juste, S., del Mar Garcés, M., Pérez, R., and Carretero, P.J. (1996). Anaphylaxis caused by linseed (flaxseed) intake. *J. Allergy Clin. Immunol.* **98**, 469–470.

Berlin, J., Wray, V., Mollenschott, C., and Sasse, F. (1986). Formation of β-peltatin-A-methyl ether and coniferin by root cultures of *Linum flavum. J. Natural Prod.* **49**, 435–439.

Bernstein, D., and Bernstein, I. (1988). Occupational asthma. In Middleton, E., Reed, C.E., Ellis, E.F., Adkinson, N.F., and Yunginger, J.W. (eds), *Allergy: Principles and Practice*, C.V. Mosby Company, St. Louis, pp. 1197–1218.

Bhatty, R.S. (1993). Further compositional analyses of flax: Mucilage, trypsin inhibitors and hydrocyanic acid. *J. Am. Oil Chem. Soc.* **70**, 899–904.

Bierenbaum, M.L., Reichstein, R., and Watkins, T.R. (1993). Reducing atherogenic risk in hyperlipiemic humans with flaxseed supplementation: A preliminary report. *J. Am. Coll. Nutr.* **12**, 501–504.

Broomhead, A.J., and Dewick, P.M. (1988). *Linum* species as a source of anticancer lignans. *J. Pharm. Pharmacol.* **40 Suppl**, 55p.

Carlsson, L., Mlingi, N., Juma, A., Ronquist, G., and Rosling, H. (1999). Metabolic fates in humans of linamarin in cassava flour ingested as stiff porridge. *Food Chem. Toxiocol.* **37**, 307–312.

Cunnane, S.C., Ganguli, S., Menard, C., Liede, A.C., Hamadeh, M.J., Chen, Z.-Y., Wolever, T.M.S., and Jenkins, D.J.A. (1993). High α-linolenic acid flaxseed (*Linum usitatissimum*): Some nutritional properties in humans. *Br. J. Nutr.* **49**, 443–453.

Dinkel, C.A., Minyard, J.A., Whitehead, E.I., and Olson, O.E. (1957). *Research at the Reed Ranch Substation.* S. Dak. Agric, Exp. Stn. Circular 135. pp. 24–26.

Elinder, C.-G., and Järup, L. (1996). Cadmium exposure and health risks: Recent findings. *Ambio* **25**, 370–373.

Ermans, A.M., Bourdoux, P., Kinthaert, J., Lagasse, R., Luvivila, K., Mafuta, M., Thilly, C.H., and Delange, F. (1983). Role of cassava in the etiology of endemic goitre and cretinism. In Delange, F., and Ahluwalia, R. (eds), *Cassava Toxicity and Thyroid: Research and Public Health Issues*, International Development Research Center, Ottawa, Canada, pp. 9–16.

Evenstad, E.T., Lamoureux, G.L., Klosterman, H.J., and Cooley, A.M. (1965). Pilot scale extraction of the antipyridoxine factor in linseed meal. *Proc. N. Dakota Acad. Sci.* **19**, 110–114.

Fan, T.W.-M., and Conn, E.E. (1985). Isolation and characterization of two cyanogenic β-glucosidases from flaxseeds. *Arch. Biochem. Biophys.* **243**, 361–373.

Gordaliza, M., Castro, M.A., Garcia-Gravalos, M.D., Ruiz, P., Miguel del Corral, J.M., and San Feliciano, A. (1994). Antineoplastic and antiviral activities of podophyllotoxin related lignans. *Arch. Pharm (Weinheim)* **327**, 175–179.

Gutzwiller, A. (1993). The effect of a diet containing cyanogenic glycosides on the selenium status and the thyroid function of sheep. *Anim. Prod.* **57**, 415–419.

Halverson, A.W., Hendrick, C.M., and Olson, O.E. (1955). Observations on the protective effect of linseed oil meal and some extracts against chronic selenium poisoning in rats. *J. Nutr.* **56**, 51–60.

Hauth, J.C., Hauth, R.B., Drawbaugh, L.C., Gilstrap, L.C., and Pierson, W.P. (1984). Passive smoking and thiocyanate concentrations in pregnant women and newborns. *Obstet. Gynecol.* **63**, 519–522.

Järup, L., Berglund, M., G., E.C., Nordberg, G., and Vahter, M. (1998). Health effects of cadmium exposure: A review of the literature and a risk estimate. *Scand. J. Work. Environ. Health* **24**, 1–52.

Jenkins, D.J., Kendall, C.W., Vidgen, E., Agarwal, S., Rao, A.V., Rosenberg, R.S., Diamandis, E.P., Novokmet, R., Mehling, C.C., Perera, T., Griffin, L.C., and Cunnane, S.C. (1999). Health aspects of partially defatted flaxseed, including effects on serum lipids, oxidative measures, and *ex vivo* androgen and progestin activity: A controlled crossover trial. *Am. J. Clin. Nutr.* 69, 395–402.

Jensen, L.S., and Chang, C.H. (1976). Fractionation studies on a factor in linseed meal protecting against selenosis in chicks. *Poultry Sci.* 55, 594–599.

Jin, T., Lu, J., and Nordberg, M. (1998). Toxicokinetics and biochemistry of cadmium with special emphasis on the role of metallothionein. *Neurotoxicol.* 19, 529–536.

Kaufmann, H.P., and Tobschirbel, A. (1959). Über ein Oligopeptid aus Leinsamen. *Chem. Ber.* 92, 2805–2809.

Klosterman, H.J. (1974). Vitamin B_6 antagonists of natural origin. *J. Agric. Food Chem.* 22, 13–16.

Klosterman, H.J., Lamoureux, G.L., and Parsons, J.L. (1967). Isolation, characterization, and synthesis of linatine, a vitamin B_6 antagonist from flaxseed (*Linum usitatissimum*). *Biochem.* 6, 170–177.

Klosterman, H.J., and Smith, F. (1954). The isolation of β-hydroxy-β-methylglutaric acid from the seed of flax (*Linum usitatissimum*). *J. Am. Chem. Soc.* 76, 1229–1230.

Kolodziejczyk, P.P., and Fedec, P. (1995). Processing flaxseed for human consumption. In Cunnane, S.C., and Thompson, L.U. (eds), *Flaxseed in Human Nutrition*, AOCS Press, Champaign, IL, pp. 261–280.

Konuklugil, B. (1996). Aryltetralin lignans from genus *Linum*. *Fitoterapia* 67, 379–381.

Kratzer, F.H., and Williams, D.E. (1948). The effect of pyridoxine upon growth of chicks fed linseed oil meal. *Poultry Sci.* 27, 671.

Kratzer, F.H., Williams, D.E., Marshall, B., and Davis, P.N. (1954). Some properties of the chick growth inhibitor in linseed oil meal. *J. Nutr.* 52, 555–563.

Lezaun, A., Fraj, J., Colas, C., Duce, F., Dominguez, M.A., Cuevas, M., and Eiras, P. (1998). Anaphylaxis from linseed. *Allergy* 53, 105–106.

Li, Y.-M., Chaney, R.L., Schneiter, A.A., Miller, J.F., Elias, E.M., and Hammond, J.J. (1997). Screening for low grain cadmium phenotypes in sunflower, durum wheat and flax. *Euphytica* 94, 23–30.

Madhusudhan, K.T., and Singh, N. (1983). Studies on linseed proteins. *J. Agric. Food Chem.* 31, 959–963.

McLaughlin, M.J., Parker, D.R., and Clarke, J.M. (1999). Metals and micronutrients – food safety issues. *Field Crops Research* 60, 143–163.

Majak, W., and Benn, M.H. (1994). Glycosides. In Hui, Y.H., Gorham, J.R., Murrell, K.D., and Cliver, D.O. (eds), *Foodborne Disease Handbook-Disease Caused by Hazardous Substances*, Vol 3, Marcel Dekker Inc., New York, pp. 311–370.

Morita, H., Shishido, A., Matsumoto, T., Itokawa, M., and Takeya, K. (1999). Cyclolinopeptides B-E, new cyclic peptides from *Linum usitatissimum*. *Tetrahedron* 55, 967–976.

Morita, H., Shishido, A., Matsumoto, T., Takeya, K., Itokawa, H., Hirano, T., and Oka, K. (1997). A new immunosuppressive cyclic nanapeptide, cyclolinopeptide B from *Linum usitatissimum*. *Bioorg. Med. Chem. Lett.* 7, 1269–1272.

Moxon, A.L. (1941). The influence of some proteins on the toxicity of selenium. Ph.D., University of Wisconsin, Madison (Cited in Halverson *et al.*, 1955).

Muthiah, R., Louthain, R., Knoll, M.T., and Kagen, S. (1995). Flaxseed-induced anaphylaxis (ANA): Another cross-reactive carbohydrate food allergen. *51st Ann. Meeting Am. Acad. Allergy and Immunology. New York, N.Y., J Allergy Clinical Immunol.* p. 370.

Nowier, M., El-Sadik, Y., El-Dakhakny, A., and Osmar, H. (1975). Dust exposure in manual flax processing in Egypt. *Br. J. Industrial Med.* 32, 147–152.

Nugent, P.J. (1971). Chemical synthesis and metabolism of linatine. Ph.D., North Dakota State University, Fargo.

Olson, O.E., and Halverson, A.W. (1954). Effect of linseed oil meal and arsenicals on selenium poisoning in the rat. *Proc. South Dakota Acad. Sci* 33, 90–94.

Oomah, B.D., Kenaschuk, E.O., and Mazza, G. (1996). Phytic acid content of flaxseed as influenced by cultivar, growing season, and location. *J. Agric. Food Chem.* 44, 2663–2666.

Palmer, I.S. (1995). Effects of flaxseed on selenium toxicity. In Cunnane, S.C., and Thompson, L.U. (eds), *Flaxseed in Human Nutrition*, AOCS Press, Champaign, Ill, pp. 165–173.

Palmer, I.S., Olson, O.E., Halverson, A.W., Miller, R., and Smith, R. (1980). Isolation of factors in linseed oil meal protective against chronic selenosis in rats. *J. Nutr.* 110, 145–150.

Schilcher, H., Schulz, V., and Nissler, A. (1986). Zur Wirksamkeit und Toxikologie von Semen Lini. *Zeitschrift für Phytotherapie* 7, 113–117.

Schulz, V., Löffler, A., and Gheorghiu, T. (1983). Resorption von Blausäure aus Leinsamen. *Leber. Magen. Darm.* 13, 10–14.

Smith, C.R., Weisleder, D., Miller, R.W., Palmer, I.S., and Olson, O.E. (1980). Linustatin and neolinustatin: Cyanogenic glycosides of linseed meal that protect animals against selenium toxicity. *J. Org. Chem.* 45, 507–510.

Tarpila, S., and Kivinen, A. (1997). Ground flaxseed is an effective hypolipidemic bulk laxative. *Digestive Disease Week and 97th Ann. Meet. Am. Gastroenterological Assn. Washington, D.C.*, Gastroenterology. p. A836.

van Uden, W., Homan, B., Woerdenbag, H.J., Pras, N., Malingré, T.M., Wichers, H.J., and Harkes, M. (1992). Isolation, purification, and cytotoxicity of 5-methoxypodophyllotoxin, a lignan from a root culture of *Linum flavum. J. Natural Prod.* 55, 102–110.

Wahlstrom, R.C., Kamstra, L.D., and Olson, O.E. (1956). *Preventing Selenium Poisoning in Growing and Fattening Pigs.* S. Dak. Agric. Exp. Stn. Bull. No. 456. pp. 10–12.

Wang, H.-K., and Lee, K.-H. (1997). Plant-derived anticancer agents and their analogs currently in clinical use or in clinical trials. *Bot. Bull. Acad. Sinica* 38, 225–235.

Wichers, H.J., Versluis-De Haan, G.G., Marsman, J.W., and Harkes, M.P. (1991). Podophyllotoxin related lignans in plants and cell cultures of *Linum flavum. Phytochemistry* 30, 3601–3604.

William, R., and Kenneth, P. (1988). Aerobiology and inhalant allergens. In Middleton, E., Reed, C.E., Ellis, E.F., Adkinson, N.F., and Yuninger, J.W. (eds), *Allergy: Principles and Practice*, C.V. Mosby Company, St. Louis, pp. 312–372.

Wolever, T.M.S., and Jenkins, D.J.A. (1993). Effect of dietary fiber and foods on carbohydrate metabolism. In Spiller, G.A. (ed.) *CRC Handbook of Dietary Fiber in Human Nutrition*, CRC Press Inc., Boca Raton, FL, pp. 111–152.

13 Traditional food and medicinal uses of flaxseed

N. Lee Pengilly

Introduction

The use of flax for food, fiber and medicine reaches back to the most remote periods of history and true to its domestic name *Linum usitatissimum* (pronounced LY-num yew-si-ta-Tiss-i-mum) has proven itself to be both the most used and the most useful of the *Linum* genus. As early peoples experimented with many plants to determine which were suitable to eat, other properties were also discovered and used to humankind's advantage. Recorded history is not likely to disclose the first utilization of flax.

Domestication

Although the *Linum* L. genus is large with species spread over the steppe belts of the northern hemisphere and the temperate Mediterranean, there is general agreement that our present day cultivated flax is most closely related to wild *L. angustifolium* (syn. *L. bienne*) (see Chapter 2 for a more detailed discussion). Whereas some wild forms of *L. angustifolium* are biennial or perennial and our modern cultivated flax is an annual, they share a commonality in having the same number of chromosomes. The two species are fully inter-fertile and intercross with ease. Distribution of *L. angustifolium* through Iran, Caucasia, North Africa, the Near East, west Europe and the Mediterranean basin, as well as its close genetic and morphological affinity to *L. angusitatissimum*, adds further evidence to substantiate the claim that it is the wild progenitor of the cultivated crop (Zohary and Hopf, 1988). Through domestication *L. angustifolium*, also known as Pale Flax, experienced primary changes: a marked increase in seed size, the development of a non-dehiscent seed boll, higher oil yield and/or a longer stem (Stuart, 1979; Zohary and Hopf, 1988). Genetically, flax responded positively to the investment of human time and energy (tending and selective pressure) which was put into the early efforts of "domestication." These important genetic changes helped flax to achieve a dominant position in the social, religious, economic and political lives of Neolithic peoples, a status that became securely entrenched. The oldest flax remains retrieved from various sites in Europe and the Near East have been traced to the wild *L. angustifolium*. However, remains found in Syria put flax into the class of *L. usitatissimum* by 6250–5950 B.C.E. indicating flax cultivation prior to that time (Zohary and Hopf, 1988).

The beginning of the Neolithic period (6800–2550 B.C.E.) saw peoples closely tied to a series of plant and animal resources on which they had come to depend through knowledge gained by observation and experimentation. Gradually that knowledge propelled humans from a subsistence existence of hunting and gathering to arable and pastoral agriculture. This practice of farming and living in settled hamlets was a seminal event of the Neolithic Revolution. Farmers were tied to the land. The seed had been invested in and it was necessary to wait for the return on that investment; the dividend, the security that sufficient food supplies afforded. This "bond (or bondage)" to the land was of great importance in societal development (Hawkes and Woolley, 1963).

Egypt

Although the cultivation and use of flax has been chronicled in several areas, one of the best documented patterns of development comes from the Fayum Depression which lies to the west of the Nile River and south of the Delta Region of modern day Egypt. By the fifth millennium B.C.E., the Fayumis fished and hunted, but their settlements also displayed the full complement of Neolithic culture. They kept domestic pigs, goats, sheep and cattle, cultivated wheat, barley, emmer, and grew flax which they wove into linen (Hawkes and Woolley, 1963). Some of the earliest specimens of coil technique basketry that have been found at the Fayum Depression were utilized to line grain storage pits. The cores of these basket fibers were of grass, wrapped with thread of flax bast. Coiled basketry remained popular in Egypt through Dynastic times. Earlier dated sites indicate that flax fibers were initially used for fishing nets (Mesolithic era) and the process of twisting fibers to make "thread" for bow strings, to sew skins together or to attach projectile points may have been used as early as Paleolithic times (Hawkes and Woolley, 1963).

Surges in population movement from various Neolithic sites precipitated by increased demands for food (soil exhaustion or game depletion), clan fights, warfare, illness, or the human need to explore advanced this early culture throughout the Near and Middle East and into Europe. Many cultural similarities between the people of the Fayum Depression and the Swiss Lake-dwellers have been documented, including coiled and twining basketry techniques utilizing flax, the use of flax twining for bags, and a similar weaving pattern is evident in their linen textiles. The Swiss Lake-dwellers had access to a flax fiber that displayed fleecy, napped properties probably used for mats, cushions or thick cloaks (Hawkes and Woolley, 1963). From these sites, the cultivation of flax continued to spread through Europe, reaching Scandinavia before the end of the Neolithic period (Hawkes and Woolley, 1963).

Although we know little of how these ancient peoples extracted oil and fiber from plants, it seems logical to assume that the oil was obtained by decantation; crushing the seed, pouring hot water over the meal and scooping the oil out after setting. Prior to crushing, the seeds may have been softened by soaking in water (Zohary and Hopf, 1988). Alternately, the seeds may have been put in water and heated over a fire, the oil removed by skimming the surface (Wilkinson, 1988). To obtain the fiber, the stalks were immersed in water for a period of time in order to soften them so that the fiber could be more easily separated from the woody plant parts, a method referred to as retting. In all likelihood, these processes predate agriculture (Zohary and Hopf, 1988).

With knowledge garnered from their ancestors and improved upon by their own experience, the farmers of Ancient Egypt harvested a profusion of products that more than provided the necessities of life. This abundance spearheaded the growth and success of their agrarian based economy. Life in Egypt was significantly influenced by the

Nile River and it was of paramount importance in agricultural development. The inundation began between the end of May and the middle of June and crops were planted after it subsided. Apart from foodstuffs, flax was the main agricultural crop grown and, as an industry, flax production came to be as important as the growing of grain (Hawkes and Woolley, 1963). The Pharaoh, who was "God Incarnate" to the people of Egypt, personally owned the largest portion of the flax-growing area. This harvest was so important that he placed it under the control of a high officer of his court. The art of weaving was highly developed and Egypt was famous for the fineness of its fabric. Although the Pharaoh held a monopoly over cloth making for trade and export, the domestic textile industry, typical within the homes of rich and poor alike, was not impacted by this policy (Hawkes and Woolley, 1963).

Harvesting and processing techniques

Harvested flax was bound in sheaves and forcibly drawn through a comb-like instrument that stripped off the bolls (rippled) which were collected (Wilkinson, 1988). Today's microscopic examination of textile remnants reveals that the harvest took place at three different stages. The first one occurred soon after the flowering was complete while the stems were still green. This produced a soft, fine fabric reserved for the exclusive use of the aristocracy. Plants pulled 30 days after flowering when the stems were turning from green to yellow would yield a stronger fiber. A final harvest took place several weeks later when the stalks were golden and the seed bolls fully ripened. This fiber was suitable for mats and ropes. The seed for the next year's crop and for linseed oil, which was employed in a multitude of ways, was also collected at this time (Heinrich, 1992).

Much of the research and discussion regarding flax usage in antiquity has focused on flax for fiber, but the large volumes of flaxseed harvested beg the question as to the uses of the oil extracted (one bushel of seed, 25.40 kilograms, produces 9.5 litres of linseed oil (Anonymous, 1992)). A variety of oilseed crops were grown and harvested, and we can assume that there were several interchangeable uses for these oils as well as plant-specific niches for the various extractions.

Oxen were generally used to tread out the seeds (Wilkinson, 1988), but heavy long-handled mallets were also used to remove and crush the bolls prior to winnowing (Heinrich, 1992). At the domestic level, stone querns were used to grind the seeds to make linseed meal, while on an industrial scale, people were engaged to pound the seeds. Two men, taking turns, raised heavy pestles and directed the falling point into the center of a mortar made of hard stone which had been partially hollowed out to form a narrow tube. When the substance had been well pounded, it was removed, run through a sieve and the larger particles returned to the mortar (Wilkinson, 1988). The type of process used to extract the oil often determined the end use, with the pressing method being utilized for coarse oil production and the skimming method more desirable for domestic, medicinal and religious applications (Wilkinson, 1988).

Industrial uses

Primarily, industrial oil was used to fuel lamps (domestic, sacred and workplace), but it was also employed in a number of other ways. For dragging stone from the quarries, placing a statue in position or moving a burial sledge, water and oil or grease were poured before the sledge to facilitate easier movement (Wilkinson, 1988). Lead soldering required the use of oil (Wilkinson, 1988). It is quite likely that the Egyptians were well aware of linseed oil's properties as a preservative. Oil use is cited in the manufacture of wooden furniture, ship building, vehicles of transportation, articles of warfare and hunting, agricultural implements, statues, musical instruments and in the intricate wooden boxes used to hold possessions and cosmetics. Oil was employed as an ingredient to tan hides and was used to finish and preserve a wide variety of leather goods and components. In all probability the Egyptians experimented with the application of linseed oil onto fabric which would have yielded the earliest forms of oilcloth. Fibers exposed to the elements of fresh and salt water, wind and sun were preserved. To spin flax fiber, the yarn had to be moistened to insure a strong, smooth, even finished product. Whether modern spinners, who often use a wetting solution prepared by boiling flaxseed in water (Heinrich, 1992) took their lead from Ancient Egyptians is a matter of conjecture.

Domestic uses for oil fulfilled a variety of human needs beyond nutrition. The Egyptians "anointed" themselves with oil prior to leaving home and it was customary that guests were attended to by servants who anointed each person with oil as a principal token of welcome (Wilkinson, 1988). Wrestling was a prime amusement and the bodies of the contestants were oiled (Wilkinson, 1988). Oil or mucilage gum was the likely ingredient women used in preparations to style their hair (Wilkinson, 1988). Cleansing oils were used extensively to heal and protect the skin from the unrelenting sun and the torment of biting insects (Aldred, 1984). The lustrations of the Egyptians, who were a highly magico-religious people, required oil in the preparation and selection of donations to the deities. Those things that they were most grateful for or found the most useful were considered accordingly acceptable. Products derived from flax would have been high in the ranking and were presented through flower, wreath, chaplet, ointment, oil, seed (alone or on baked goods) and fabric. Ointments and oils were presented according to the festival or ceremony being celebrated. The statues of deities were anointed, a process which was carried out with the little finger of the right hand. Sweet scented ointments were liberally offered at every shrine (Wilkinson, 1988). When animals or birds were sacrificed, their bodies were stuffed with a variety of odoriferous substances, saturated with oil, then burnt over a fire (Wilkinson, 1988). High priests had oil poured upon their heads after they had completed their entire dress and adornment (Wilkinson, 1988); during "*Services for the Dead*," anointing with oil was part of the ceremony (Wilkinson, 1988). To be anointed with "oil of gladness" was not restricted to kings and priests, but was applied to the ordinary occurrences of the day (Wilkinson, 1988). Embalmers, who were part of the medical profession, made extensive use of linen fabric for mummy cloth, and during the "*Prayers for the Dead*," offerings similar to those presented to the deities were donated. The ingredients used to embalm and cleanse the body and internal organs as well as the bandaging (as much as 1000 yards) required scented oils and gummy mucilage (Wilkinson, 1988). Mummified bodies were anointed with oil or ointment during the time that elapsed between embalming and the funeral which was often several months later (Wilkinson, 1988).

Medical uses

The Egyptians paid great attention to their health, studying both medicine and surgery. Through the elaborate process of embalming, much opportunity was afforded to examine the human body, although it was fundamentally misunderstood by today's reckoning. Pharmacopoeia and drug inventories of Egypt, Mesopotamia and India show such similarity that there must have been continual discovery and exchange among the health professionals (Griggs, 1981). Knowledge of these "simples" (Hawkes and Woolley, 1963) was the cumulative result of many generations of experience – experience that had proven the efficacy of certain remedies. Linseed was among the favorites: "the delicate, blue-flowered plant from whose seeds comes (the) gentle oil used . . . since time immemorial" (Griggs, 1981). Recorded prescriptions indicate how closely these recipes interconnected common sense, medicine, magic and incantation. The curative value of an herb was often equated with a magical virtue inherent in the plant credited perhaps to the place, timing or ritual of its cultivation, growth or harvest. Ingredients were carefully prepared and administered in the form of poultice, ointment, pill or potion, often followed with or accompanied by an incantation; the manual rite and the oral rite.

Early doctors recognized the importance of diet. Modifications were often recommended as it was believed that a great number of diseases were caused by over-eating and poor digestion. Digestive problems may have required the induction of vomiting and the importance of free action of the bowels had been recognized with oil enemas administered when necessary. In both Egypt and Mesopotamia, records indicate that doctors specialized in various complaints and sometimes developed an impressive reputation which was not necessarily confined to their own country. Doctors would travel great distances to treat important patients and medical books were circulated freely from one country to another. It would appear that both the medical profession and the pharmacopoeia upon which it relied were highly internationalized (Hawkes and Woolley, 1963).

Those who took care of sick animals and had knowledge of veterinary practices held a higher social status than those involved in other forms of animal husbandry. Although not as privileged as those in the medical profession, they utilized many of the same herbal remedies with equivalent results.

The Greeks and Romans

Medical and food uses

Much of the foundation for flax usage had been laid during Egyptian times, with the Greek and Roman civilizations continuing to praise its merits as a food, a fiber and a medicine (Adkins and Adkins, 1997). The ancient Greeks followed two lines of thought in their medical practice. Hygeia was the natural approach to healing dependent upon the use of air, light, water, nutrition, herbs, massage and other natural non-invasive remedies. This school of thought, embraced by Hygeia, the goddess of health and beauty, was an attempt to provide the body with the tools that it needed to heal itself. The Greek god, Aesculapius, embodied the second school of thought which used invasive and often toxic methods to treat or suppress symptoms or infections. Healers subscribed primarily to Hygeia, relying upon Aesculapius only in times of crisis and for short term treatment (Erasmus, 1986).

The Roman armies made a substantial contribution to the spread of information regarding the healing arts of both medicine and surgery. Legions traveling with their own doctors engaged the services of local doctors when and where capable ones were found. The regular army doctors took supplies of the most important drugs (herbals) with them and "physic" gardens were tended and guarded around many of the army camps (Griggs, 1981).

Through antiquity, authors began to document the medical-botanical information previously handed down by oral tradition. Flax use has been described in detail in many of the classical writings of the Egyptians, Hebrews, Greeks and Romans. It was believed that appropriate food and drink were useful not only to cure disease, but also to guarantee good health, making it necessary to understand the individual properties of the associated product. Recipes that encouraged good eating habits were recorded. Flaxseed (linseed) was used in combination with barley, spices and salt to make a flavorful and nutritious polenta, mixed with wheat to make a variety of breads and with other grains, fruits and seeds to create grain-paste mixtures (puls or porridge) (Gozzini Giacosa, 1992). In many regions, the practice of adding flaxseed to foods has survived and retains its place as an essential ingredient for certain festival dishes.

Oil was an extremely precious commodity in the ancient world and it was in demand everywhere. All plants that yielded oil were prized, but flax, because of its multitude of uses, was particularly valued. Pliny the Elder, the great Roman naturalist (A.D. 23–79), is often quoted, "What department is there to be found of active life in which linseed is not employed? And in what production of the Earth are there greater marvels than this?" (Grieve, 1995). He did, however, caution early agriculturalists of the exhaustive nature of flax upon the soil, indicating that it "scorched" (Grieve, 1995) the earth, which is in agreement with Talmudic literature which notes that flax cultivation "impoverishes" the soil and should be grown in the same field only once every three or seven years (Anonymous, 1971).

Flax and Jewish law

The numerous Biblical and Talmudic references to flax cultivation and oil usage in ancient times add further to the reputation of this valued crop. Oil was symbolic of light as it was used to fuel the lamps in temples and shrines. As light, in turn, symbolized God and law, oil was regarded as having come from the "highest realm of the universe" (Lamsa, 1964). Kings and priests were anointed with oil as a symbol of their worth and capability of leading and guiding their subjects (Lamsa, 1964). In another context, oil used to refresh the body was of substantial importance where water was scarce and bathing was a rare or unknown occurrence. Oil preserved the skin and relieved painful sores on the feet, hands and body (Lamsa, 1964). Ointments and oils loosened dirt and cleansed the skin. "Dip his foot in oil" is an Eastern expression meaning, "Let him become very prosperous." This reference is from Biblical times when oil was used as a medium of exchange or loaned, as was silver or gold (Lamsa, 1964).

The importance of flax cultivation is stated in the Bible. In Exodus (9:31), reference is made to the crop damaged by a hailstorm and in Isaiah (19:5–9), the withering of the flax crop and decimation of the associated linen industry is described as a consequence of the Nile drying up (Anonymous, 1971). The Torah prohibits the wearing of garments spun from both flax and wool, a reference to the story of Cain and Abel, in which Cain is said to have brought an offering of flaxseed, and Abel, of sheep's wool; a probable precursor of the antagonism between farmer and shepherd (Anonymous, 1971). Talmudic literature tells of flax being attacked by plant diseases and of subsequent offerings of public prayers for a healthy and bountiful harvest (Anonymous, 1971).

Middle Ages

By the Middle Ages, flax was grown throughout Europe, with its diverse uses applauded: linen for clothing, sails, fishnets, thread, rope, candle wicks, string, measuring lines, sheets, bandages, sacks, bags and purses and mention is made of the tow being "used for stuffing into the cracks in ships . . . with none herbe . . . so needfull to so many dyurrse uses to mankynde as is the flexe" (Grieve, 1995).

Paper

The increased popularity of transferring information from oral tradition to written form saw the evolution of another utilization for flax fiber with invention and advancement coming out of China. Thin strips of bamboo were stitched or laced together with flax, silk or animal hair contributing to the advent of paper. From China, papermaking spread to Central Asia, Persia and India. In the Middle East, flax fiber was one of the predominant ingredients used by paper makers as many of the Chinese materials were unavailable to them. The Europeans were slow to accept the use of paper and it was well into the Middle Ages and the invention of the printing press before paper became more popular than parchment or vellum (Williams, 1999).

Medical uses

The compendiums of herbal medicine that had come down from ancient times were protected in the monasteries during the Dark Ages and the monks provided continuity to the framework in which medieval herbalism was practiced. Practical reference books were compiled outlining the various actions herbs had on the body, their mode of activity, the range of healing properties and their innumerable applications. By virtue of the timeless nature of much of this information, it continues to be invaluable.

Flaxseed's medicinal popularity, documented since earliest times, was in part due to the many forms in which it could be utilized. Generally referred to as linseed, flax was known as an emollient, useful for balms, salves, ointment and unguents. It was frequently put into layered poultices (cataplasms) to treat tumors, inflammations and abscesses (imposthumes). Cold-pressed linseed was recommended for diseases of the chest and lungs such as asthma, coughs, tuberculosis (consumption), pneumonia, and pleurisy. Taken by mouth or given as an enema (clyster), flax was reported to be useful in treating colic or "stones" (Culpeper, 1654).

Later herbalists continued to add to the virtues of *L. usitatissimum*, citing the benefits of linseed tea which was prepared by boiling a tablespoon of linseed for fifteen minutes in one quart of water. After straining, and adding honey, those suffering from coughs, coughs accompanied by blood, bladder complaints, including stones, as well as intestinal and other internal inflammations could have their symptoms relieved by consuming two cups per day (Aloysius, 1992). Pregnant women were encouraged to drink one cup daily commencing six weeks prior to their "confinement" (Aloysius, 1992). For throat infections, sore throats, hoarseness or painful swallowing, a tea made by boiling one tablespoon of flaxseed in two cups of water, cooled and taken one tablespoon at a time was said to bring comfort (Aloysius, 1992).

Flaxseed meal poultices placed on the lower part of the abdomen were reported to be useful to treat bladder or stomach inflammation (Aloysius, 1992). Ulcers were treated with hot poultices or compresses along with a daily dose of linseed tea (Aloysius, 1992).

A stewed leek–linseed meal poultice placed on the buttocks, abdomen or legs was said to encourage the sores associated with smallpox to manifest on those site-specific areas (Aloysius, 1992).

Fresh linseed oil was rubbed onto the sore joints of those suffering from rheumatism (Aloysius, 1992). One tablespoon of oil by mouth was a cure for piles (hemorrhoids) and if the anus had been torn, linseed oil was smeared on the injured area (Aloysius, 1992). Children with diarrhea had hot, raw linseed rubbed on their abdomens (Aloysius, 1992).

A "sugar spoon" of linseed ground to a powder and mixed with a small amount of honey taken four to six times per day was beneficial in treating patients with dropsy (edema) or chest complaints (Aloysius, 1992). Chronic "pain in the side" required an embrocation of linseed meal, honey and lard which was applied as plaster on the area three times per day (Aloysius, 1992).

The alliance between medicine and botany that had evolved with humankind encouraged patience with the healing process. It was common knowledge that nature could not be rushed. Medicinal plants were expected to do their work gently and thoroughly but not necessarily quickly.

Aggressive interventions such as purging and bloodletting, the increased study of alchemy and subsequent development of elaborate and costly medicines intensified the divide between the professionally trained doctor and the traditional healer, whether village wise woman, member of a religious order, medicine man or housewife. Many who had inherited the knowledge of the efficacy of medicinal herbs, where to find them and how to store, prepare and administer them were accused of witchcraft or sorcery, often paying a terrible price for practicing their traditional healing skills.

Industrial uses

Environmental consequences of the industrialization of the linen industry were manifested when fish deaths in the freshwater rivers of Europe and the British Isles were attributed to the retting process. Animals and humans were also poisoned through ingestion of food or water from these contaminated waterways. In some areas, legislation was enacted to prevent retting in "rivers, streams, brooks and other running waters" (Heinrich, 1992).

Artistic skills were revived during the European Renaissance and the Old Masters depended in part upon the drying properties of linseed oil combined with tree resins and balsams to create their special effects (Anonymous). The benefits of adding cold pressed linseed oil to pigments had been known and written about centuries earlier. It was reputed that the best linseed oil came from the Baltic Sea region. In order to purify the oil, it was frozen together with snow for one week and then sun dried in a covered container until the oil became thick (Jusko, 1997). The carefully formulated varnishes developed during this period have withstood the test of time and are seeing a modern day resurgence. These same oil based products were used to finish musical instruments, furniture and to paint murals; traditions that live on.

Flax in the New World

L. usitatissimum had arrived in the Americas with the first Europeans. A sowing of two acres of flax was deemed sufficient to support a family's linen requirement, but these pioneers also depended on flax as a medicine (Heinrich, 1992).

When illness struck, families relied upon personal and local knowledge of herbs in addition to information garnered from the "doctor books" that they had brought with them. Later they benefitted from information published in their newly adopted country. If the ingredients cited were unavailable, these pioneers had no choice but to improvise.

For abdominal pain, chewing and swallowing a teaspoon of flaxseed was recommended. Through mastication, the thick mucilage of the seed coat was released which helped to protect the lining of the stomach and intestine from the painful irritation of bile or acid. This same teaspoon of flaxseed would act as a bulk laxative used to treat constipation. Another method of taking flaxseed for its laxative properties was as an ingredient in a breakfast combination consisting of bran and flaxseed covered in boiling water and steeped for five minutes. Honey-sweetened and served with cream, it was both tasty and nutritious (Kerr, 1981). A single moistened flaxseed put into an eye helped remove irritants such as a speck of dirt, a glass shard or chaff. Because the irritant would adhere to the flaxseed, it would come out when the flaxseed was removed (Kerr, 1981).

Cold-pressed linseed oil was used to soothe the pain of a burn and to protect and aid in the healing process. An effective enema was derived from the oil. The itch of poison ivy was relieved with linseed oil and applications on boils, sties, burns and hemorrhoids alleviated the associated discomfort. A complicated remedy to eliminate pinworms called for a teaspoon of linseed oil mixed with molasses or sugar which was given morning, night and morning (three times), then missed three times and repeated three times until the patient had received the dosage nine times (DuBose and Micheletti, 1998). For rheumatism sufferers, an application of linseed oil mixed with equal amounts of spirits of turpentine and vinegar was said to lessen the pain. The same combination of ingredients was cited as an excellent furniture polish (Kerr, 1981)!

Linseed tea made of flaxseed steeped in water and sweetened with honey was soothing for a cough, cold, sore throat or a related respiratory disorder. Adding a slice of lemon or lemon juice was recommended if available. Mucilage from boiled flaxseed mixed with lemon and alcohol made a popular sore throat syrup.

Flaxseed poultices were applied to the chest to nurse a cold and to the throat to relieve painful swallowing or hoarseness. The emollient action of a poultice, made by grinding the seeds, boiling them in water and placing the resulting mash on clean cloths to retain the heat and protect the skin would bring boils to a head and ease sore skin. A similar poultice formulation when allowed to remain on the skin for several days was cited as saving an injured person from blood poisoning (Kerr, 1981).

Human resourcefulness has always resulted in a myriad of uses for nature's bounty. Women knew well of the mucilage properties of flaxseed that had been soaked in water. This jelly-like substance came to be an effective and inexpensive hair-setting gel (Vickers, 1980). Mixed with honey, it was also employed as a cosmetic to remove spots from the face (Grieve, 1995). Flax straw had been utilized as thatching on buildings since Biblical times and when supplies permitted, it was used as a fuel for the fire. When appropriate footwear was unavailable to those experiencing the harshness of winter weather, flaxseed proved to be an excellent insulator when a layer of seed was placed between foot and footwear.

Linseed oil was used as a principal ingredient in recipes for cleaning, water-proofing and preserving leather goods and the first explorations into commercial paints and varnishes were dependent upon the drying properties of the oil. The essential fatty acids in flaxseed oil when exposed to oxygen cause it to dry in a thin film. Wood, leather and cloth surfaces benefit from applications of linseed oil which protects against excessive wear caused by exposure to the elements. The word *linoleum* comes from the Latin words *linum* (flax) and *oleum* (oil) (Barnhart and Barnhart, 1991) and the early attempts at painting linseed oil onto canvas ultimately resulted in both table and floor coverings for homes and industrial application.

Veterinary applications

Flax has entered the veterinary pharmacy in varied forms; logic would tell us that animals would benefit from its nutritional and medicinal properties in capacities comparable to humans. The general health of pets and livestock may be improved with a dietary supplement of flax, which enhances the immune system, leaving the animal less vulnerable to infectious and contagious diseases. It is said that flaxseed fed to pregnant cows aids in the development of healthy offspring that are easily birthed. The jelly-like mucilage formed by boiling the seeds can alleviate scours when fed to calves and the general disposition of animals is likely to improve (Erasmus, 1986). Unrefined linseed oil may be used to induce vomiting in sheep and horses and its purgative properties can be safely employed to treat young animals (Grieve, 1995). Livestock pulmonary problems (coughs, pneumonia, tuberculosis and bronchitis) can be treated with a dose of medicinal linseed brew which is made of linseed oil, water and molasses. To eliminate worms, it is claimed that adding small amounts of turpentine and ginger to the above concoction is effective, the strength of the dose being determined by the size of the animal (de Bairacli-Levy, 1973). It has long been common knowledge that flax taken internally or applied externally helps to keep animal coats shiny and glossy. The addition of spruce fir shoots to linseed oil is reputed to make a valuable rubbing lotion (de Bairacli-Levy, 1973). Poultices made with flaxseed, water and olive oil are effective in healing pulmonary complications, carbuncles, abscesses, boils, sprains and injured ligaments when applied to the ailing or injured area (de Bairacli-Levy, 1973).

Human health

Regarding human health, traditional remedies that have been handed down over the centuries are experiencing a modern day revival, becoming a prime component of the new "integrative" medicine. The divergent opinions held within the health profession since the time of the ancient Greeks have been coerced into a fragile alliance precipitated by an educated public willing to assume more personal responsibility for their health and well-being. Personal and professional investigations are attesting to the benefits of a combination of conventional medicine, herbal remedies, exercise and nutrition all interfaced with attributes of the healing traditions drawn from a global network of human evolution.

The modern day herbalist continues to hold flax in high regard within the pharmacopoeia of medicinal plants and herbs, broadly classifying its healing properties as demulcent, emollient, aperient, purgative and, occasionally, a vermifuge (Mairesse, 1981; Hoffman, 1988).

A demulcent rich in mucilage is effective in soothing and protecting irritated and inflamed internal tissue (Hoffman, 1988). In that capacity, flax can be used in all chest related infections. An infusion made by pouring a cup of boiling water over two or three teaspoons of flaxseed and steeped for ten to fifteen minutes should be drunk morning and evening (Hoffman, 1988). The addition of honey and lemon may make the remedy more efficacious. In difficult or chronic cases (emphysema, pneumonia, cough accompanied with blood, bronchitis, pleurisy) the amount consumed may be increased. This same infusion can be used to soothe and heal the stomach, inflamed intestines (gastritis, enteritis, colitis and diverticulitis), kidneys, and urinary tract (Heatherley, 1998). It is cautioned that ingesting whole flaxseed can cause gas and it is contraindicated for use in patients with internal obstructions (DuBose and Micheletti, 1998). The pain and discomfort of swollen joints due to arthritis, gout or rheumatism has been relieved through ingestion or external application of cold pressed linseed oil.

An emollient is essentially a demulcent used topically to soften, soothe and protect the skin (Hoffman, 1988). Applying linseed tea directly to dry skin or flaxseed oil as an addition to bathwater allows the skin to become more supple and eases tension. Mixed with scented herbs, a refreshing bath oil is created. A topical treatment of linseed oil is beneficial to treat burns or scalds (Grieve, 1995) and may lessen the discomfort of psoriasis (Duke, 1997).

Aperients are mild and gentle laxatives made effective by the action of the bulk seeds which not only improve the digestive process but also stimulate bowel movement (Hoffman, 1988). Each herbalist seems to have a personal purgative recipe with flaxseed being one of the most common laxative ingredients. When including flaxseed in the diet it is necessary to increase water consumption in order to keep the bulk moving through the system. The prussic acid content of flaxseed is an important consideration when consuming flax and amounts in the range of one hundred grams (three and one-half ounces) have been known to cause poisoning (Mairesse, 1981). Symptoms of an overdose (immature seeds grown in warm climates seem to be most toxic) include increased respiratory rate, gasping, staggering, excitement, weakness, paralysis and convulsion (Spoerke, 1980).

Although not cited as often, flax oil has been recommended as a vermifuge, a medicine used to expel worms from the intestine (Mairesse, 1981).

The age-old act of applying poultices to relieve the symptoms of numerous disorders is also experiencing a revival. Flaxseed has maintained its reputation as a most efficacious ingredient because of its anodyne and anti-inflammatory properties (Mairesse, 1981). Alone or in combination with other herbs, flaxseed poultices are placed on boils and carbuncles, infections and abscesses to encourage suppuration while reducing pain. Chronic coughs, pleurisy, pneumonia and emphysema are aided by linseed poultices and the discomfort of those suffering from psoriasis, shingles and other skin disorders is said to be lessened.

Cold pressed linseed oil taken internally is used to treat constipation. However, the benefits of using a bulk laxative are not achieved. Gallstones are reputedly dissolved when one and one-half to two tablespoons of the oil are taken daily (Mairesse, 1981). In more recent times, linseed oil has been recommended to help leach heavy metals such as lead or aluminum from the body (Mabey, 1988; Heatherley, 1998).

One of the contributing factors adding to the renewed popularity of herbal remedies comes from advanced chemistry and pharmacognosy technologies. The research community has been able to provide society with increased accuracy in the scientific analysis of the active chemical constituents that evoke medicinal action. This growing body of evidence appears to be substantiating the claim that the secret to wellness lies in nutraceuticals and phytochemicals that are found in wholesome foods. It is necessary to be fair to both tradition and modern science when evaluating the data on natural products and, as in most cases, the standardization of quality and purity remains an unresolved issue. Measuring the nuances between the synergistic effects of a whole plant relative to the unique attributes of a single constituent continues to be a challenge.

Linum usitatissimum has been a plant of considerable importance for centuries and today, traditional healers and modern scientists alike share a curiosity as to what discoveries future investigation will hold. Research shows promise in affirming many of the traditional uses, as well as targeting several modern day afflictions. Cancer, heart and immune system diseases, high blood pressure, depression, migraine headaches, HIV and various nutritional disorders continue to be investigated to determine flaxseed's contribution in potential treatment. The exploration of the constituent parts will continue to determine its overall role in human health. Cognisance of the ethnobotanical uses of flaxseed may well be the best approach to directing the future research which will not only validate some of the historical uses, but inevitably lead to a future of expanded utilization.

Folklore

Whether it be fable or fact, the people of linseed-growing regions have flax lore woven into their history. Holda (Hulda), a Germanic goddess often described as an old woman with long, tangled hair, was the protector of the household and patron of the housewife and mother. She was also the patron of the spinner and was said to tour villages inspecting the handiwork of women. The best spinners were rewarded with gifts, usually a distaff of the finest flax or golden thread. Occasionally a woman would awaken to find that her work had been completed, a gift from Holda. Those deemed lazy were punished and found their work tangled or destroyed. With the arrival of Christianity, Holda's influence held, as working over the Christmas period, on Sunday, or a Saint's Day was likely to arouse her wrath. Lightning was an indicator that Holda was working her flax. She was most associated with winter, because spinning was conducted indoors at that time (Thorn, 1996).

In one particular folktale, Holda appeared as a beautiful queen with handmaidens in attendance. She lured a peasant farmer into a magical cavern high on a mountain top. Holda told the farmer he might choose anything he desired from the room filled with gold and precious stones. Overcome by the surrounding beauty, he asked only for the flowers that Holda held in her hand. As the legend goes, the blooms were of flax, a plant not yet known to mankind. True to her promise, the beautiful queen gave the man a bag of flaxseed which he planted when he returned home. When the plants ripened, Holda returned to the farmer and his wife, teaching her how to make linen cloth from the plants. The family's prosperity was ensured from that time on (Thorn, 1996).

In days gone by society deemed it respectable for unmarried girls and women to fill their time with spinning and by the seventeenth century, any woman who remained unwed after the usual age for marriage was referred to as a *spinster*. The word acquired a hint of the immoral in England when wayward ladies were put into correction houses and became *spinsters* as a means to earn their keep (Funk, 1950). A *distaff* which held the bunch of flax that women were to spin came to be a symbol for women and their realm of responsibility; the phrase *the distaff side* meant the *female side* of the family (Funk, 1950).

The *large linen squares* that diners required in the days when table cutlery was rudimentary, rare or nonexistent, were called *little nape* which is an Old French term meaning, *little tablecloth*. *Napery* was *table linen*, the root of which is *nape* which once meant *table cloth*. An *aperon* was originally called *a naperon*, but because of a very old transcribing error we moved from *a naperon* to *an aperon (apron)*. These words all go back to a Latin source, *mappa* which meant *napkin* or *cloth* (Funk, 1950).

In seventeenth century France, Croatian mercenaries arrived wearing linen scarves tied around their necks. It wasn't long before the French men and women created their own version of the *cravate* – linen scarves trimmed with lace and tied with long flowing ends. Today, *cravat* is a term for *necktie*, which makes perfect sense as *cravate* means *Croatian* in the French language (Funk, 1950).

It was once believed that flax flowers afforded protection from sorcery and in Bohemia it was said that seven year old children who danced among the flax flowers would grow up to be beautiful (Grieve, 1995).

In Greece, the *Kallikantzaroi*, inhabitants of the underworld, were claimed to emerge from the bowels of the earth for twelve days each year. One of the ways to protect the household from the *Kallikantzaroi* was to hang a tuft of tangled flax over the door which would cause them to stop so that they could untangle the mass. Upon completion of that task, they would commence counting the threads which ensured their capture before sunrise (MacDonald, 1992).

In Flanders, a winter of heavy snowfall gave the promise of a bountiful flax harvest the following year. During planting, crushed hard-boiled eggs were mixed with the seed, the yellow of the egg imparting the wish for golden seed bolls and the egg white symbolizing the desired color of the linen fabric (Heinrich, 1992). On Ash Wednesday, ashes garnered from the burned remains of branches and discarded liturgical linens were mixed and planted with the seed to bestow favorable blessing upon the crop, and prayers were offered preceding the planting (Heinrich, 1992).

The Swedes claimed that placing flaxseed in one's shoes made the wearer more fertile and the act of throwing flaxseed behind a funeral ensured that the dead would not return to haunt the living (Heinrich, 1992).

Lithuanians celebrated the tying of the last flax sheaf of the harvest with a feast. The farmer's wife sat in a chair decorated with flowers and the harvest workers placed a wreath made of flax upon her head and lifted her chair into the air. The best flax puller also received a wreath which she was allowed to wear only after the lead harvester had drunk a designated cup of beer. Flax breaking often commenced at dusk and continued through the night. Of the many games that accompanied the task, *Kursis* carrying survived until the twentieth century. The *Kursis* was a straw effigy said to represent the evil spirits that had dried the grain. Towards the end of the session, some of the young men from one barn would take the *Kursis* and throw it into a neighboring barn challenging the workers to "Take the *Kursis*." They would then run off to hide with the flax-breakers who had received the *Kursis* in pursuit. If unable to locate the challengers, the pursuers were humiliated by the entire village. The groups later joined together to tear the *Kursis* figure apart or burn it at the stake (Ambrazevicius, 1996).

Numerous fairy tales make reference to flax with variations of Sleeping Beauty, Little Briar Rose, Sun, Moon and Talia and The Ninth Captains Table all having a heroine who "would fall to the floor as if dead," should she come into close contact with flax (Ashliman, 1999).

Since ancient times, people have used flowers as symbols and emblems. The "language of flowers" is thought to have originated in Turkey with harem women wishing to communicate with their lovers, who were far removed from them in the outside world. Eventually, the concept reached Paris where the first "Language of Flowers Dictionary" appeared. Within the art of floriography, flax came to be defined as *domestic industry; fate; utility* and was used to connote the phrase, "*I feel your kindness.*" This custom spread to England and America during Victorian times and was popular with those in genteel society (Marsh, 1978).

Flax and language

Several words in common usage today can trace their roots back to ancient times attesting to the importance of flax fiber and oilseed production. The word *flax* is from a Latin word meaning *flail* referring to the beating of the fibers so that they can be separated prior to spinning (Martin, 1987).

Christ/Christian is from the Greek word *christos* meaning *anointed*; and from this source, comes the Greek and Latin word *chrisma, the sacred oil* (Funk, 1950).

When a Roman politician campaigned, he made certain that his (linen) toga was impeccably white guaranteeing he made the best possible impression. *Candidate*, from the Latin word *candidatus*, signified a *person clad in white*, but later took the meaning of *one who is seeking office* (Funk, 1950).

When a speaker is being *heckled*, someone is verbally *harassing* or *annoying* him; the Middle English word *hekel* from which *heckle* is derived is *an instrument used to card flax* or hemp; literally to clean out the tangles by pulling through wire teeth. To *tease* someone would be from a similar context as flax and hemp fabrics were teased *to raise the nap of the fiber* (Funk, 1950).

Lint, which is the *soft downy material obtained by scraping linen fabric* and used to dress wounds, is from the Latin word *linum* (flax) (Barnhart and Barnhart, 1991).

Liniment is from the Late Latin word *linimentum*, related to the Latin, *linere, to smear or anoint* and *line* is from the Latin *linea* which traces its roots to *linen thread* from *linum* (flax). *Line, lineage, linear*, and *lingerie* all share this common Latin root (Barnhart and Barnhart, 1991).

References

Adkins, L., and Adkins, R.D. (1997). *Handbook to Life in Ancient Greece*. Facts on File, New York. 472 p.

Aldred, C. (1984). *The Egyptians*. Thames and Hudson, London. 216 p.

Aloysius (1992). *Comfort to the Sick*. Samuel Weiser, York Beach, ME. 434 p.

Ambrazevicius, R. (1996). Lithuanian Roots: An Overview of Lithuanian Traditional Culture – Flax Pulling and Flax Breaking. Lithuanian Folk Culture Center, http://www.lfcc.lt/publ/roots/node1.html

Anonymous. Alchemist Paints and Varnishes. What was the Secret of the Old Masters?, http://www.amberalchemy.com/secret.html

Anonymous (1971). *Encyclopedia Judaica*. Keter Publishing House, Jerusalem. Vol. 6, pp. 1338–1339.

Anonymous (1992). *The World Book Encyclopedia*. Scott Fetzler, Chicago. Vol. 7, p. 231.

Ashliman, D.L. (1999). Sleeping Beauty (Folklore and Mythology Electronic Texts). University of Pittsburgh, http://www.pitt.edu/~dash/folktexts.html

Barnhart, C.L., and Barnhart, R.K. (eds) (1991). *The World Book Dictionary*, Scott Fetzer, Chicago. 2430 p.

Culpeper, N. (1654). Culpeper's Complete Herbal: Consisting of a comprehensive description of nearly all herbs with their medicinal properties and directions for compounding the medicines extracted from them. In., W. Foulsham, London, pp. 146–147.

de Bairacli-Levy, J. (1973). *Herbal Handbook for Farm and Stable*. Faber and Faber, London. 320 p.

DuBose, F., and Micheletti, E. (eds) (1998). *North American Folk Healing: An A-to-Z Guide to Traditional Remedies*, Reader's Digest, Pleasantville, NY. 408 p.

Duke, J.A. (1997). *The Green Pharmacy: New Discoveries in Herbal Remedies for Common Diseases and Conditions from the World's Foremost Authority on Healing Herbs*. Rodale Press, Emmaus, PA. 508 p.

Erasmus, U. (1986). *Fats that Heal, Fats that Kill: The Complete Guide to Fats, Oils, Cholesterol and Human Health*. Alive Books, Burnaby. 456 p.

Funk, W.J. (1950). *Word Origins and their Romantic Stories*. Bell Publishing, New York. 432 p.

Gozzini Giacosa, L. (1992). *A Taste of Ancient Rome*. The University of Chicago Press, Chicago. 231 p.

Grieve, M.A. (1995). A Modern Herbal: The Medicinal, Culinary, Cosmetic and Economic Properties, Cultivation and Folklore of Herbs, Grasses, Fungi, Shrubs and Trees with their Modern Scientific Uses, http://www.botanical.com/botanical/mgmh/f/flax—23.html

Griggs, B. (1981). *Green Pharmacy – A History of Herbal Medicine*. The Viking Press, New York. 379 p.

Hawkes, J., and Woolley, L. (1963). *Prehistory and the Beginnings of Civilization: History of Mankind, Cultural and Scientific Development*. Allen and Unwin, London. 873 p.

Heatherley, A.N. (1998). *Healing Plants: A Medicinal Guide to Native North American Plants and Herbs*. Harper Collins, Toronto, ON. 252 p.

Heinrich, L. (1992). *The Magic of Linen: Flaxseed to Woven Cloth*. Orca Book, Victoria, B.C., 231 p.

Hoffman, D. (1988). *The Herbal Handbook: A User's Guide to Medical Herbalism*. Healing Arts Press, Rochester, VT. 240 p.

Jusko, D.A. (1997). Academic Realism with Impressionistic Shortcuts, http://mauigateway.com/~donjusko

Kerr, S. (1981). *Early Prairie Remedies*. Barker Gifts, Calgary, AB. 48 p.

Lamsa, G.M. (1964). *Old Testament Light: The Indispensable Guide to the Customs, Manners and Idioms of Biblical Times*. Harper and Row, San Francisco. 976 p.

Mabey, R. (ed.) (1988). *The New Age Herbalist: How to Use Herbs for Healing, Nutrition, Body Care and Relaxation*, Macmillan, New York, NY. 288 p.

MacDonald, M.R. (ed.) (1992). *The Folklore of World Holidays*, Gale Research, Detroit. 739 p.

Mairesse, M. (1981). *Health Secrets of Medicinal Herbs*. Arco Publishing, New York, NY. 178 p.

Marsh, J. (1978). *The Illuminated Language of Flowers*. Macdonald and Jane's, London. 78 p.

Martin, L.C. (1987). *Garden Flower Folklore*. Globe Pequot Press, Chester, CT. 273 p.

Spoerke, D.G. (1980). *Herbal Medications*. Woodbridge Press, Santa Barbara, CA. 192 p.

Stuart, M. (ed.) (1979). *Color Dictionary of Herbs and Herbalism*, Van Nostrand Reinhold, New York. 89 p.

Thorn, T. (1996). Thorshof: Holda, Nidram Co, http://www.thorshof.org/holdapic.htm

Vickers, J. (1980). *Let's Talk Basic Herbs and Simple Remedies*. NuYu Enterprises, Calgary, AB.

Wilkinson, J.G. (1988). *A Popular Account of the Ancient Egyptians*. Bonanza Book, New York. 438 p.

Williams, R.C. (1999). The Invention of Paper: Early Papermaking in China, Institute of Paper Science and Technology, Atlanta, GA, http://www.ipst.edu/amp/museum_invention_paper.htm

Zohary, D., and Hopf, M. (1988). *Domestication of Plants in the Old World: The Origin and Spread of Cultivated Plants in West Asia, Europe and the Nile Valley*. Clarendon Press, Oxford. 249 p.

Supplemental literature

Anonymous (1991). *Encyclopedia Americana. International Ed. Deluxe Library ed.* Grolier Inc. Danbury. Vol. 17.

Carper, J. (1988). *The Food Pharmacy: Dramatic New Evidence that Food is your Best Medicine.* Bantam Books, Toronto. 367 p.

Chang, K.C. (ed.) (1977). *Food in Chinese Culture: Anthropological and Historical Perspectives*, Yale University Press, New Haven. 429 p.

Chevallier, A. (1996). *The Encyclopedia of Medicinal Plants.* Reader's Digest Assn (Canada), Montreal. 336 p.

Fowler, C., and Mooney, P.R. (1990). *Shattering: Food, Politics and the Loss of Genetic Diversity.* The University of Arizona Press, Tucson, AZ. 278 p.

Hastings, J., Selbie, J.A., Mathews, S., and Lambert, J.C. (eds) (1909). *Dictionary of the Bible*, Charles Schribner's Sons, New York. 992 p.

Heiser, C.B. (1973). *Seeds to Civilization: The Story of Man's Food.* W.H. Freeman, San Francisco.

Hoffman, M.A. (1979). *Egypt before the Pharaohs: The Prehistoric Foundations of Egyptian Civilization.* Alfred A. Knopf, New York, NY. 391 p.

Lee, N.E. (1960). *Harvests and Harvesting Through the Ages.* Cambridge University Press, Cambridge. 208 p.

Leung, A.Y., and Forster, S. (1966). *Encyclopedia of Common Natural Ingredients used in Food, Drugs and Cosmetics.* John Wiley and Sons, New York. 649 p.

McIntyre, A. (1994). *Folk Remedies for Common Ailments.* Key Porter Books, Toronto, ON. 96 p.

Mindell, E. (1994). *Earl Mindell's Food as Medicine.* Simon and Schuster, New York. 393 p.

Null, G. (1992). *No More Allergies: Identifying and Eliminating Allergies and Sensitivity Reactions to Everything in your Environment.* Villard Books, New York. 421 p.

Phillips, P.A. (1980). *The Prehistory of Europe.* Indiana University Press, Bloomington, IN. 314 p.

Pyke, M. (1968). *Food and Society.* Murray, London. 178 p.

Reno, L., and Devrais, J. (1995). *Allergy Free Eating: Key to the Future.* Celestial Arts Publishing, Berkeley, CA. 507 p.

Riotte, L. (1995). *Carrots Love Tomatoes: Secrets for Companion Planting for Successful Gardening.* Garden Way Publishing, Pownal, VT. 226 p.

Ritiche, C.I.A. (1981). *Food in Civilization: How History has been Affected by Human Tastes.* Beaufort Books, New York. 192 p.

Romain, E., and Hawkey, S. (1996). *Herbal Remedies in Pots.* Little, Brown (Can.), Toronto, ON. 96 p.

Sharon, M. (1989). *Complete Nutrition: How to Live in Total Health.* PRION, London. 222 p.

Tannahill, R. (1973). *Food in History.* Eyre Methuen, London. 448 p.

Turner, L. (1996). *Meals that Heal: A Nutraceutical Approach to Diet and Health.* Healing Arts Press, Rochester, VT., 235 p.

14 Use of flaxseed in animal diets to create consumer products with modified fatty acid profiles

Sheila E. Scheideler

Introduction

Flaxseed has a long history of use in animal diets, primarily in its by-product form as linseed meal in ruminant rations (Ensminger *et al.*, 1990; McDonald *et al.*, 1996). Linseed meal is a by-product of flaxseed extraction by either a mechanical (cold press) or solvent extraction process. Mechanical crushing of flaxseed produces a high oil containing product, which is preferred by horse caretakers as a result of its ability to enhance the glossiness of the hair coat. Linseed meal is utilized as both a fiber (mucilage) and protein source in other ruminant rations. Linseed meal is considered a good source of thiamin, riboflavin, nicotinamide, pantothenic acid and choline in ruminant diets. However, linseed meal is not as valued in monogastric rations due to its high mucilage content and deficiency in the essential amino acids lysine and tryptophan. The use of linseed meal in poultry diets has actually been avoided as a result of the negative effects of cyanogenic glycosides on growth rate and anti-vitamin B_6 factors (Kratzer and Williams, 1948a; Kratzer and Williams, 1948b). Very little research has been reported in poultry since the 1940s when Kratzer reported feeding chicks linseed meal processed by the "old mechanical" extraction procedures.

Recent interest in ω-3 fatty acid deposition in poultry meat and eggs has renewed interest in utilizing flaxseed as a ω-3 fatty acid source in poultry, swine and ruminant diets. While linseed meal continues to be utilized in ruminant rations as a protein and mucilage source, the value of the oil in flaxseed is being pursued as a feed ingredient utilizing either whole flaxseed or whole ground flaxseed, depending on the species of animal. For example, birds have a gizzard muscle capable of grinding seeds, thus enabling the chicken to consume whole flaxseed with high digestibility. Simple stomached and ruminant animals like pigs and dairy cows must be fed the ground flaxseed to maximize digestibility. Poultry, beef, pork, milk and eggs can be modified by the dietary lipid profile fed to the animal, thereby allowing the producer to produce high quality food products with added ω-3 fatty acids for the discerning consumer. Efficiency of ω-3 fatty acid deposition from flaxseed does vary by species of animal and some attributes of altered fatty acid profiles in pork and beef may develop undesired qualities such as soft fat. Nevertheless, modified animal proteins containing enhanced ω-3 fatty acid offer consumers another choice of food products to meet their essential fatty acids requirement, along with a high quality protein product (Wiseman, 1990).

Poultry products

Poultry meat

The chicken is a unique animal in how it digests and deposits fat into meat, adipose tissues and the egg. The chicken is able to elongate and desaturate the essential fatty acid C18:3 – linolenic acid to eicosopentanoic acid (EPA) and docohexanoic acid (DHA) in the liver and then deposit all three ω-3 fatty acids in substantial quantities into adipose tissue, eggs and to a lesser degree in phospholipids in muscle tissue (Yau *et al.*, 1991; Hargis and Van Elswyk, 1993). Deposition of ω-3 fatty acids into meat tissue is more difficult to achieve than into adipose fractions of subcutaneous fat or eggs. Fatty acid deposition into phospholipids in the muscle tissue membranes is limited as a result of the functional role of these membranes (Smith, 1991). Olomu and Baracos (1991) conducted trials feeding 1.4 to 6.0 percent flaxseed oil to broiler chicks and found significant accumulation of ω-3 fatty acids (C18:3, C20:3, C20:5, C22:5 and C22:6) in thigh muscle lipids after 21 days. The quantity of ω-3 fatty acid deposition increased linearly as dietary flaxseed oil increased. As ω-3 fatty acid deposition increased, ω-6 fatty acid deposition (C20:2, C20:3 and C20:4) decreased in the muscle lipids, indicating an enzymatic pathway preference for the elongation and desaturation of ω-3 fatty acids when the precursor C18:3 is present in substantial amounts in the diet. This observation is also supported by the findings of Phettleplace and Watkins (1989). Adding flaxseed oil to a commercial broiler diet raises concerns about oxidation potential, especially in the southern warm climate of the broiler industry in the United States. Feeding of the whole flaxseed removes the concerns of stability and oxidation, offering more flexibility for the commercial poultry producer. Born reported substantially increased ω-3 content and reduced monounsaturates and saturates per 50 gram serving size of broiler meat in chickens fed a diet rich in flaxseed (Table 14.1) (Born, 1998). The "Born 3" chicken product is presently available for consumers in Canada and has been authorized by Health and Welfare Canada to label the product as having 50 percent less fat and 29 percent fewer calories than regular chicken. Little to no sensory work has been reported for ω-3 enriched poultry meat at this time. Earlier work (Klose *et al.*, 1951; Klose *et al.*, 1953) indicated some fish flavor in turkey meat from turkeys fed 2–5 percent linseed oil. Some of the negative oxidation effect on taste may be alleviated by feeding of the whole flaxseed rather than the oil.

Table 14.1 Nutrition content of a Born 3 chicken per 50-gram serving

	Born 3 chicken	*Regular chicken*
Energy, cal/kJ	75/314	106/444
Protein, g	9.9	8.7
Fat, g	3.9	7.9
ω-6 polyunsaturates, g	0.7	0.7
ω-3 polyunsaturates, g	0.7	0.2
Monounsaturates, g	1.7	4.4
Saturates, g	0.9	2.6
Cholesterol, mg	33.0	39.0
Carbohydrates, g	0.1	0.1

Source: Born (1998).

Eggs

Enriching the egg with ω-3 fatty acids is probably the most efficient of all the land based animal protein sources. The egg contains 5–6 g of fat which is a combination of triglycerides and phosopholipids. The fatty acid profile of egg lipids is directly affected by dietary lipid composition. A linear relationship between level of fatty acid supplementation in the diet and level of fatty acid deposition in the egg has long been established for the long chain ω-3 and ω-6 polyunsaturated fatty acids (Cruickshank, 1934). This linear relationship does not exist for saturated fats.

Flaxseed can be utilized quite efficiently by the laying hen with little or no detrimental effects on rate of egg production and egg size (Scheideler and Froning, 1996). Flaxseed can be fed either whole or ground, although whole is recommended to avoid oxidative products during feed storage. A typical poultry diet may contain between 5 and 20 percent flaxseed which linearly affects the amount of ω-3 fatty acid deposition into the egg lipids. A recent patent describes a feed ration containing flaxseed that results in eggs with enriched levels of ω-3 lipids (Scheideler, 1999). The patent is a specific program to produce ω-3 enriched eggs. Table 14.2 shows the composition of egg yolk lipids from hens fed 10 percent flaxseed in their diets (Scheideler and Froning, 1996). Others have also reported the success of altering egg yolk lipid composition by feeding flaxseed to laying hens (Caston and Leeson, 1990; Jiang *et al.*, 1992; Hargis and Van Elswyk, 1993; Aymond and Van Elswyk, 1995). One concern about ω-3 fatty acid enriched eggs has been their sensory quality characteristics. There has been some indication of increased oxidation in eggs with high levels of ω-3 fatty acids during storage and cooking (Aymond and Van Elswyk, 1995). These reports are mixed with some reports indicating little difference in taste (Scheideler *et al.*, 1997) to others detecting a "fishy taste" especially when ω-3 fatty acid deposition is accomplished through supplementation of dietary fish oil (Van Elswyk *et al.*, 1992). Public acceptance of eggs as a source of ω-3 fatty acids is increasing as awareness of ω-3 fatty acid requirements in the human diet increases. Marshall *et al.* (1994) conducted a survey of consumer acceptability for ω-3 enriched eggs and found 65 percent surveyed were willing to

Table 14.2 Enriched egg composition from hens fed a 10 percent flaxseed diet

Egg weight (g)	Average (%)	Average (g/egg)
Yolk	27.5	16.9
Egg yolk lipids	31.6	5.34
Palmitic acid (C16:0)	27.30	1.45
Palmitoleic acid (C16:1)	3.59	0.19
Stearic acid (C18:0)	8.64	0.22
Oleic acid (C18:1)	37.2	1.99
Linoleic acid (C18:2)	14.2	0.760
Linolenic acid (C18:3)	5.61	0.299
Arachidonic acid (C20:4)	0.90	0.048
DPA (C22:5)	0.21	0.011
DHA (C22:6)	1.85	0.099
Total ω-6		808 mg
Total ω-3		409 mg
ω-6:ω-3 Ratio		2:1

Source: Scheideler and Froning (1996).

purchase a ω-3 fatty acid enriched egg and of these, 71 percent were willing to pay an additional $US0.50 per dozen. There is a growing number of designer eggs enriched with ω-3 fatty acids available in the commercial marketplace in the USA and Canada. These would include Omega Eggs, by PDI/Hy-Vee, Ankeny, Iowa; Eggs Plus, by Pilgrims Pride, Dallas, Texas; Gold Circle Farms, Omegatech, Boulder, Colorado; Golden Premium Eggs – Rose Acre Farms, Seymour, Indiana and Born-3 Eggs, Born-3 Marketing Corporation, Abottsford, British Columbia, Canada and Sumas, Washington, USA.

Pork products

Limited research has been reported on feeding flaxseed to pigs to alter pork fat composition. An acknowledged trend in the pork industry worldwide has actually been to decrease the amount of fat in pork products such as ribs, chops and hams. This has been successfully done through intensive genetic selection for lean muscle gain in modern breeds of pigs. Enhancing the quality of fat left in a product being bred to reduce overall fat content is not an easily accomplished goal. Romans *et al.* (1995a, 1995b) found that feeding ground flaxseed to pigs at rates of 5 to 15 percent of the diet did not negatively affect production or carcass traits typically measured. No processing problems as a result of decreased fat firmness were noted. Amounts of C18:3 – linolenic acid, and C20:3 – EPA were increased in backfat and other fat deposits in the pig. Sensory tests for bacon products from flax fed pigs did reveal some negative response to flavor of the bacon despite no measurable changes in thiobarbituric acid reactive substances (TBARS) in bacon from test versus control pigs. Howe (1998) preferred using fish meal products to enhance the ω-3 fatty acid content of pork products. Howe (1998) reported no off-flavor in fresh pork products from pigs fed fish meal, but some unfavorable flavor was detected in frozen pork products.

It would seem, given the limited data available, that ω-3 fatty acid enrichment of pork products has limitations at this time for consumer acceptability.

Milk

A unique challenge exists when feeding flaxseed to ruminants to alter the fatty acid composition of milk or tissue fats. As flaxseed is metabolized in the rumen, rumen bacteria hydrogenate long chain fatty acids so that what is fed may not be what is deposited. A mechanism to protect long chain polyunsaturated fatty acids from ruminal degradation is needed to make this process of ω-3 fatty acid deposition in ruminants more efficient. Khorasani and Kennelly (1994) fed dairy cows either whole flaxseed, rolled flaxseed or a combination of rolled flaxseed/canola seed and reported lower milk production for cows fed the whole flaxseed diet, but no differences in milk fat, protein, lactose percentage or yield. Supplementation of rolled flaxseed increased monounsaturated and polyunsaturated fatty acids in milk (Table 14.3) while saturated fatty acid content decreased. The percentage increase of C18:1, C18:2 and C18:3 was much greater for cows fed the rolled versus whole flaxseed, indicating a need for processing flaxseed before feeding to ruminants to optimize deposition in milk. Wright *et al.* (1998) reported a linear increase in milk DHA content as intake of the dietary rumen non-degradable protein supplement containing fish meal (source of DHA) increased. Levels of DHA equivalent to human breast milk (North American) were achieved in this trial with no detrimental effects on milk yields. The potential for enriching dairy

Table 14.3 Fatty acid composition of milk fat for cows fed control, whole flaxseed (WFS), rolled flaxseed (RFS), and a 50:50 mixture of RFS and rolled canola seed (RCS)

Fatty acids	Control	Oilseed			SEM
		WFS	RFS	RFS-RCS	
Short-chain*	12.85[a]	11.76[ab]	10.92[b]	11.64[ab]	0.66
Medium-chain[†]	40.13[a]	38.86[b]	35.49[c]	36.97[bc]	0.97
Long-chain[‡]	36.34[b]	44.47[a]	47.14[a]	46.16[a]	1.38

* C4:0, C6:0, C8:0, C10:0, C12:0.
[†] C14:0, C14:1, C15:0, C16:0, C16:1, C17:0.
[‡] C18:0, C18:1, C18:2, C18:3, C20:0.

Source: Khorasani and Kennelly (1994).

products, especially those that still contain a quantity of lipids, is good if the dietary source of ω-3 fatty acids can be protected from ruminal degradation. The potential of the cow to elongate and desaturate C18:3 to C22:6, however, is quite limited compared to the chicken egg model.

Red meat

Research on feeding of flaxseed to beef cattle to alter the ω-3 fatty acid composition of red meat is extremely limited (Mandell *et al.*, 1998). Ashes *et al.* (1992, 1998) stated that dietary C18:3 fatty acids will be hydrogenated by microorganisms (bacteria) in the rumen producing trans-monounsaturated fatty acids and stearic acid. The authors suggested that EPA and DHA be fed to beef cattle as these very long chain polyunsaturated fatty acids are not as readily degraded in the rumen. The feasibility of feeding flaxseed to beef cattle has not been tested or reported to our knowledge at the time of this publication. The data from Khorasani and Kennelly (1994) in dairy cows would indicate that some of the C18:3 reaches lipid tissues for deposition. However, this has not been proven in red meat tissue except in forage range animals. Foraging animals can obtain C18:3 by consuming fresh forage and there is some reference (Larick and Turner, 1990) that long chain fatty acid deposition in forage fed beef compared to grain fed beef will vary and that this can negatively affect the flavor of the meat. Forage finishing of beef in North America is basically non-existent, but is common in Australia and the United Kingdom. Finishing beef with flaxseed in the ration has not been tested or reported to date and warrants some consideration.

Summary

Deposition of ω-3 fatty acids into livestock protein products such a poultry meat, eggs, pork, milk and red meats by feeding flaxseed varies in efficiency considerably by species of animal tested. The chicken appears to be the most efficient in processing ω-3 fatty acids from the diet into the egg and into poultry meat. The chicken would be followed by the pig (also a monogastric animal) in efficiency of ω-3 fatty acid deposition. Dairy cows and beef animals are the least efficient in depositing ω-3 fatty acids from flaxseed due to the disruptiveness of the rumen microflora. Animal proteins can provide another source of ω-3 fatty acids to the health conscious consumer.

Note

Born 3 Marketing Corp., 141 Ross Road, Abbotsford, B.C., Canada V4X 2M6 and P.O. Box 8000 – 343, Sumas, Washington U.S.A. 98295.

References

Ashes, J.R., Gulati, S.K., and Scott, T.W. (1998). Utilization of ω3 fatty acids in ruminants. In Simopoulos, A.P. (ed.) *The Return of ω3 Fatty Acids into the Food Supply. I. Land-Based Animal Food Products and Their Health Effects, World Rev. Nutr. Diet.*, Vol. 83, Karger, Basel, pp. 223–234.

Ashes, J.R., Siebert, D.B., Gualati, S.K., Cuthbertson, A.Z., and Scott, T.W. (1992). Incorporation of 3 fatty acids of fish oil into tissue and serum lipids of ruminants. *Lipids* 27, 629–631.

Aymond, W.M., and Van Elswyk, M.E. (1995). Yolk thiobarbituric acid reactive substances and n-3 fatty acids in response to whole and ground flaxseed. *Poultry Sci.* 74, 1358–1394.

Born, F. (1998). ω3 products: From research to retail. In Simopoulos, A.P. (ed.) *The Return of ω3 Fatty Acids into the Food Supply. I. Land-Based Animal Food Products and Their Health Effects, World Rev. Nutr. Diet.*, Vol. 83, Karger, Basel, pp. 166–175.

Caston, L., and Leeson, S. (1990). Research note: Dietary flaxseed and egg composition. *Poultry Sci.* 69, 1617–1620.

Cruickshank, E.M. (1934). Studies in fat metabolism in the fowl. I. The composition of the egg fat and depot fat of the fowl as affected by the ingestion of large amounts of different fats. *Biochem. J.* 28, 965–977.

Ensminger, M.E., Oldfield, J.E., and Heinemann, W.W. (1990). *Feeds and Nutrition Digest.* Ensminger Pub. Co., Clovis, CA. 794 p.

Hargis, P.S., and Van Elswyk, M.E. (1993). Manipulating the fatty acid composition of poultry meat and eggs for the health conscious consumer. *World's Poultry Sci. J.* 49, 251–264.

Howe, P.R.C. (1998). ω3-enriched pork. In Simopulos, A.P. (ed.) *The Return of ω3 Fatty Acids into the Food Supply. I. Land-Based Animal Food Products and Their Health Effects, World Rev. Nutr. Diet.*, Vol. 83, Karger, Basel, pp. 132–143.

Jiang, A., Ahn, D.U., Ladner, L., and Sim, J.S. (1992). Influence of feeding full-fat flaxseed and sunflower seeds on internal and sensory qualities of eggs. *Poultry Sci.* 71, 378–382.

Khorasani, G.R., and Kennelly, J.J. (1994). Influence of flaxseed on the nutritional quality of milk. *Proc. of the Flax Institute of the United States. Fargo, ND,* Flax Institute of the United States. pp. 127–134.

Klose, A.A., Hanson, H.L., Mecchi, E.P., and Anderson, J.H. (1953). Quality and stability if turkey as a function of dietary fat. *Poultry Sci.* 32, 82–88.

Klose, A.A., Mecchi, E.P., Hanson, H.L., and Lineweaver, H. (1951). The role of dietary fat in the quality of fresh and frozen storage turkeys. *J. Am. Oil Chem. Soc.* 28, 162–164.

Kratzer, F.H., and Williams, D.E. (1948a). The improvement of linseed oil meal for chick feeding by the addition of synthetic vitamins. *Poultry Sci.* 27, 236–238.

Kratzer, F.H., and Williams, D.E. (1948b). The relation of pyridoxine to the growth of chicks fed rations containing linseed oil meal. *Poultry Sci.* 27, 297–305.

Larick, D.K., and Turner, B.E. (1990). Flavour characteristics of forage-grain-fed beef as influenced by phospholipid and fatty acid compositional differences. *J. Food Sci.* 55, 313–317.

McDonald, P., Edwards, R.A., and Greenhalgh, J.F.D. (1996). *Animal Nutrition.* Longman Scientific and Technical, England and John Wiley and Sons, New York.

Mandell, I.B., Buchanan-Smith, J.G., and Holub, B.J. (1998). Enrichment of beef with ω3 fatty acids. In Simopoulos, A.P. (ed.) *The Return of ω3 Fatty Acids into the Food Supply. I.*

Land-Based Animal Food Products and Their Health Effects. World Rev. Nutr. Diet., Vol. 83, Karger, Basel, pp. 144–159.

Marshall, A.C., Kubena, K.S., Hinton, K.R., Hargis, P.S., and Van Elswyk, M.E. (1994). n-3 fatty acid enriched table eggs: A survey of consumer acceptability. *Poultry Sci.* 73, 1334–1340.

Olomu, J.M., and Baracos, V.E. (1991). Influence of dietary flaxseed oil on the performance, muscle protein deposition and fatty acid composition of broiler chicks. *Poultry Sci.* 70, 1403–1411.

Phetteplace, H.W., and Watkins, B.A. (1989). Effects of various n-3 lipid sources on fatty acid compositions in chicken tissues. *J. Food Comp. Anal.* 2, 104–117.

Romans, J.R., Johnson, R.C., Wulf, D.M., Libal, G.W., and Costello, W.J. (1995a). Effects of ground flaxseed in swine diets on pig performance and on physical and sensory characteristics and omega-3 fatty acid content of pork: I. Dietary level of flaxseed. *J. Animal Sci.* 73, 1982–1986.

Romans, J.R., Wulf, D.M., Johnson, R.C., Libal, G.W., and Costello, W.J. (1995b). Effects of ground flaxseed in swine diets on pig performance and on physical and sensory characteristics and omega-3 fatty acid content of pork. II. Duration of 15% dietary flaxseed. *J. Animal Sci.* 73, 1987–1999.

Scheideler, S.E. (1999). Feed to produce omega-3 fatty acid enriched eggs and method for producing such eggs. Patent # 5897890.USA.

Scheideler, S.E., Froning, G., and Cuppett, S. (1997). Studies of consumer acceptance of high omega-3 fatty acid-enriched eggs. *J. Appl. Poultry Res.* 6, 137–146.

Scheideler, S.E., and Froning, G.W. (1996). The combined influence of dietary flaxseed variety, level, form, and storage conditions on egg production and composition among vitamin E-supplemented hens. *Poultry Sci.* 75, 1221–1226.

Scheideler, S.E., Jaroni, D., and Froning, G. (1998). Strain and age effects on egg composition from hens fed diets rich in n-3 fatty acids. *Poultry Sci.* 77, 192–196.

Smith, S.B. (1991). Dietary modification for altering fat composition of meat. In Haberstroh, C., and Morris, C.E. (eds), *Fat and Cholesterol Reduced Foods: Technologies and Strategies*, The Portfolio Publishing Company, The Woodlands, TX, pp. 75–97.

Van Elswyk, M.E., Sams, A.R., and Hargis, P.S. (1992). Composition, functionality, and sensory evaluation of eggs from hens fed dietary menhaden oil. *J. Food Sci.* 57, 342–344.

Wiseman, J. (1990). Broiler production: Market trends, meat quality and nutrition in the light of changing consumer requirements. In Wiseman, J., and Lyons, T.P. (eds), *6th Conference on Biotechnology in the Feed Industry*, Alltech Technical Publications, Nicholasville, KY, pp. 119–134.

Wright, T., McBride, B., and Holub, B. (1998). Docosahexaenoic acid-enriched milk. In Simopoulos, A.P. (ed.) *The Return of ω3 Fatty Acids into the Food Supply. I. Land-Based Animal Food Products and Their Health Effects, World Rev. Nutr. Diet.*, Vol. 83, Karger, Basel, pp. 160–165.

Yau, J.C., Denton, J.H., Bailey, C.A., and Sams, A.R. (1991). Customizing the fatty acid content of broiler tissues. *Poultry Sci.* 70, 167–172.

15 Current market trends and economic importance of oilseed flax

John R. Dean

Introduction

There is a long history of oilseed flax production which has been noted elsewhere in this book. There are therefore traditional uses and markets, which are rooted in the commerce and culture of several nations. Attention is often focused on international trade, particularly on the volume and value of exports and imports in the countries involved based on the publication of trade figures. In addition, however, substantial domestic trade can also take place, and such is the case for a sizeable portion of the annual production of oilseed flax. For such internal trade, official records may be less precise.

In this chapter, the extent and changing pattern of international trade in oilseed flax over the past decade, and the major players involved in that trade, will be discussed. The value of this business and the relationship of prices within the context of the vegetable oilseeds and oils market will be examined. Finally, the impact of this commerce on both the farmers who produce the grain, the marketing mechanisms which provide orderly price discovery and risk management, and the processors who manufacture products from the oil, meal and whole seed will be noted.

Global oilseed flax (linseed) production

Production of oilseed flax for the decade 1987/88 to 1998/99 ranged from 1.97 to 2.88 million metric tonnes (mmt), with an average of 2.4 mmt. Global production of all oilseeds in 1999/2000 was forecast to be 300 mmt (Oil World Annual, 1999) – so oilseed flax was less than 1 percent of the total. The distribution between the major producing countries (Figure 15.1)[1] shows that while Canada is the largest producer, and also exporter in recent years, substantial and consistent production is found in India (325,000 t) and China (450,000 t). India and China use all their production domestically, with additional supply from imports.

Two other trends are worth noting from Figure 15.1. Production from Argentina was substantial in the late 1980s at 500,000 t, when Argentina was one of the two major exporters along with Canada. Since then production in Argentina has decreased dramatically to less than 100,000 t. This was due to the impact of severe disease (fusarium and rust), the loss of Russia as a key importer, and the switch to more profitable soybean production. The second trend was the substantial production in Europe (particularly France and the UK) from 1990/91 to 1993/94. This was due to a steep decline in world supply, based on a decrease in production in Canada and Argentina, and attractive subsidies within the EC for the production of linseed. However, after

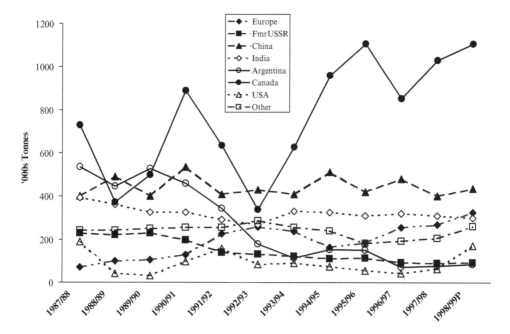

Figure 15.1 Global oilseed flax production 1987/88 to 1998/99P.

Sources: Canada Grains Council, Statistical Handbook, 99; Oil World Annual, 1999.

1993/94, Canadian production soared to the one million tonne range where it has remained until 1999/00. Production in the EU has also taken a jump in recent years, with supply soaring to an estimated 500,000 t in 1999/00, stimulated by the final years of high subsidies for linseed before the introduction of reduced subsidy levels under Agenda 2000.

Price discovery and price protection for oilseed flax

Trade determines price, and visible markets such as the Winnipeg Commodity Exchange (WCE), which is the premium market for canola/oilseed rape and oilseed flax, make that price broadly available to buyers and sellers. The WCE, as with other such exchanges like the CBOT (Chicago Board of Trade – the premium soybean and soyoil market), provide the means for sellers and buyers to conclude transactions for physical grain, and also provide a public forum to determine the value of that grain. The Exchanges provide a meeting place for sellers and buyers to cover risk through "hedging" in a future trading month. This is done for a series of trading months throughout the year, such that transactions may take place today for settlement in the future. This means that price risk may be off-set to a large degree by hedging, a process whereby a position can be taken in the market in the future which can be liquidated nearer to the time of its maturity. This provides someone holding physical stocks of grain (a farmer or a grain company) with a mechanism to off-set that same amount of grain in a future month. When the physical grain is actually sold, the market may have changed – moved up or down in value. This price movement will be reflected in the futures month which was used as a hedge. Liquidating the position in the futures month should off-set any price swing in the cash transaction, thereby protecting the value of the grain. To take a position without such off-setting protection is called speculation. Price risk management is important because of market volatility, and because there is often a significant delay between production and supply, and the physical transaction.

Hedging – an example

A seller may take a position in the market to protect an acceptable price by taking an opposite position in a futures month until physical delivery of the commodity. The off-set to cover gains or losses in value is not necessarily exact, but generally the future months will reflect the current (cash) value with the addition of carrying charges.

An example of a "hedge" for a flaxseed seller (e.g. a Canadian exporter) is given in Table 15.1, based on an example for soybean oil published by the Chicago Board of Trade (Chicago Board of Trade, 1996).

In the examples in Table 15.1, when the November futures price went up between June and September, the futures price was 50 cents per tonne less than the selling price of the physical flaxseed. When the futures hedge was bought back, it actually yielded a small profit by making the effective selling price 50 cents more. So the hedge yielded an additional 50 cents per tonne once both transactions were completed. The original target price of $250 was met and the futures transaction provided a 50 cent bonus. Of course, if the November futures had been more than $260, then the target of $250 would not have been met. However, any loss would likely have been minimal. One objection to the use of the futures hedge at all in this scenario would be that since the actual sale realized $260, by using the futures an opportunity for an additional $10 on the target price was lost. However, this is a speculative approach as the market could have gone the other way, resulting in a loss. This is shown in the second example when the futures market dropped to $240. The hedge in the futures

Table 15.1 The effect of futures on the return on the effective selling price of flaxseed

Cash	Futures (up)		Futures (down)
	WCE*		WCE*
June			
Objective:			
To sell 2000 t of	Sell 100 Nov flaxseed		Sell 100 Nov flaxseed
flaxseed at $250/t	contracts at $250.00/t		contracts at $250.00/t
Sept			
Sell 2000 t of flaxseed:	Buy 100 Nov flaxseed		
– at an avg. price of $260/t	contracts at $259.50/t		
– at an avg. price of $240/t			Buy 100 Nov flaxseed
			contracts at $240.00/t
Result			
Sell physical flaxseed @	$260.00/t		$240.00/t
Futures gain ($250–259.50)	–$9.50/t	($250–240)	$10.00/t
Effective selling price	$250.50/t		$250.00/t

Note
* WCE, the Winnipeg Commodity Exchange, is the only flaxseed futures market in the world. One flaxseed contract equals 20 tonnes. Nov is the abbreviation for the November futures.

market provided a gain of $10, which brought the total return to $250, so meeting the target price. Without the hedge it would have been $240.

Basis

The value of oilseed flax to the farmer selling his seed to a grain company in western Canada (at a price referred to as the "street" price) is determined as a "basis" to the nearby futures month. The basis reflects costs, such as handling, storage, insurance, etc of the grain and hopefully a reasonable margin. The demand for that grain in the short term also affects the basis. Thus, a wider basis will reflect less real and anticipated demand, while a tighter basis will mean stronger demand. As with any market, changes to the supply and demand situation will affect the value of the futures price and thus the street price.

In the case of oilseed flax, there are two main types of market – the global export market reflected by the WCE, and local domestic/regional markets. Domestic/regional markets include traditional markets, which may be significantly affected by traditional customs and government protection (such as in Russia, China and India); and larger markets within major trading blocs such as the European Union.

The most transparent market for oilseed flax is the WCE futures market. Farmers, grain companies, processors, exporters and importers all participate in this market. Production and export demand are the key contributors to the WCE price of Canadian oilseed flax (which is quoted as "basis Thunder Bay" – the main eastbound grain port on Lake Superior). In Europe, prices are generally more or less in line with the WCE, but from time to time may diverge, depending on short-term conditions and specific demand at various times of the year. Thus, if the WCE and European prices are in line, the price in Europe will reflect the Thunder Bay price in Canada, plus the cost of handling and unloading at the port, ocean freight to NW Europe, insurance, etc. The prices of oilseed flax and linseed oil for Europe are usually quoted as cif NW Europe (cif stands for "cost of goods, insurance, freight") for seed, or ex-tank Rotterdam for oil. The reason for choosing NW Europe or Rotterdam as a reference is because Rotterdam is the largest port in the world where huge quantities of oilseeds are trans-shipped, or off-loaded for crushing in Dutch plants. Similarly, other NW European ports such as Antwerp in Belgium, and Hamburg in Germany, are major centers for oilseed processing. This pricing basis therefore refers to the physical point at which the product value is quoted – delivered to a port unloading facility for seed, or unloaded from a storage tank in Rotterdam for oil. The price for seed or oil at other locations in Europe will reflect the NW Europe (seed) or Rotterdam (oil) price, plus a freight adjustment.

Similarly, the price to other export destinations from Canada (such as Japan) will reflect the cost of shipping from the west coast of Canada (Vancouver) rather than through Thunder Bay. The price of oilseed flax may be quoted fob ("free on board") at Vancouver port facility, or c&f ("cost of goods and freight") Japan port, depending on which party is looking after the freight. There are other arrangements but the above discussion covers the most common methods for pricing.

Global supply and trade

It has been noted earlier that Canadian production of oilseed flax has increased significantly over the past few years to over one million tonnes. However, this supply

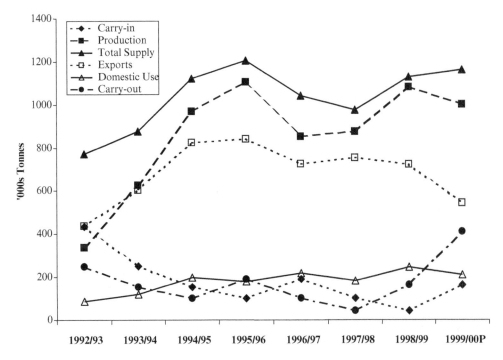

Figure 15.2 Canadian oilseed flax supply and demand 1992/93 to 1999/00P.

Sources: Oil World Annual, 1999; Statistics Canada.

was balanced by export demand, and carryout stocks were reduced to very low levels by the end of the 1997/98 crop year (Figure 15.2). Large Canadian crops in 1998 and 1999, and a surge in EU production, significantly reversed this trend, so that carry-out stocks in Canada are projected to be around 400,000 t by the end of the crop year (July 31) in 2000. A significant reduction in the area seeded to flaxseed in Canada, the EU and the USA in 2000 should lead to a drawing down of stocks in 2000/01.

Exporters

According to *Oil World*, global trade in oilseed flaxseed is currently about 900,000 t, of which Canada exported 760,000 t, in 1997/98 based on Canada Grains Council statistics (Canada Grains Council, 1999). The next largest exporters were the USA and the former USSR with about 8500 t, or just over 1 percent of Canada's total. However, in the context of global oilseed trade, oilseed flax is a small player when compared to 39 mmt of soybeans, five mmt of rapeseed (canola), and three mmt of sunflower seed in export channels in 1997/98. In addition to this, oil exports accounted for an additional 7.5 mmt for soyoil, 2 mmt for rapeseed oil (canola) and 3 mmt for sunflower oil. Linseed oil was about 134,000 t, mostly from western Europe. The greatest disappearance (manufacturing plus exports) of linseed oil during the same period was in western Europe (160,000 t) and China (130,000 t), followed by India (90,000 t), the USA (70,000 t), and Canada and Japan at about 30,000 t each.

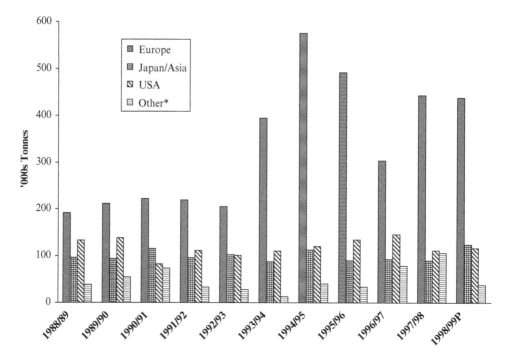

Figure 15.3 The major markets for Canadian oilseed flax, 1988/89 to 1998/99P.

Source: Canada Grains Council, Statistical Handbook 99.

* Other: container and truck shipments to various countries.

Importers

If Canada is the major exporter of oilseed flaxseed, who are the importers? The largest importing region is Europe, with most of the seed going to Belgium, Germany and The Netherlands. Imports by European countries have increased from about 200,000 t in 1992/93, to 473,200 t in 1997/98, after reaching 576,000 t in 1994/95, with almost the entire amount supplied from Canada (Figure 15.3). Japan imports most of its requirement of about 80,000 t from Canada. The USA imports about 200,000 t from Canada, made up of shipments through licensed elevators, and also direct from "unlicensed" local cleaning plants.

Linseed oil trade

In addition to trade in seed, linseed oil is exported by producing countries (Canada, Argentina, USA), and also re-exported by seed importers such as The Netherlands, Belgium, the UK and Germany. As noted above, total world trade in linseed oil was projected to be 134,000 t in 1998/99, having been in the range of 115–183,000 tonnes since 1993/94. The largest importers of linseed oil during this time were East Europe (about 20,000 t), China (14–95,000 t), and Egypt 10,000 t. Many countries engaged in such trade at low levels – including the major exporter, Canada. This reflects natural patterns of trade and crushing capacity, based on geography and logistics, and therefore cost.

Value of oilseed flax products

Commercial seed (grain)

The value of oilseed flax grain sold by the farmer is dependent on the value of the two main products from processing – oil, which forms about 40 percent of the seed, and the residual meal, at about 60 percent. The precise oil content of the seed will vary with variety, environment and climate. The yield of oil and meal after processing will depend on the method of oil extraction. There are traditional uses for both oil and meal, which require expelling the oil by pressing without further extraction using a solvent. Expeller processes leave between 8 to 15 percent of the oil in the meal. The oil portion is of more value per tonne than the meal. If more oil is left in the meal, then the value of the meal must increase to compensate for the loss of revenue from oil. Traditional users of high oil meal (often called oil cake) will therefore pay a premium for the product. Whether this premium fully compensates for the loss of revenue from oil will be a reflection of the prevailing market conditions for both oil and meal.

The value of oilseed flax on the futures market makes interesting reading in relation to the supply and demand picture discussed above. In the late 1980s, Canada and Argentina combined produced about 800,000 tonnes of seed. At the same time, the cash prices in Winnipeg (WCE), were in the $400–450 range (Figure 15.4) – reflecting strong demand and limited supply. In 1990/91, the cash price dropped $175 as production in the two main exporting countries rose to 1.25 mmt – an increase of 63 percent. Production then dropped dramatically and by 1992/93 was 500,000 tonnes combined from Canada and Argentina (Figure 15.1). However, the huge production from the previous years had built a carryover of 400,000 tonnes in Canada,

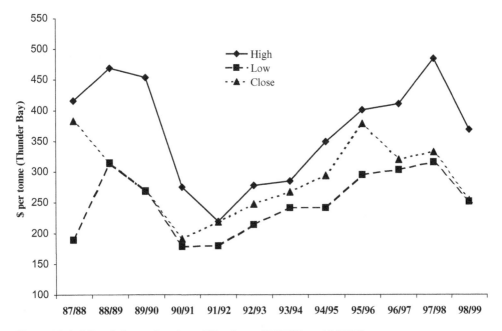

Figure 15.4 Oilseed flax cash prices, Winnipeg, 1987/88 to 1998/99.

Source: Canada Grains Council, Statistical Handbook 99; Winnipeg Commodity Exchange.

282 John R. Dean

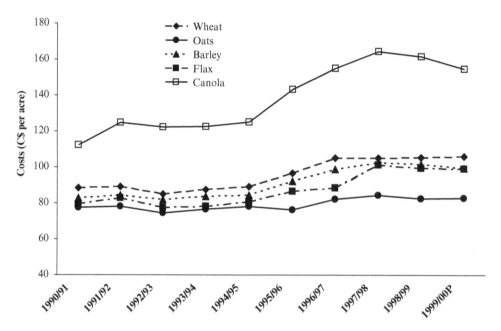

Figure 15.5 Operating costs for five major crops in Manitoba, 1990/91 to 2000/01P.

Source: Manitoba Agriculture, Market Analysis and Statistics, 2000.

so that total supply was close to 800,000 tonnes (Figure 15.2). As stocks began to drop so prices gradually climbed, eventually breaking through the $400 level again in 1995/96 (Figure 15.4). By this time, as we have seen, only Canada was a major exporter. The potential volatility of the flax price is also illustrated in Figure 15.4 where, as noted above, the cash price dropped nearly $175 per tonne in one year (between 1989/90 and 1990/91).

The effect of these changes in the global pattern of production, on farmers in western Canada, was to re-establish oilseed flax as the valuable rotation crop that it is. Lacking diseases in common with either cereals or canola/oilseed rape, flax has always played a valuable role as an alternative crop in rotations. Although lacking the yield potential of canola, its hardiness in tolerating extremes of weather has also contributed to its attractiveness. In addition, the much lower costs of production (Figure 15.5) for oilseed flax versus canola, are a major incentive and will be discussed later.

While production was falling from 1990/91 in Canada and Argentina, it was rapidly building in Europe. The lack of a winter variety and the lower yield compared to even spring oilseed rape, required the highest subsidies of all grain and oilseed crops in Europe, to ensure production in the European Union. By 1992/93 the supply of linseed from domestic production in Europe had reached more than 250,000 tonnes (Figure 15.1), only to gradually fall back again as other higher yielding crops were preferred by farmers in the UK, France and Germany, the main producing countries in the EU. It is worth noting that under the EU Agenda 2000 proposals, the area subsidy for oilseeds will be gradually reduced until it is equal to cereals. This is expected to result in an increase in cereal production on account of the higher yield potential for cereals, and a decrease in the production of oilseeds, especially linseed. However, further modification of support programs, agronomic considerations, and of course market prices, will all continue to impact production levels in Europe.

Linseed oil and meal

A range of valuable products is produced from linseed oil, including: oil based paints and varnishes; wood preservatives; floor coverings such as linoleum; environmentally friendly adhesives; concrete preservative; printing inks. Although other vegetable oils can substitute for linseed in some of these applications, the additional cost from slower drying (lower Iodine Value) or other functional concerns, means linseed oil has few if any substitutes in its many applications. Virtually all the linseed oil produced in North America, Europe and Japan is used for such purposes. As noted earlier, however, use by traditional producers such as China, India, and Russia has also included some human consumption.

This lack of substitutability, and the dependence on one major exporter, means that a certain degree of in-elasticity and therefore potential price volatility is associated with linseed oil. This is illustrated in Figure 15.6, where the NW European price for linseed oil lacks a consistent relationship with the major vegetable oils, sometimes at a huge premium (1989/90) and sometimes at a significant discount (1993/94).

Taking a snapshot of prices to illustrate pricing relationships, on June 7, 2000 the NW Europe crude vegoil prices per tonne were quoted as: soybean oil $323.84, rapeseed oil $341.04, sunflower oil $382.50, linseed oil $425.00. These values are substantially lower for all oils than in 1998 and 1999, as shown in the calculation made in the following example of product values.

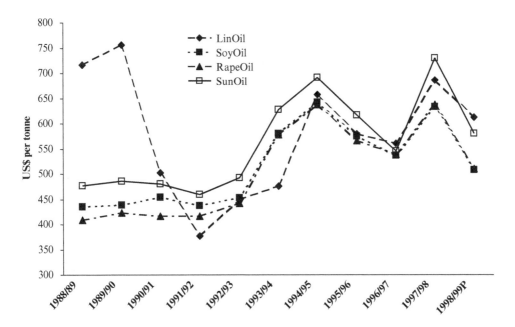

Figure 15.6 Crude vegoil prices NW Europe 1988/89 to 1998/99P.

Source: Oil World Annual, 1999.

Table 15.2 Equivalent seed value of flaxseed (linseed), soybean and rapeseed/canola based on their component values

	Average price ($/tonne)§		% of seed¶	Contribution ($/tonne)		Contribution (% of value)	
	1998	1999		1998	1999	1998	1999
Linseed oil	729	540	38	227.02	205.20	75	72
Linseed meal (36%)	147	128	62	91.14	79.36	25	28
seed equivalent value*				**368.16**	**284.56**		
Soybean oil	671	452	18	120.78	81.36	44	40
Soybean meal (44%)	187	152	82	153.34	124.64	56	60
seed equivalent value				**274.12**	**205.64**		
Rapeseed/canola oil	673	452	40	269.20	180.80	77	75
Rape/canola meal (34%)	133	99	60	79.80	59.40	23	25
seed equivalent value				**349.00**	**240.20**		

Notes
* Seed equivalent value is the value of seed based on component value after crushing and therefore includes costs of handling, storage, crushing, etc, and should not be confused with the value of commercial seed. The relationship of the cost of seed to the seed equivalent value will vary depending on a number of factors and will differ between oilseeds.
§ Prices based on published values for oils and meal for May 1998, and May 1999 (Oil World Annual, 1999).
¶ Total oil extracted, which equals the seed oil content less the residual oil remaining in the meal.

For oil-rich seeds such as oilseed flax, sunflower and canola (which contain 40 percent or more oil), the meal provides a smaller portion of the value, particularly following solvent extraction. In comparison, soybeans contain only about 20 percent oil. In the case of oilseed flax, the following equivalent seed value based on component values in May 1999 versus 1998 may be derived, and compared with soybean and rapeseed (canola) oil (Table 15.2).

Thus, at the prices used in Table 15.2, the contribution to the value of the seed from oil was up to 75 percent for linseed, slightly more for rapeseed, but only 40–44 percent for soybeans. On the meal side, 25 percent of the value was from meal for linseed, but more than 55 percent for soybeans. These relative values reflect mostly the lower oil content of soybeans, and to some extent the lower value of soybean oil.

The value of soybean meal is higher than linseed meal, mostly on account of the higher protein level (44% versus 36%), its versatility for use with all classes of livestock, and its place as the standard among protein meals. When solvent extraction is not used, the increased level of residual oil from the expeller press increases the value of linseed meal. Certain sectors of the livestock industry, such as dairy farmers in Europe, are willing to pay for this additional energy as they see the benefits in animal productivity. The cyanogenic glucosides in linseed meal are of no concern in ruminant feeding, but a reduced level of feeding for growing pigs is recommended. Plant breeding is underway to reduce the level of the cyanogenic glycosides to very low levels, thereby enhancing the value of the meal for pig feeds. Additional breeding developments, which will reduce the level of soluble fiber (gums), will enhance the value of linseed meal for use in poultry feed.

While the oil component commands most of the value in linseed, the bulk of the product after processing is meal (about 60 percent). There are no long term statistics for linseed meal disposition or prices, since linmeal is not used as the major meal component of livestock rations. There are specialized markets for particular classes of livestock such as dairy cattle where linmeal, especially expeller high oil cake, is favored, but these uses are relatively small and regional. The latter are premium markets, but generally linseed meal is valued between soymeal and rapeseed meal.

Whole seed flax for human consumption

A small, but growing, market is the use of whole seed oilseed flax for human consumption. This is a premium market, but again no statistics exist as to the precise values or volumes that are used in this way. Flax is receiving increased attention by nutritionists for the valuable fiber, lignans, and the omega-3 in the oil. In spite of its nutritional value, government recommendations exist in some countries restricting the maximum level of flax included in foods. The need for these restrictions is questioned by some, especially when the seed is included in baked products.

The largest market for whole seed consumption is Germany. No official statistics exist for this market, but estimates suggest 80,000 tonnes is used for food in baked products, mueslis, and packaged for sale through health food stores. The total whole seed consumption in Europe has been estimated at 100,000 tonnes or more annually, including human consumption and specialized feeds for horses, show animals and in pet foods. In North America the level of consumption is much smaller, but it is growing as elsewhere in the world. This growth will be based on the real health benefits of flax, namely the soluble and insoluble fiber; the lignan SDG (a valuable phytoestrogen, which is at a very high level in flax); and the omega-3 fatty acid in the oil.

Economic importance in producing and importing countries

Producing countries

Oilseed flax production is generally not the major crop revenue earner for farmers in any of the producing countries. However, it provides an excellent break in cereal/oilseed crop rotations in temperate climatic areas. This is because it has no diseases in common with the major cereal crops (wheat, oats, barley, corn, sorghum) or with the main oilseed crops, particularly soybeans, canola/oilseed rape, or sunflowers. The decision to produce oilseed flax will be determined by a number of factors. As has been noted above, production and price swings can be considerable, and are generally more volatile than with the major vegetable oils (Figure 15.6). Price alone will not determine a farmer's desire to grow flax – field choice, a particular management problem, time of seeding, contracting opportunities, cash flow, support programs, and even tradition, will all have an impact on the decision.

The value of oilseed flax to farmers in western Canada is illustrated in Figure 15.7. As an oilseed, flax plays second fiddle to canola in terms of total crop value. In 1998, the total crop value for oilseed flax ranked fifth of the major crops produced in western Canada, as shown in Table 15.3.

In Canada, the cereal grains, which are the best adapted crops to the vast prairie region, dominate. However, the increasing adaptability of modern varieties, with competitive prices, has led to the sizeable production of oilseeds within which flax

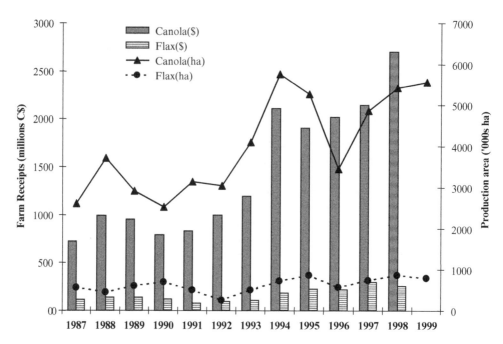

Figure 15.7 Farm receipts (C$) and production ('000s ha), canola and flax, Canada, 1987–1999.

Source: Canada Grains Council, Statistical Handbook 99.

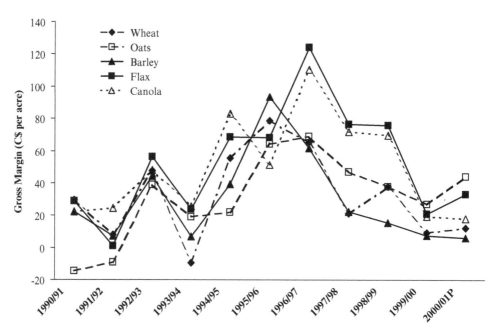

Figure 15.8 Gross margin for five major crops in Manitoba, 1990/91 to 2000/01P.

Source: Manitoba Agriculture, Market Analysis and Statistics, 2000.

Table 15.3 Total farm cash receipts for grain and
oilseed crops grown in western Canada, 1998

Crop	Millions C$
Canola	2,703
Wheat	2,428
Durum	905
Barley (malt and feed)	583
Oilseed flax	254
Oats	219

Source: Canada Grains Council, Statistical Handbook, 99.

Table 15.4 Commodity prices for vegetable oils in Mumbai (India) and Rotterdam

Product	Price/tonne (Rupees)	Price/tonne (US$)	Rotterdam price (US$)
Linseed oil	33,500	782.53	540.00
Soybean oil	27,500	642.37	450.00
Rapeseed oil	29,500	689.09	450.00
Sunflower oil	27,300	637.70	590.00
Soybean oilcake	6,500	151.83	152.00
Linseed oilcake	6,900	161.18	128.00

plays a valuable role. In terms of the value of the crop to individual farmers, oilseed flax remains one of the most profitable of the grain and oilseed crops (Figure 15.8) in Manitoba. Compared to canola in particular, oilseed flax has a much lower cost of production (Figure 15.5), in most years more than off-setting its lower yields.

In the European Union, very attractive subsidies, and the isolation of oilseed flax from the production-restrictive oilseeds agreement (Blair House) between the USA and the EU, have led to increased production in some years. In non-exporting countries such as India and China, oilseed flax has a traditional place with a balance between production and local markets, tending to isolate the crop from the vagaries of the global market. In addition, an import duty in India of 21.16 percent (in 1999) on vegetable oils in bulk serves to further protect the domestic market. Commodity prices in Mumbai (Bombay) for Indian vegetable oils and meals for May 18, 1999 (from India Mart, 1999) provide some comparisons (Table 15.4).

If the price of oil in Mumbai (Bombay) fairly reflects the farm price for seed, then Indian farmers were receiving a good price for oilseed flax compared to the Rotterdam price, including taking freight and import duty into account. It is also interesting to note the relative value of oils in India and Europe. In Europe, sunflower oil was trading at a considerable premium – whereas in India it was at a discount to soy and rape oils in May 1999. This reflects tradition and culture – sunflower has always been the premium oil in Europe, whereas rapeseed is favored in India. Linseed oil commanded a significant premium over soybean and rapeseed oils in Mumbai as in Rotterdam. It should be noted that these prices reflect one point in time – normal changes in supply and demand will affect the price levels and relationships.

Importing countries – value in processing

The most common use for linseed oil is in paints, varnishes and wood preservatives. This traditional use is based on the characteristic of the oil to polymerize during oxidation of the fatty acids, producing a smooth textured product with excellent durability, color retention, and waterproof properties. The coatings industry is huge, encompassing paints and coatings for dwellings, industrial (automotive, transportation, furniture, cans, containers, etc), and special purpose uses. Global statistics for linseed oil are not available, but to illustrate the size of the coatings industry, sales in the USA in 1997 were $16.4 billion. Of this, about 37 percent (or $6.1 billion) was used in the so-called "Architectural Coatings (house paints)" sector, and within this the solvent-borne (oil) based products were included. The impact of durable, attractive paints is a basic but vital and positive part of home-ownership and neighborhood ambiance, within which linseed oil plays its essential part.

The other major use is in the manufacture of linoleum floor covering, where linseed oil forms about 30 percent of the raw materials used. In the UK, for example, the total floor coverings industry in 1997 was £1.6 billion, of which 80 percent was carpeting and 12 percent vinyls and linoleum. The linoleum portion was about £48 million or 3 percent of the total floor coverings industry in the UK. As an interesting historical note, long before vinyl flooring was developed, linoleum had been in widespread use, being made from a process developed by Frederick Walton in England in 1863, and later improved by Michael Nairn in Scotland (National Paints and Coatings Association, 1999).

The main use of linseed is in coatings in all countries where the oil is available, and from an economic activity point of view, it plays a part in supporting the manufacturing, distribution and retailing industries. The manufacture of linoleum has seen a resurgence in recent years as its environmentally friendly properties have been appreciated, along with a desire to move away from non-renewable sources of raw materials. The linoleum industry is now concentrated in a few large plants in Europe, and boasts use of the product in a variety of high profile public spaces worldwide, such as airports, hospitals, and universities. It is also popular in modernistic designs for corporate offices and private homes.

Future economic importance of oilseed flax

It is expected that the traditional uses of oilseed flax in coatings, linoleum, wood preservatives, etc, will continue strongly in the future, especially in light of constantly increasing environmental awareness and the need for biodegradable products. In addition to these traditional uses, there are a number of developing markets which may significantly increase in value in the future.

Increased use of flax oil in foods

Many nutritional studies have emphasized the need for increased consumption of omega-3 fatty acids. Flax oil is the richest source of omega-3 from vegetable oils. Where consumption of fish is lower than meats, as in North America, there is a significant need for a higher level of intake of omega-3 to narrow the ratio with omega-6. Improvements in the processing of flax oil in the future may well lead to the incorporation of the oil in food products, when in the past it was assumed the oil's oxidative instability would not permit such use.

New types, new uses

Plant breeding has for many years ensured a steady introduction of improved crop varieties. These have either provided farmers with better yields and disease resistance, or have supplied the consumer with more nutritious and safer food. For example, the transformation of rapeseed to canola through the virtual elimination of erucic acid, led to the introduction of a whole new crop for farmers and a new product for vegetable oil processors and the retail market.

While rapeseed/canola has received a lot of attention from plant breeders, oilseed flax has tended to remain in the shadows. An exception is a low linolenic flax type (designated as solin in Canada), which was first developed in Australia by CSIRO. The characteristics of linseed type flax which permit its highly successful use for industrial applications have prevented it in the past from being used widely in food products. The rapid oxidation of the high linolenic oil means the rapid onset of rancidity in food applications.

Low linolenic flax (solin) has a linolenic content of only 2 percent compared to more than 50 percent in regular flax, and is yellow rather than brown seeded for ease of identification. In fact, the fatty acid profile of solin is similar to sunflower oil, so that since its commercial introduction in 1995, the oil has been used successfully in premium quality margarines and spreads in Canada and Europe. The value of solin as a premium vegetable oil is thus closely related to the major oils, especially sunflower oil. The use of solin in the baking industry has provided an additional product to regular brown flax. The interesting yellow color is of value for cosmetic purposes, and where long-term storage is needed, the low linolenic characteristic ensures oxidative stability. In all respects, other than the supply of omega-3 fatty acid, solin provides the same nutritional benefits including lignans, and soluble and insoluble fiber.

Production of solin is centered in Canada with more than 100,000 hectares, and smaller production areas in the UK and Australia. This new flax is already providing yet another alternative crop for farmers, and a consistent quality of a polyunsaturated oil for processors from a new geographic location. Such initiatives generate new economic activity, enhancing the opportunities for diversification and better risk management for farmers, and also strengthen overall business opportunity in the food grains and oilseeds sector, as well as in food processing and manufacturing.

Nutraceuticals, functional foods and extractable components

In the same way, the very new but rapidly developing interest in nutraceuticals and functional foods may be expected to enhance value and provide new opportunities for development of both better products and hopefully better health through food.

Flax, whether the traditional or the new low linolenic oilseed, is a strong candidate to enter this market. Traditional oilseed flax has a number of vital nutritional benefits – soluble and insoluble fiber; omega–3 (linolenic) fatty acid; lignans. The use of flax as an ingredient in baked products is seen as an excellent way to increase consumption of these valuable nutrients. In addition, extraction of the soluble fiber (gums), and in separate patented developments, the lignans and protein, could lead to the production of purified ingredients or encapsulated products. Such developments could lead to the production of a whole range of new products with significant economic spin-offs.

At this point, component extraction is still in the development phase, so it is not yet possible to put a value on this for the flax, food processing, or health food industries. However, the identification of such value within the seed of the once lowly flax plant points to a bright future and increased value for this ancient and durable crop.

Biotechnology

Much has been said about the benefits which biotechnology will bring in the future. Examples of such benefits include: pharmaceutical products supplied in association with the oil in an oilseed, and separated during processing; enhanced levels of the more valuable components such as protein and oil in a feed grain; modified properties to increase processing efficiencies; herbicide and insect resistance in many different crop species.

To date, the products of biotechnology have been introduced mostly at the field production end of the chain with the benefit of herbicide and insect resistance going to the producer. In some countries this has caused concern in that no benefits are provided to the consumer, and yet the consumer is required to purchase food from genetically modified crops. The farmer gains an economic benefit, in this scenario, while the consumer does not. Concerns continue to be expressed in very public debates about the long-term safety of the technology. In other countries, a much higher level of trust in government regulatory systems has satisfied such concerns for the most part, so that consumers accept that the new technology will in time not only definitively prove to be safe, but will also bring significant benefits.

The economic impact of these developments remains to be seen. There are claims of reduced cost in the process of production where less chemical for weed and pest control is used, at the same time providing an environmental benefit. Crops which have enhanced nutritional value have been introduced for use in livestock feeding, thereby packing in more nutrients for lower incremental cost. The use of oil in oilseed crops such as oilseed flax as a carrier for beneficial pharmaceuticals has been demonstrated in the lab and if commercialized could provide valuable drugs at cheaper cost. There is also the claim that crops capable of significant productivity in poor soils will allow depressed regions in the world to produce more of their own food.

All of these possibilities as products of biotechnology are generating significant economic activity in development and the initial introduction phase. The full benefits from commercialization for the most part have yet to be seen. There is great confidence on the part of the developers, expectation on the part of producers and processors, and a mixture of opposition, skepticism and anticipation on the part of consumers. In the final analysis, measurable benefits at little or no additional cost from safe and nutritionally enhanced products will build confidence in consumers and generate the expected economic value.

Acknowledgments

The author wishes to thank G. Pownall (Agricore United, Winnipeg), D. Frith (Flax Council of Canada), C. Gunvaldsen (Manitoba Agriculture and Food) and R. Gupta (Canadian High Commission, New Delhi, India) for their insight and for access to unpublished data.

Note

1 In all charts the letter P denotes projections for the year which may be subsequently revised.

References

Canada Grains Council (1999). *Canadian Grains Industry Statistical Handbook 99*. Canada Grains Council. Winnipeg, Canada.

Canadian Grain Commission (1999). *Exports of Canadian Grain and Wheat Flour*. Winnipeg, Canada.

Chicago Board of Trade (1996). *Taking Control of Your Bottom Line*. Market and Product Development Department, Chicago Board of Trade. Chicago, IL.

India Mart (1999). Current Commodity Prices, India Mart Market Watch, http://finance.indiamart.com

National Paints and Coatings Association (1999). Products of the Paints and Coatings Industry, National Paints and Coatings Association, Washington, DC, http://www.paint.org/ind_info/types/htm

Oil World Annual (1999). *Oil World*. Mielke GmbH. Hamburg, Germany.

16 Current regulatory status of flaxseed and commercial products

Alister D. Muir and Neil D. Westcott

Introduction

As with any products that enter the commercial world, there is an inevitable pro-liferation of regulations to ensure that the products are described correctly, that they are safe, and if claims are made, that there is some verifiable basis for these claims. Patents and other forms of intellectual property protection are another inevitable development as businesses seek to protect their investment in research and develop-ment and as researchers and organizations funding research attempt to recover some of their investment. In this chapter we will review the current regulatory status of flaxseed and products derived from flaxseed and provide an overview of the scope of patents related to flaxseed. Standards for the preparation of flax oil for food use are not well defined at this time. The ISO standard "specifies flaxseed requirements for the manufacture of oil for industrial use including foodstuffs" (ISO, 1982; Kolodziejczyk and Fedec, 1995). However, the standard for minimum Iodine Value varies with specification and ranges from 170 for the ISO standard to 177 for the American Society for Testing and Materials (ASTM).

Pharmacopoeias

In spite of a long history of use as a food and for its medical properties (see Chapters 1 and 13), there is relatively little formal documentation of flax in pharmacopoeias. For example, flax is not included in the US Pharmacopoeia.

British Pharmacopoeia

The British Pharmacopoeia describes flax (linseed), whole or in powdered form, for use as a demulcent. The specification calls for a swelling index of not less than 4 for the whole seed and 1.5 for the powdered form, not more than 1.5 percent foreign matter and not more that 6 percent sulphated ash (Anonymous, 1993). The British Pharmacopoeia also describes linseed oil with an acid value of not more than 4.0, Iodine Value not less than 175 and a saponification value of not less than 188 for use as a pharmaceutical aid.

Monographs of Commission E

A monograph (*B.Anz.* No 228) was published on May 12, 1984, describing the use of flaxseed for internal use as a laxative (Anonymous, 1998b), where the principal active ingredient is presumed to be the gum or mucilage component of the seed coat. An external use as a cataplasm for local inflammation is also described. A revised monograph was published in 2000 (Blumenthal *et al.*, 2000).

ESCOP monographs

The European Scientific Co-operative on Phytotherapy (ESCOP) was formed in 1990 to organize, collate and harmonize therapeutic data on herbal drug products. Flaxseed is described in Fascicule 1 of this series (Anonymous, 1996). The ESCOP monograph describes the constituents of flaxseed and describes the accepted dosages and methods of administration for internal use as a laxative and for external use as a poultice or compress. Potential delays in the absorption of drugs and delays in glucose absorption associated with the use of whole flax as a laxative are noted but no adverse effects during pregnancy or lactation were noted.

Physicians Desk Reference

The PDR for Nutritional Supplements (Hendler and Rorvik, 2001) contains an entry for flaxseed oil but no information on the seed meal. Contraindications and precautions reported, including advice against pregnant women consuming flaxseed oil on account of the theoretical possibility that these lignan-containing substances might induce menstruation, are not supported by any experimental or epidemiological evidence (Blumenthal *et al.*, 2000). Since the PDR incorrectly identifies the lignan SDG as a component of the oil, the credibility of these recommendations is suspect. The authors of this report also appear to have incorrectly attributed the laxative effect of the dietary fiber to the oil content. The reported possible antithrombotic activity attributed to the lignan content of the oil has not been substantiated by any adverse reaction reports (Lininger, 1999). The PDR for Herbal Medicines contains an entry for flaxseed which lists a contraindication for ileus, stricture of the esophagus and acute inflammatory illnesses of the intestine, esophagus and stomach entrance (Gruenwald *et al.*, 2000). The possible delayed absorption of drugs taken simultaneously is identified as precautionary advice.

"GRAS" status

Under the U.S. Federal Food, Drug, and Cosmetic Act, a substance can be legally added to a food if it is a food additive approved for use, or if the substance is deemed to be "Generally Recognized As Safe" (GRAS). A substance can be designated as GRAS if it has a long history of common safe use in food prior to 1958 (Vanderveen, 1995).

Flaxseed

Flaxseed as a food ingredient does not have food additive status nor has it been affirmed as GRAS (Vanderveen, 1995). However, the FDA is not currently challenging the status of flaxseed as an ingredient in bread and other cereal foods. In a communication to Harvest Foods Ltd., of Saskatoon, Saskatchewan, the FDA stated "Foods containing up to 12 percent flaxseed (whether or not defatted or heat treated) are not deemed unsafe and we (FDA) wouldn't object to their use" (Vanderveen, 1995). We are not aware of any submission to the FDA for GRAS status for either flaxseed or flax oil. Flaxseed is approved for use in food in France at levels not exceeding 5 percent (Bruneton, 1999).

Solin (Linola™)

In addition to high α-linolenic acid oil obtained from traditional oilseed flax, solin oil with less than 5 percent α-linolenic acid is now commercially available and this oil has received an informal GRAS status in the USA (Anonymous, 1998a). The request for GRAS status for solin oil (GRAS Petition No GRP5G0416) originally filed on June 30, 1995 and subsequently converted to a notification, was accepted without objection by the FDA (GRAS Notice No. GRN00002/Docket No 98S-0104) (FDA, 1998) on May 27, 1998. This GRAS status ruling applies to the solin oil. In a ruling by William Price (Deputy Director, Division of Animal Feeds, Center for Veterinary Medicine, FDA, Rockville, MD), the solin meal was ruled to fall within the Association of American Feed Control Officials (AAFCO) definition for flax meal. In a letter dated Nov 10, 1994, Mr. Price states "Since seed composition appears to be unaffected by the mutation except for the alterations in the concentrations of linoleic and linolenic acids, it appears that low linolenic acid linseed should be considered another linseed variety which does not require a new AAFCO definition for meal products."

The term solin is recognized by the Canadian Grain Commission as the official name for yellow seed coat flax with low α-linolenic acid. The term Linola™ is a trademark held by the Commonwealth Scientific and Industrial Research Organization (CSIRO) in Australia and refers specifically to low α-linolenic acid flaxseed. Linola™ is synonymous with solin but the reverse is not true.

Patents

In keeping with the focus of this book, only patents related to food, animal feed, nutraceutical, functional food and pharmaceutical applications will be reviewed. The patents identified will in general be the first filing for each particular technology or the first filing when an English summary is available. It is beyond the scope of this chapter to identify all the patents related to the original filing. However, the reader should be aware that some of these patents have been granted in many countries. In addition, there are patents pending for a number of applications.

Food and food-related uses

Along with the growing number of uses for flaxseed, there have been a considerable number of patents filed on various subjects related to flax. Flaxseed or flaxseed components have potential for incorporation into food products by virtue of their functional properties, i.e. viscosity properties of the gums (Hanaoka and Nogami, 1980). As medical opinion and the regulatory environment changes to recognize functional claims for gums, proteins and secondary plant compounds, some of these patents may assume even greater significance in terms of influencing future product development. US patents have been granted for use of flaxseed in beverages (Lauredan, 1979; Stitt, 1989), production of protein hydrolysates (Stahel, 1985), and for incorporation into foodstuffs such as ground flaxseed (Stitt, 1989). Patents have been granted for the preparation of proteins and mucilage for use in baked goods (Nikkila, 1965;

Kankaanpaa-Anttila and Anttila, 1999) and for the roasting of flaxseed for incorporation into food products (Fukuda *et al.*, 1997). A US patent has also been granted for the use of a combination of flaxseed and microflora that accumulate long chain omega-3 fatty acids as a food (Barclay, 1999).

Patents have also been granted for the use of flax oil in food (Martin, 1964) and beverages (Durst, 1977).

Dietary supplements, nutraceutical and pharmaceutical use

Patents have been granted in the USA and in other countries on processes to isolate the lignan SDG from flax for nutraceutical and pharmaceutical use (Westcott and Muir, 1998; Westcott and Muir, 1999; Empie and Gugger, 2001; Westcott and Paton, 2001). Patents have also been granted for the use of SDG as an antioxidant (Prasad, 1998), and for the treatment of lupus nephritis (Clark and Parbtani, 1998).

In Europe and elsewhere, patents have been granted for the extraction, purification and use of flaxseed polysaccharides as saliva substitutes (Attström *et al.*, 1993) and as a component in topical medications and cosmetics (O'Mullane and Hayter, 1995). Flaxseed in combination with other ingredients is claimed as a composition for natural digestion regulation (Kowalsky and Scheer, 1982) and for inhibition of colon cancer (Alabaster, 1999). A European patent was recently granted for the use of crushed flaxseed as a drug composition for the regulation of bowel activity (Kolari *et al.*, 1999).

Medical uses of flax oil include patents for a hemorrhoid remedy (Gros, 1986), acne treatment (Yeo, 1994), reduction of platelet adhesiveness (Martin, 1977), and as a carrier in a treatment for the symptoms of menopause (Maxson *et al.*, 1990). Flax oil in combination with a number of different herbs is described in a US patent (Emanuel-King, 1993) for the relief of symptoms of gingivitis, gum disorders, cold sores, acne and other conditions.

Animal feed

In addition to patents for human use, US patents have also been granted for use of whole flaxseed in animal feed (Stitt, 1991; Yeo, 1993), defatted meal for cows and turkey polts (Stahel, 1985), for the use of flaxseed in animal rations to increase live births (Stitt, 1992). US Patent 3246989 (Biehl, 1966) discloses the use of a fermented flaxseed meal product for calves. A German patent (Albrecht, 1995) discloses the use of flax oil as a supplement to promote muscle growth in horses.

The practice of feeding whole or partially defatted flaxseed to hens to produce eggs enriched in α-linolenic acid (Omega-3 eggs) is increasing in North America (see Chapter 14) and patents have been granted for feed formulations containing either flaxseed (Scheideler, 1999) or flax oil (Weiss and Schwartz, 1990; Ise, 1992; All *et al.*, 1994).

References

Alabaster, O. (1999). Dietary fiber composition. Patent # EP0954984A1.European Patent.

Albrecht, I. (1995). Use of natural cold-pressed linseed oil. Patent # DE4411499.Germany.

All, T., Terada, F., Muraoka, M., Tsurusaki, M., Ono, H., Kojima, Y., Murakami, T., Matsuzaki, M., Hayasawa, H., Shimizu, T., Ishida, S., and Nakamura, T. (1994). Feed for edible poultry and breeding of edible poultry by the same feed. Patent # JP7255387A.Japan.

Anonymous (1993). *British Pharmacopoeia*. London. 385–386 p.

Anonymous (1996). Lini semen (linseed). In *ESCOP Monographs*, Fascicule 1, ESCOP Secretariat, Exeter, UK, pp. 1–5.

Anonymous (1998a). FDA accepts solin as GRAS, *INFORM*, AOCS, pp. 761–762.

Anonymous (1998b). Flaxseed (Lini semen) (English Translation). In Blumenthal, M., Busse, W.R., Goldberg, A., Gruenwald, J., Hall, T., Riggins, C.W., and Rister, R.S. (eds), *The Complete German Commission E Monographs. Therapeutic Guide to Herbal Medicines*, Vol. B. Anz. No. 228, American Botanical Council, Austin, TX, p. 132.

Attström, R., Glantz, P.O., Hakansson, H., and Larsson, K. (1993). Saliva substitute. Patent # 5260282.USA.

Barclay, W.R. (1999). Food product containing thraustochytrium and/or schizochytrium microflora and an additional agricultural based ingredient. Patent # 5908622.USA.

Biehl, H. (1966). Method for the production of an improved animal feed. Patent # 3246989.USA.

Blumenthal, M., Goldberg, A., and Brinckmann, J. (eds) (2000). *Herbal Medicine. Expanded Commission E Monographs*, Integrative Medicine Communications, Newton, MA. 519 p.

Bruneton, J. (1999). *Pharmacognosy, Phytochemistry, Medicinal Plants*, 2nd ed. Hatton, C.K. (Trans). Lavoisier Publishing, Paris. 1119 p.

Clark, W.F., and Parbtani, A. (1998). Method for treatment of lupus nephritis. Patent # 5837256.USA.

Durst, J.R. (1977). Preparation of fat-containing beverages. Patent # 4031261.USA.

Emanuel-King, R. (1993). Herbal dietary supplement. Patent # 5248503.USA.

Empie, M., and Gugger, E. (2001). Method of preparing and using isoflavones. Patent # 6261565.USA.

FDA (1998). GRAS Notice No. GRN00002/Docket No. 98S-0104.

Fukuda, N., Sato, T., and Nakahara, R. (1997). Processed food containing linseed and its production. Patent # JP9252750A.Japan.

Gros, C.P. (1986). Remedy for hemorrhoids. Patent # 4626433.USA.

Gruenwald, J., Bendler, T., and Jaenicke, C. (eds) (2000). *PDR for Herbal Medicines*, Medical Economics Company, Inc., Montvale, CA., 858 p.

Hanaoka, J., and Nogami, Y. (1980). Linseed mucilage, having novel properties and preparation of the same. Patent # JP165902A.Japan.

Hendler, S.S., and Rorvik, D. (eds) (2001). *PDR for Nutritional Supplements*, Medical Economics Company, Inc, Montvale, NJ. 575 p.

Ise, S. (1992). Method of producing commercially useful poultry products with increased concentrations of omega-3 polyunsaturated fatty acids. Patent # 5133963.USA.

ISO (1982). *Linseed for Manufacture of Oil-Specification*. International Organization for Standardization. Geneva, Switzerland. Vol. 5513-1982, 3 p.

Kankaanpaa-Anttila, B., and Anttila, M. (1999). Flax preparation, its use and production. Patent # 5925401.USA.

Kolari, P., Koskimaa, P., Gröhn, P., Kivinen, A., and Tarpila, S. (1999). Linseed drug. Patent # EP889729A1.European Patent.

Kolodziejczyk, P.P., and Fedec, P. (1995). Processing flaxseed for human consumption. In Cunnance, S.C., and Thompson, L.U. (eds), *Flaxseed in Human Nutrition*, AOCS Press, Champaign, IL, pp. 261–280.

Kowalsky, H., and Scheer, H. (1982). Dietetic composition for natural digestion regulation. Patent # 4348379.USA.

Lauredan, B. (1979). Nutritive seamoss composition and method for preparing same. Patent # 4180595.USA.

Lininger, S.W. (ed.) (1999). *A-Z guide to Drug-Herb-Viatmin Interactions*, Prima Health, Rocklin, CA. 436 p.

Martin, W. (1964). Producing edible oil from grain. Patent # 3163545.USA.

Martin, W. (1977). Process for reducing platelet adhesiveness. Patent # 4061738.USA.

Maxson, W.S., Hargrove, J.T., and Delk, J.H. (1990). Novel pharmaceutical composition containing estradiol and progesterone for oral administration. Patent # 4900734.USA.

Nikkila, O.E. (1965). Egg white substitutes. Patent # 34558.Finland.

O'Mullane, J.E., and Hayter, I.P. (1995). Linseed mucilage. Patent # EP6745221A1.European Patent.

Prasad, K. (1998). Purified SDG as an antioxidant. Patent # 5846944.USA.

Scheideler, S.E. (1999). Feed to produce omega-3 fatty acid enriched eggs and method for producing such eggs. Patent # 5897890.USA.

Stahel, N.G. (1985). Method of treating oilseed material. Patent # 4543264.USA.

Stitt, P.A. (1989). Stable nutritive and therapeutic flaxseed compositions, methods of preparing the same and therapeutic methods employing the same. Patent # 4857326.United States.

Stitt, P.A. (1991). Therapeutic and nutritive flaxseed composition and methods employing the same. Patent # 5069903.United States.

Stitt, P.A. (1992). Method of increasing live births to female animals and animal feed blend suitable for same. Patent # 5110592.USA.

Vanderveen, J.E. (1995). Regulation of flaxseed as a food ingredient in the United States. In. Cunnane, S.C., and Thompson, L.U. (eds), *Flaxseed in Human Nutrition*, AOCS Press, Champaign, Il., pp. 363–366.

Weiss, H.S., and Schwartz, C.S. (1990). Method and composition for increasing the conentration of omega-3 polyunsatutated fatty acids in poultry and poultry eggs and poultry and eggs resulting therefrom. Patent # 4918104.USA.

Westcott, N.D., and Muir, A.D. (1998). Process for extracting lignans from flaxseed. Patent # 5705618.USA.

Westcott, N.D., and Muir, A.D. (1999). Process for extracting and purifying lignans and cinnamic acid derivatives from flaxseed. Patent # 702410.Australia.

Westcott, N.D., and Paton, D. (2001). Complex containing lignan, phenolic and aliphatic substances from flax and process for preparing. Patent # 6264853.USA.

Yeo, Y.K. (1993). Feed composition for breeding pigs with n-3 fatty acid-accumulated pork. Patent # 5234699.USA.

Yeo, Y.K. (1994). Pharmaceutical composition for treating acne. Patent # 5312834.USA.

Index